AF840581

RHEOLOGY and FRACTURE MECHANICS of FOODS

TON VAN VLIET

RHEOLOGY and FRACTURE MECHANICS of FOODS

CRC Press
Taylor & Francis Group
Boca Raton London New York

CRC Press is an imprint of the
Taylor & Francis Group, an **informa** business

Cover design, reprinted with permission from Elsevier, includes a figure from J. F. V. Vincent, Application of fracture mechanics to the texture of food, *Engineering Failure Analysis*, 11, (2004): 695–704.

CRC Press
Taylor & Francis Group
6000 Broken Sound Parkway NW, Suite 300
Boca Raton, FL 33487-2742

First issued in hardback 2019

© 2014 by Taylor & Francis Group, LLC
CRC Press is an imprint of Taylor & Francis Group, an Informa business

No claim to original U.S. Government works

ISBN-13: 978-1-4398-2703-1 (hbk)

This book contains information obtained from authentic and highly regarded sources. Reasonable efforts have been made to publish reliable data and information, but the author and publisher cannot assume responsibility for the validity of all materials or the consequences of their use. The authors and publishers have attempted to trace the copyright holders of all material reproduced in this publication and apologize to copyright holders if permission to publish in this form has not been obtained. If any copyright material has not been acknowledged please write and let us know so we may rectify in any future reprint.

Except as permitted under U.S. Copyright Law, no part of this book may be reprinted, reproduced, transmitted, or utilized in any form by any electronic, mechanical, or other means, now known or hereafter invented, including photocopying, microfilming, and recording, or in any information storage or retrieval system, without written permission from the publishers.

For permission to photocopy or use material electronically from this work, please access www.copyright.com (http://www.copyright.com/) or contact the Copyright Clearance Center, Inc. (CCC), 222 Rosewood Drive, Danvers, MA 01923, 978-750-8400. CCC is a not-for-profit organization that provides licenses and registration for a variety of users. For organizations that have been granted a photocopy license by the CCC, a separate system of payment has been arranged.

Trademark Notice: Product or corporate names may be trademarks or registered trademarks, and are used only for identification and explanation without intent to infringe.

Library of Congress Cataloging-in-Publication Data

Vliet, Ton van.
Rheology and fracture mechanics of foods / Ton van Vliet.
pages cm
"A CRC title."
Includes bibliographical references and index.
ISBN 978-1-4398-2703-1 (alk. paper)
1. Food texture. 2. Food--Composition. 3. Rheology. I. Title.

TX531.V45 2013
664'.07--dc23 2013006493

Visit the Taylor & Francis Web site at
http://www.taylorandfrancis.com

and the CRC Press Web site at
http://www.crcpress.com

Contents

PART III Experimental Evaluation

PART V Relationships among Food Structure, Mechanical Properties, and Sensory Perception

Preface

The main theme of this book is the rheological and fracture behavior of foods. The focus is on the basic principles of this behavior, the determination of rheological and fracture properties, and the relation between these properties and the structure of model food products.

People working in the field of food production have realized for quite some time that the mechanical properties of food are essential quality characteristics of these products. These properties play an important role during the manufacture of food and intermediate products, their storage, handling, and last but not least during consumption.

Initially, mainly empirical approaches were used for determining the mechanical properties of food. This was especially the case with respect to getting instrumentally an impression of humans' perception of these properties during consumption and meal preparation. Less than a century ago, scientists generally considered sensory perception of the mechanical properties of food during handling and consumption so subjective and unreliable that it could not be studied seriously in a profound way.

On the other hand, process engineers started to realize quite soon the great value of determining the rheological properties of liquid and liquid-like food products for improving existing processes of food production and developing new ones. For this purpose, they could make use of series of measuring methods and sets of equations describing the flow of liquids under various conditions that were developed outside food science and technology. However, many liquid and liquid-like foods are structurally very complex. The description of the flow behavior of these products turned out to be more complex than, for instance, that of simple polymer solutions. Moreover, the relation between the rheological properties of foods and their sensory properties turned out to be very complicated. Pioneering work on the rheology of food products was carried out by the famous British scientist, Scott Blair, and his coworkers.

It was the valuable contribution of Alina Szczesniak and coworkers and many other scientists that highlighted the essential role that the mechanical (both rheological and fracture) properties of food (food texture) play with respect to the appreciation of solid and solid-like food products. Moreover, after 1960 more and more methods became available for determining these properties in such a way that the results obtained were relevant for estimating the texture perception of food products. In addition, there was a growing awareness of the importance of many more aspects of the fracture behavior of food products for texture perception of solid and solid-like foods than originally thought.

Initially, the fracture behavior of brittle (hard solid) food products was characterized relatively simply by determining the stress at which the product fractured. However, for soft solid products such as margarines, butter, gels, various cheeses, jellies, dough, meat and meat products, fracture behavior was observed to be much more complex than that of materials usually studied in the field of fracture mechanics. Fracture behavior was observed to depend on certain factors such as speed and the way of deformation. It turned out that for a profound characterization of the mechanical properties of soft solid products, concepts developed in the fields of rheology and fracture mechanics are required.

There are several good books available that discuss the rheology of foods. Mostly they are limited to liquid and liquid-like food products. Furthermore, there are several good books that discuss food texture and how to measure it using instruments and/or human panels. In some of these books, some aspects of fracture mechanics are also discussed. In addition, there are various books on fracture mechanics of engineering materials and/or polymer systems. However, as discussed above, for a profound analysis of the mechanical properties of many food products, one needs a basic understanding of rheology and fracture mechanics as well as the ability to combine concepts from both science fields. With this book, I have attempted to provide basic knowledge of both rheology and fracture mechanics to enable the readers to gain a profound understanding of the mechanical behavior of various classes of food products. Moreover, a discussion is given on the relation between mechanical properties and the structure of food products at various length scales. In the last chapter, a short discussion is given on the relation between rheological and fracture characteristics of foods and texture perception.

The book can be used as a course book for food science majors and for students with a minor in food engineering. I realize that for both purposes, a selection has to be made of the topics discussed to let the content fit better with the background of the students following the course. The book is primarily written as a textbook; a second aim is that it can be used as a reference book by people working in the field of food research and development. I have tried to keep the mathematics at a relatively low level (no tensor calculus). In places where a derivation is given for a relatively complicated equation, I have tried to do so in such a way that one can understand the most important steps in this derivation without being able to follow the mathematics in detail.

For several chapters, I used parts of the chapter on rheology that I wrote together with Hans Lyklema for volume IV of his book, *Fundamentals of Interface and Colloid Science.* In addition, extensive use was made of the numerous lectures that I have given as part of the courses on rheology and fracture behavior of foods for undergraduate, graduate, and postdoctoral students at Wageningen University and abroad and for industries.

As mentioned above, I have made no attempt to cover all aspects of rheology and fracture mechanics of foods. Had I done so, the size of this book would have doubled at least. Self-evidently, the selection of the topics is colored by my experience and opinions and the extent to which topics have been discussed in the literature. This is especially the case for the choice of examples illustrating the relation between food structure and their rheological and fracture behavior.

Acknowledgments

Several colleagues have contributed to this book in a direct or indirect way. First of all I am indebted to the late Prof. Pieter Walstra for the many discussions we had on rheology and fracture mechanics of foods. I also want to especially thank Prof. Tiny van Boekel, Wageningen University, who has read and commented on all chapters. I am likewise grateful to Prof. Allen Foegeding, Raleigh, North Carolina, USA, who scrutinized several chapters and provided good suggestions that improved the manuscript. Without the contributions of my PhD and postdoctoral students and coworkers to the development of various parts of the knowledge discussed in this book, I would never have been able to write it.

I am greatly indebted to Marlies Meijer for all the computer work she did, which was necessary to enhance the figures from the basic ones I prepared. And last, but not least, I am very grateful to my wife, Willemien, for her understanding, love, and support.

Author

Ton van Vliet graduated with a degree in food technology from Wageningen University in 1973. He earned his PhD in physical chemistry and colloid sciences from the same university in 1977. After a postdoctoral fellowship at the Weizmann Institute of Science in Israel, he became an assistant professor at Wageningen University in the Food Physics Group, and in 1984 he became an associate professor. From 1998 until his retirement in 2010, he worked as a project leader at the Top Institute Food and Nutrition (TIFN).

His main research interest is rheology and fracture mechanics of food systems, in terms of their composition, structure, and physical stability. In his research, a range of rheological techniques were used in addition to techniques such as microscopy, permeametry, x-ray diffraction, and pulse nuclear magnetic resonance (NMR). He tries to bridge the gap between fundamental and predominantly curiosity-driven research and research resulting from problems in the food industry.

At TIFN, Dr van Vliet was the project leader of large projects on the stability of emulsions and foams, crispy/crunchy behavior of cellular solid foods, and effects of water migration in solids.

During his career, he has worked with scientists from various disciplines, and has published about 300 scientific papers on many topics in the field of food physics in a wide variety of journals. He taught at the university holding lectures for undergraduate and graduate students, and was responsible for establishing experimental courses, and thesis supervision for many MSc and PhD students. He also conducted various post-doctoral courses at the university as well as in industry on a contractual basis both in the Netherlands and abroad.

Dr. van Vliet served as the editor of the *Journal of Texture Studies* for 14 years. In 2008, he received the Carl Wilhelm Brabender award from the American Association of Cereal Chemists (AACC) for his outstanding contribution to the field of rheology as applied to the milling and baking industry (this award is presented every 3 years), is listed by Thomson ISI as a highly cited researcher, and is an elected member of the International Academy of Food Science and Technology (IAFoST).

I

Introduction

1

Rheology and Fracture Mechanics in Food Science and Technology

The mechanical properties of food constitute essential characteristics with respect to their behavior during consumption, meal preparation, and production. Often, they are so characteristic for a food that consumers will not buy the product if these characteristics do not meet certain minimum requirements or even decline to eat the food when it is offered to them. Moreover, during the (industrial) production of food products, storage, and use of these products for the preparation of a meal, these mechanical properties play an essential role.

The mechanical properties of different food products vary drastically. Liquid food products such as water, milk, and beer flow rapidly under a low force, whereas a large force is needed for obtaining fast flow of honey or sugar syrup. So-called semisolid food products such as tomato ketchup, mayonnaise, and many desserts only flow if the applied force is higher than a certain characteristic value. Below this force, they do not flow at all or only do so very slowly. With increasing force they yield, and the mechanical behavior changes from solid-like to liquid-like. Solid products such as candies, bread, chocolate bars, and many types of cheese do not flow at all, but fracture when they are subjected to a large force. These examples illustrate that food scientists and technologists should be familiar with both the flow and yielding and with the deformation and fracture behaviors of foods.

The flow, yielding, and deformation behaviors of solid materials subjected to a low stress are studied in *rheology*, whereas the fracture behavior of solid materials is studied in *fracture mechanics*. The latter field is primarily developed in relation to the construction of buildings, apparatus, means of transport, etc.

As will be discussed in later chapters, for a thorough understanding of the large deformation and fracture behavior of many food products (e.g., cheese, puddings, pies), one has to combine concepts developed in rheology and fracture mechanics.

Rheology is defined as *the science of deformation and flow*. The first use of the word "rheology" is credited to E.C. Bingham (about 1928). The rheological properties of a material are noted when a force is exerted on it, and as a result of which it deforms or flows. As indicated above, the extent to which a material deforms under a certain force depends strongly on its properties. Water flows already even under a low force, whereas a hard candy only deforms to a negligible extent under a low or moderate force. On consumption, it has to dissolve in the saliva present in the mouth. In this book, we will not make a distinction between hydrodynamics and rheology. Hydrodynamics deals with the flow behavior of simple liquids such as water, which can be characterized by a constant ratio between applied stress and flow rate (constant viscosity). In rheology, the flow and deformation behavior of more complicated materials are studied. The flow behavior of simple liquids will be dealt with as part of the discussion on rheological properties.

Fracture mechanics is *the field of mechanics concerned with the study of the formation of cracks in a material*. As indicated by its name, fracture mechanics deals with fracture phenomena and events. It has been developed with the goal of improving the performance of engineering materials. Its importance for food products is often underestimated. Not everybody realizes that, for instance, sensory characteristics such as crispiness and brittleness are primarily determined by the fracture behavior of the product and can only be understood by using an approach based on fracture mechanics. Also, during (industrial) processing and meal preparation (e.g., during cutting, grinding, and crushing), the fracture properties of the product are essential.

In daily life, everybody is aware of the fact that some materials only deform if a force is applied (*elastic* behavior), whereas others start to flow (*viscous* behavior). For still other materials, the response to a force is less unequivocal. For example, most cheeses and bread- and cake dough will only deform temporarily under a force of short duration, but will exhibit flow behavior over a longer course of time, implying permanent deformation. These materials react *viscoelastically* to an applied force; so do most paints. Their reaction to an applied force is partly elastic and partly viscous. Typically, time is an important parameter determining their reaction to a force. To stress the importance of time in rheology, one could include this notion in a descriptive definition of rheology: "Rheology concerns the study of the relations between forces exerted on a material and the ensuing deformation as a function of time." The function of time with respect to the rheological behavior of materials will be discussed extensively in **Chapter 4**. Bingham, who introduced the word rheology was also the person who proposed as motto of this science field "panta rhei" (παυτα ρει)—from the work of Heraclitus, a Greek philosopher from about 500 B.C.—meaning "everything flows" (Reiner 1964). This motto is more realistic than most people think, but before one can appreciate this, one should understand the role of time in determining the rheological behavior of a material.

As indicated above, rheology and fracture mechanics play an important role in food science and technology and in the industrial and household application of food science and technology for various reasons:

1. Knowledge of rheological and fracture properties is essential for designing and constructing machines and for manufacturing and handling of food products, raw materials, and intermediate products. A machine for cutting sugar cane should be constructed in a different way than a coffee grinder.
2. Process control. During manufacturing of foods, their rheological and/or fracture properties often change gradually (e.g., thickening of milk during preparation of evaporated milk) or more abruptly (e.g., whipping of cream). This change in rheological or fracture properties can be used to follow the manufacturing process. Because fracture experiments are inherently destructive, the determination of rheological properties is often preferred, although not always. An example from the kitchen is the determination whether boiled potatoes are well cooked by sticking a fork in them.
3. Quality control of the end product. This concerns the properties determining whether products are suitable for further processing, practiced by catering or at home, as well as sensory characteristics. Examples of important mechanical characteristics regarding further use include spreadability of margarine and peanut butter, cutting properties of bread and tomatoes, flow behavior of fruit juices, and fracture characteristics of chocolate bars. Sensory characteristics of foods are often related to their rheological (e.g., thickness of liquids and semisolid foods) or fracture (e.g., hardness of solid foods) properties or a combination of both (e.g., spreadability of soft solids and semisolids).
4. Rheology and fracture mechanics help to obtain information on the structural properties of food systems. For a profound characterization of structural properties of foods, usually a combination of different methods is required. Understanding the relationship between structure and rheology helps in designing products with specific rheological properties.

For the first three reasons given above, it is sufficient to determine some phenomenological rheological properties, but for the last one the relationship between the structural properties of food products and rheological and fracture properties must be known. To a certain extent, the measurement of rheological properties is preeminently suited for this, because—in contrast to techniques such as microscopy or turbidimetry—it provides information on the interaction forces between the structural elements in the products. However, to that end, theoretical relationships between structure and mechanical properties based on models are required, and the combination of results from rheological and fracture experiments with those of other measurements is often essential for obtaining complete understanding.

The difference between the two approaches in using rheology and fracture mechanics is related to the two distinct ways in which one can interpret rheological and fracture properties, viz. phenomenological or in relation to the molecular and/or mesoscopic (colloidal) structure of the material. In the former case, no structure-based interpretation is offered, whereas in the latter case such an interpretation is the very goal of the study.

1.1 Structure of this book

In this book, the deformation, flow, and fracture behaviors of food systems will be discussed. Phenomenological aspects, the determination of rheological and fracture parameters, as well as the relation between these parameters and the structure of food products at mesoscopic and macroscopic scale will be discussed. In addition, the relation between mechanical characteristics and sensory characteristics will be briefly discussed at a general level. The rheological, flow, and fracture behaviors of interfacial layers (e.g., adsorbed protein layers) will not be discussed.

The book is aimed at providing understanding; hence, it gives primarily principles and some basic theory. The chapters on the relation between mechanical characteristics and the structure of food products contain various examples primarily intended to show how general principles can be applied in the context of a specific food material. Often, people are afraid of rheology, because of the tensor calculus used in a profound analysis of the deformation and flow behavior of a material due to an applied force. In this book, we will refrain from using tensor calculus, although very briefly some background information will be given on why it is often used. During the derivation of equations, it is assumed that readers have a basic understanding of calculus.

This book consists of 18 chapters, which are divided into five main **Parts (I–V)**. **Part I (Chapter 1)** contains the introduction, **Part II (Chapters 2–5)** deals with the phenomenological aspects, **Part III (Chapters 6–8)** discusses the experimental evaluation of rheological and fracture characteristics, **Part IV (Chapters 9–17)** deals with the relation between material structure and mechanical properties, and **Part V (Chapter 18)** concentrates on the relation between structure, mechanical properties, and sensory perception.

In **Part II**, the phenomenological aspects of rheology and fracture mechanics are discussed together with the amount of basic theory required to adequately perform rheological experiments and to understand the meaning of the determined parameters. In **Chapters 2 and 3**, we discuss the basic notions needed for an in-depth description of the deformation and flow behaviors of materials. **Chapter 4** is devoted to descriptive rheology and modeling of rheological behavior by simple models. The basics of fracture mechanics are discussed in **Chapter 5**. The function of time, with respect to the rheological and fracture behavior of materials, is discussed in parts of **Chapters 4 and 5**, respectively. A distinction will be made between the effect of the time a material is deformed at a certain speed and the effect of the time scale of the deformation, which is related to the deformation speed. The time scale of the deformation affects whether a viscoelastic material will react, relatively speaking, in a more elastic or a more viscous manner on a force.

A thorough understanding of the meaning of determined rheological and fracture characteristics requires a basic understanding of the experimental evaluation methods of these characteristics. The basic principles of the most common experimental procedures are discussed in **Part III**. **Chapter 6** provides a general introduction on how to come to a suitable instrumental method to determine the relevant rheological and fracture properties of foods and identify the major pitfalls when performing such measurements. An overview of various measuring experimental methodologies is provided in **Chapter 7**. The general principles of measuring equipment that can be used to determine rheological and fracture parameters are discussed in **Chapter 8**. In the last part of that chapter, empirical tests are discussed.

The relationship between the molecular and colloidal structure of food products and rheological and fracture properties is discussed in **Part IV**. In addition to basic knowledge of rheology and fracture mechanics, this requires a profound knowledge of the material considered, especially with respect to its structure at molecular and mesoscopic (colloidal) length scale. No extensive enumeration of rheological and fracture characteristics of food products will be given. In this part, the emphasis will be on a discussion of the relationship between the general structure characteristics of model systems for foods and their rheological and fracture behavior. Subsequently, the following topics will be discussed: dilute and concentrated dispersions (**Chapter 10**), dilute and concentrated macromolecular solutions (**Chapter 11**), general aspects of solid and solid-like materials (**Chapter 12**), gels (**Chapter 13**), composite food products (**Chapter 14**), gel-like closed packed materials (**Chapter 15**), cellular materials (**Chapter 16**), and hard solids (**Chapter 17**). In some of these chapters, examples are provided to illustrate the relation between food structure and their rheological and fracture behavior.

In **Part V**, we discuss the texture perception of liquids and semisolid products, soft solids, and hard solids (**Chapter 18**).

Throughout the book, SI units are used, unless stated otherwise. An overview of the used SI units is given in Appendix B, together with a short table of conversion factors and most common prefixes to indicate a multiplication factor. In addition, some mathematical symbols used are defined. Appendix A gives an overview of the frequently used symbols for physical quantities.

This book is primarily written as a textbook, and no full account of literature sources is provided. At the end of each chapter, several references are given. These primarily concern articles and books suited for further reading and publications dealing more extensively with examples discussed in the main text. Numerous cross-references are given in the book to illustrate the relation between the various chapters.

Reference

Reiner, M. 1964. The Deborah number. *Phys. Today* 17: 62.

II

Phenomenology

2

Basic Notions

As mentioned in the preceding chapter, in rheology one deals with the relationship between forces and deformations as a function of time. In this chapter and the next one, the discussion will be limited to the relation between forces and deformations. The basic notions stress and strain will be discussed. Different simple and well-defined ways of deforming a material and definitions of accompanying rheological quantities are discussed in **Chapter 3**. The role of time in rheology will be discussed extensively in Sections 4.1 and 4.2. In **Chapter 5**, yielding and fracture behavior of materials will be discussed.

An example of a simple rheological experiment is shown in **Figure 2.1**. It is immediately apparent that the ensuing compression of the material due to the placement of a weight on it (or the application of a *force* f (N), being weight (kg) times acceleration due to gravity g (m s^{-2})) will depend on its size. To correct for variations in size and shape of the test piece studied, forces have to be normalized to forces per unit area, resulting in what is known as *stress* (f/A (in N m^{-2}), where A is the area on which the force acts). Similarly, the accompanying deformation has to be taken relative to the size of the specimen. For the experiment shown in **Figure 2.1**, $\Delta H/H_0$, where H_0 is the initial height of the test piece and ΔH is the change in height due to the compressive force. For the case in which the relative deformation of every volume element of the test piece in **Figure 2.1** is the same, it is equal to what is known as *strain* (dimensionless).

In the next chapter, we will discus several relatively simple relations between stress and strain that hold for well-defined deformations. It should be realized that for an arbitrary deformation, relations between stress and strain may become very complicated for reasons discussed below. To obtain well-defined rheological parameters, the tensor character of stress and strain has to be considered explicitly.

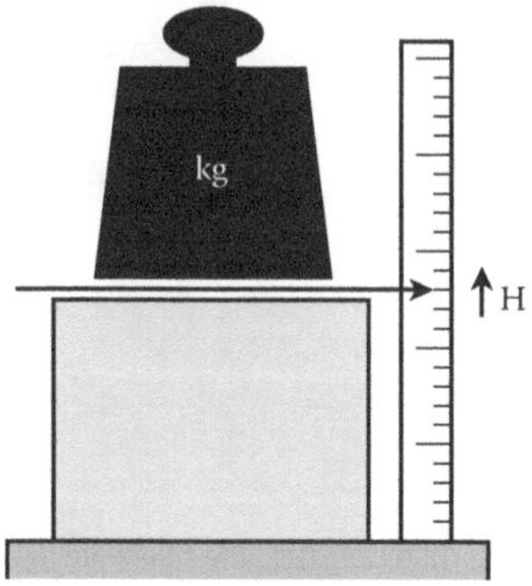

Figure 2.1 *Essence of determination of mechanical/rheological properties of a material.*

Forces are characterized by both a specified magnitude and a direction in space; they are so-called vectors. For an unambiguous characterization, the direction has to be fixed relative to a coordinate system and characterized by decomposing the force into three components, one normal to the surface on which the force acts and two in the plane of the surface perpendicular to each other. The first component gives rise to the so-called *normal stresses* and the other two to the so-called *shear stresses*. In addition, one should also realize that the plane on which the force acts has a certain magnitude (area) and spatial direction. These properties can be characterized by a vector perpendicular to the plane, its length as the measure of the area and its direction as the measure of the spatial direction of the plane.

So, a stress is given by the relationship between two vectors, force per unit area. Such a relationship is given by a second-order tensor, which is called the *stress tensor*.

To describe all stresses acting on a material, it is useful to consider a cube-shaped infinitesimal volume element of the material in a linear orthogonal (Cartesian) coordinate system (**Figure 2.2**). On each face, a force may act that can be decomposed into x, y, and z components. Each face of the volume element is characterized by the coordinate axis perpendicular to that plane. Then, nine stress components can be distinguished, each characterized by

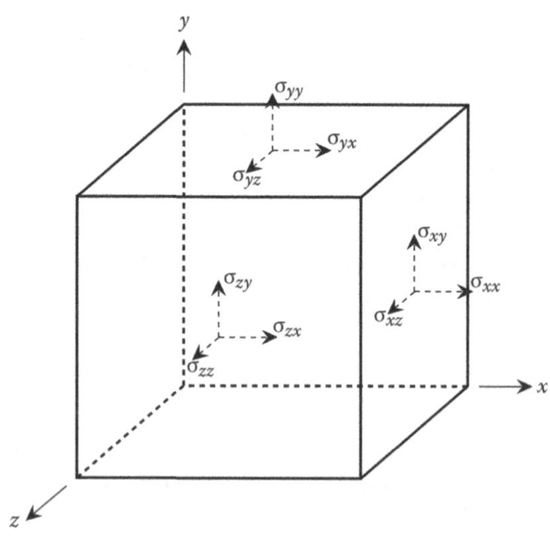

Figure 2.2 *Notation of stress components. For clarity, stress components acting on back faces of the block are not shown.*

two indices, *i* and *j*, where *i* denotes the direction of the vector normal to the plane on which the force acts and *j* denotes the direction of the force component. If the volume element is small enough to allow to neglect gravity and inertial forces, nine components are needed to completely describe the stress(-es) acting on the volume element. These nine components together constitute the stress tensor $\tilde{\sigma}$. In Cartesian coordinates,

$$\tilde{\sigma} \equiv \begin{bmatrix} \sigma_{xx} & \sigma_{xy} & \sigma_{xz} \\ \sigma_{yx} & \sigma_{yy} & \sigma_{yz} \\ \sigma_{zx} & \sigma_{zy} & \sigma_{zz} \end{bmatrix} \tag{2.1}$$

These are shown in **Figure 2.2**. In Equation 2.1, the three diagonal components (σ_{xx}, σ_{yy}, σ_{zz}) are the normal stresses acting perpendicular to the six faces of the cube. A pressure or isotropic stress applied on a material is characterized by three equal, normal stress components. Application of such a stress to an isotropic material will only lead to volume compression or expansion. For an essentially incompressible material, the absolute values of the normal stresses are of no importance for the deformation of the material. Deformation will be the case if one or two of the normal stresses are larger than the other one(s).

The other six components of the stress tensor, known as *shear* or *tangential stresses*, act parallel to—and in the plane of—the six faces of the cube. Based on the law of conservation of momentum, the components at the two sides of the diagonal terms are equal to each other, so $\sigma_{xy} = \sigma_{yx}$, etc. Moreover, the isotropic pressure can always be chosen in such a way that one of the normal stress components becomes zero, leaving two *normal stress differences*, where the first normal stress difference is

$$\sigma_{xx} - \sigma_{yy} = N_1 \tag{2.2a}$$

and the second normal stress difference is

$$\sigma_{yy} - \sigma_{zz} = N_2 \tag{2.2b}$$

This leads to five independent components in the stress tensor: two normal components and three shear components. By deforming a test piece in an arbitrary manner, all five components of the stress tensor may be affected in a manner that the relation between them cannot be described in a manageable way. If this happens to be the case, the only solution is to consider the full stress tensor and even then it may be questionable whether all components can be calculated. It is not necessary to consider the full stress tensor when well-defined deformations, as described in the next chapter, are applied and the interpretation is limited to the determination of relatively simple rheological or fracture characteristics.

As stated above, if a stress is applied to a material, it will deform. The extent of it will not only depend on the properties of the material and the magnitude of the stress, but also on the size and shape of the material. To obtain a quantitative measure of the deformation, this quantity is always taken relative to a characteristic length of the material—in **Figure 2.1**, the original height of the test piece. Both the deformation and this characteristic length are characterized by a size and a direction, so they are vectors. The relative deformation is given by the quotient of their absolute values. For the case where the material is inhomogeneous, the relative deformation will vary with the position in the material, and one can only define the relative deformation unambiguously for an infinitesimal length. The relationship between the spatial vector, which defines the position of a point and the deformation vector, is given by the deformation gradient or strain tensor.

There are various ways to define strain. Here, we will follow the classical definition of infinitesimal strain, which is, strictly speaking, only valid for small-amplitude deformations. For more general deformations, we refer to textbooks (e.g., Lodge 1979). Consider two points, A and B, in a volume element of material at a distance l apart and situated in a Cartesian coordinate system (**Figure 2.3**). Consider the case where, as a result of an applied force, the volume element is deformed, resulting in a displacement of point B with respect to point A (B moves to B'). This displacement involves both a change in distance and a rotation of the points with respect to each other. As will be discussed further in **Chapter 3**, rotation implies movement of the material in the coordinate systems without a change in mutual distance, and does not depend on the rheological properties of the material. However, the relative change in distance between A and B does. Considering only the elongational component, the relative deformation of the material between points A and B is given by $(AB' - AB)/AB = \Delta l/l$. For the present case, which only involves a change in length, the *strain* ε is defined as the infinitesimal relative elongation of l, or in general, as the relative deformation gradient, so

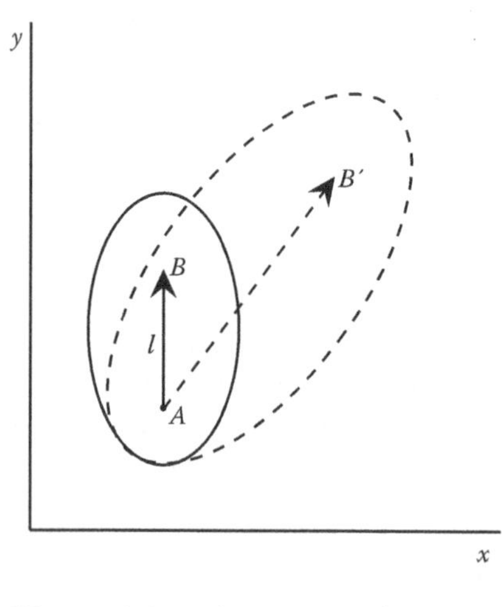

Figure 2.3 *Deformation of a material resulting from the displacement of point B to B′. l, distance between the points A and B. For further explanation, see text.*

$$\varepsilon = \frac{\partial \Delta l}{\partial l} \tag{2.3}$$

In an analogy to the stress tensor defined in Equation 2.1, the *strain tensor* describing the deformation of a material resulting from an applied stress consists of nine components. The same applies for the flow of a material described by a tensor that gives the change in strain per unit time (*rate of strain tensor*). In formal rheology, relationships between the stress, the strain, and the rate of strain tensor are formulated. This requires tensor calculus that we consider beyond the scope of this book. Therefore, in the following, we shall restrict ourselves to the main types of stress–strain relations, whereby the tensors can be read "term by term." As argued above, this requires a well-defined, relatively simple deformation of the test material. For a more extensive discussion, we refer to textbooks (e.g., Darby 1976; van Vliet and Lyklema 2005).

Because, in most cases, it is clear which tensor components of the stress tensor are meant in the text or in an equation, here they will not be written in full, but will be denoted by σ for a shear stress and by σ_e for a tensile or uniaxial compression stress except at those places where it may lead to confusion. The same applies for components of the strain and strain rate tensor.

References

Darby, R. 1976. *Viscoelastic Fluids: An Introduction to Their Properties and Behavior*. New York: Marcel Dekker.

Lodge, A.C. 1979. *Body Tensor Fields in Continuum Mechanics with Applications to Polymer Rheology*. London: Academic Press.

van Vliet, T., and J. Lyklema. 2005. Rheology. In *Fundamentals in Interface and Colloid Science,* Vol. IV, *Particulate Colloids*, ed. J. Lyklema, 6.1–6.88. New York: Academic Press.

3

Rheological Quantities, Types of Deformation

For the definition of rheological parameters, it is useful to distinguish between two extremes of rheological behavior based on whether a material deforms under an applied stress, or flows. To a first approximation, two extreme, ideal types of rheological behavior can be distinguished phenomenologically, viz., ideally fluid or *viscous* behavior and ideally solid or *elastic* behavior. Viscous materials flow at a certain rate when a stress is applied, and after removal of the stress they retain the shape they had at the moment the stress was taken away. Elastic materials deform instantaneously to a certain extent when a stress is applied and regain their original shape after the stress has been removed.

3.1 Well-defined types of deformation

In the following, three main types of deforming a material and the accompanying rheological parameters characterizing the relationships between stress and strain will be discussed. A common characteristic of these types of deformation is that, during experiments, well-defined conditions hold throughout the test piece allowing the determination of real material properties, that is, results that do not depend on the size or shape of the test piece. The discussion in Sections 3.1 and 3.2 will be limited to small deformations characterized by a constant ratio between stress and strain for elastic materials and a constant ratio between stress and strain rate (change in strain in time) for fluids. If appropriate, a distinction will be made between viscous and elastic materials. Situations in which the relation between stress and strain (rate) is not constant will be discussed in **Chapter 4**. Several definitions of stress and strain for large deformation of solid materials will be presented in Section 3.4.

3.1.1 All-sided, or isotropic, compression

For isotropic materials, this type of deformation will only lead to a change in the volume of the material. The shape of a homogeneous material will not change. The relative change in volume or volume strain, ε_v, is given by

$$\varepsilon_v = \frac{\Delta V}{V} \tag{3.1}$$

The ratio between the stress (exerted pressure p on the material) and the relative change in volume is called the *compression modulus K* (unit N m^{-2}) whereby

$$p = \sigma_{xx} = \sigma_{yy} = \sigma_{zz} = -K\varepsilon_v \tag{3.2}$$

For small ε_v, K is a constant that characterizes the material. For most solids and liquids, it has a value of about 10^9 Pa (N m^{-2}). For gases, K is about 10^5 Pa. This large difference in K between condensed matters and gas can be explained based on the molecular background of the material involved. For a gas, the resistance to a volume decrease stems from a lowering of its entropy (freedom of the molecules to divide themselves over the space), whereas for a condensed material it is a result of an increase in enthalpy owing to short range repulsion between the molecules. This difference allows users to determine the gas content of a given heterogeneous material by measuring its compressibility.

3.1.2 Uniaxial compression or extension

By uniaxial compression or extension, only one side of the material is subjected to a normal compressive or tensile stress (**Figure 3.1**). No external stress is applied to the other two sides (except atmospheric pressure), so toward these sides the material is allowed to adjust itself to the internal stresses. If

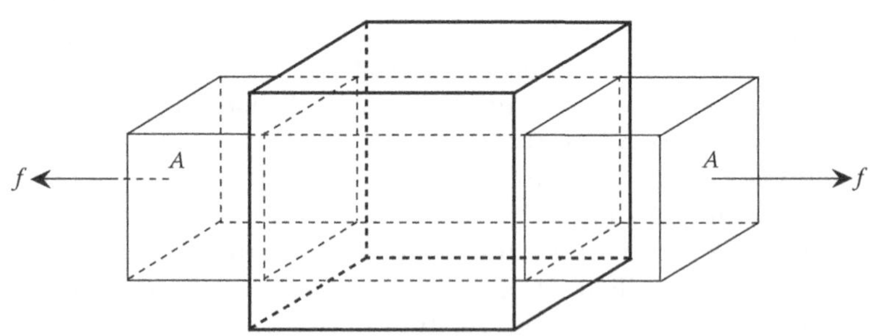

Figure 3.1 *Uniaxial extension as result of a tensile force, f. A area on which the tensile force acts.*

the volume of the material stays constant during the uniaxial deformation, the sum of all strains in the three perpendicular directions should be zero

$$\varepsilon_x + \varepsilon_y + \varepsilon_z = 0, \qquad (3.3)$$

where ε_x is the strain in the direction of the stress, and ε_y and ε_z denote those in the direction perpendicular to it. For a homogeneous deformation of the material, the two strains in the tangential direction to the applied stress will be equal, so

$$\varepsilon_x = -2\varepsilon_y = -2\varepsilon_z \qquad (3.4)$$

The material is uniaxially compressed and biaxially extended, or the other way around. The strain in the direction of the force is two times the strain in the directions perpendicular to the direction of the applied force. As a first approximation, this holds for condensed materials, which have a high compression modulus. However, for systems containing gas, a volume change may occur.

For uniaxial compression or extension of a homogeneous material, the *tensile or compressive strain* ε is given by $\mathrm{d}l/l_0$, where l_0 is the initial length of the test piece and $\mathrm{d}l$ is the change in length. The rheological quantity characterizing ideally elastic material behavior in uniaxial elongation or compression is the *Young's modulus E*, which is defined as the ratio of the exerted normal (tensile or compressive) stress $\sigma_e(f/A)$ over the tensile or compressive strain ε.

$$\sigma_e = E\varepsilon \qquad (3.5)$$

For a fluid, the material will start to flow at a certain flow rate $\mathrm{d}l/\mathrm{d}t$ in the direction of the stress applied or in terms of an elongational strain $\mathrm{d}\varepsilon/\mathrm{d}t$. The relationship between the normal stress and the so-called *elongational strain rate* $\dot{\varepsilon}(\equiv \mathrm{d}\varepsilon/\mathrm{d}t)$ reads as

$$\sigma_e = \eta_e \frac{\mathrm{d}\varepsilon}{\mathrm{d}t} = \eta_e\dot{\varepsilon} \qquad (3.6)$$

where η_e is the *elongational viscosity*. For a so-called Newtonian fluid, η_e does not depend on the elongational strain rate.*

It should be noted that, in practice, various terms are used to denote uniaxial deformation of a test piece. For solids, one speaks mostly about a tensile or (uniaxial) compression experiment and in accordance with that of a tensile or compressive stress and strain and for fluids about an extension or elongational flow experiment. For the resulting viscosity, both the terms elongational and extensional viscosity are used.

3.1.3 Shear

During a shearing deformation, usually abbreviated as shear, parts of the material are shifted parallel to other parts. It occurs when a shear stress is applied;

* Be aware that in rheology, it is common practice to denote the derivative of a physical quantity against time with a dot above the symbol for the physical quantity.

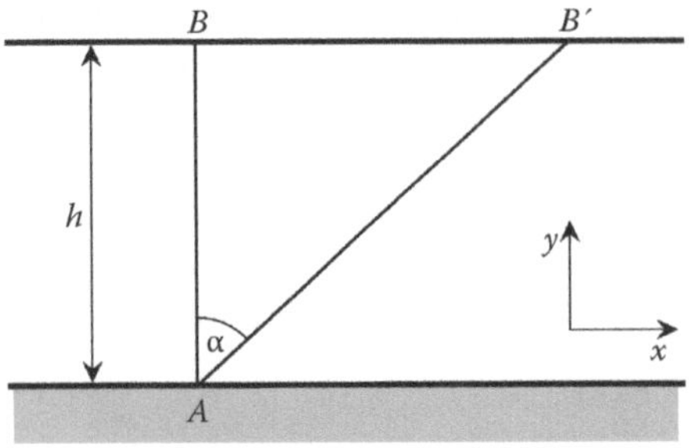

Figure 3.2 Shear between two parallel planes.

the force works in the plane of the surface on which it acts (**Figure 3.2**). For this type of deformation holds that only the shape of the material changes, whereas the volume of the material stays the same. As a result of an applied shear force, material will be displaced and deformed, resulting in a change in distance between points A and B (B moves to B'). If inertial forces can be neglected, and for a material that is homogeneous down to a size ≪ volume element considered, each volume element will be subjected to the same relative deformation $(B - B')/AB$. A common measure of the *shear strain* γ is

$$\gamma = \tan\alpha = \Delta x/\Delta y, \tag{3.7}$$

where α is the angle of shear (**Figure 3.2**).

If a shear stress is applied to an elastic material, it will deform to some extent, leading to a shear strain γ (often abbreviated to shear). For ideally elastic materials, the rheological quantity characterizing the material is the *shear modulus* G, which is defined as the ratio of the shear stress σ (force f divided by the area A on which it acts) over the shear strain γ.

$$\sigma = G\gamma \tag{3.8}$$

Shear flow is characterized by a parallel displacement of parts of the fluid with respect to other parts; layers of liquid appear to slide over each other. Shear flow will occur when a homogeneous fluid is situated between two parallel walls, and both walls are moved parallel with respect to each other. Consider a fluid confined between two parallel plates at a distance h of each other, in which the lower one is fixed (**Figure 3.3**). On application of a shear stress, σ (or more precisely, σ_{yx}), on the upper plate it will move with a velocity $v_x = \mathrm{d}x/\mathrm{d}t$ in the x direction.

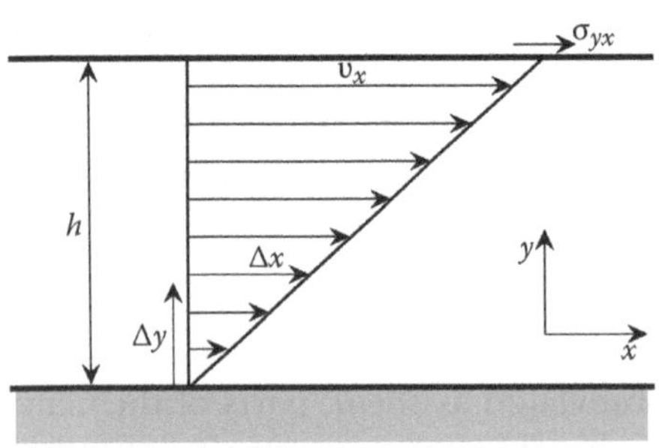

Figure 3.3 Laminar linear flow between two parallel plates. Symbols are explained in the text.

Under steady-state conditions ($\partial v_x/\partial t = 0$ throughout the fluid), σ is uniform throughout the fluid, and as the resistance to flow is also assumed to be uniform (homogeneous material), v_x will change linearly with y. The arrows in **Figure 3.3** give the velocity as a function of distance y from the lower plate. The derivative $\mathrm{d}v_x/\mathrm{d}y$ is constant. The deformation rate of the fluid is characterized by a *rate of shear* or *shear rate*, $\dot{\gamma} = \mathrm{d}\gamma/\mathrm{d}t$ (just as for $\dot{\varepsilon}$, the dot above γ

Table 3.1 Shear Rates Typical of Some Familiar Materials and Processes

Situation	Typical Range of Shear Rates (s^{-1})	Application
Sedimentation of fine particles in a liquid	10^{-6}–10^{-4}	Particles in several drinks, spices in salad dressing
Mixing and stirring	10^{1}–10^{3}	Manufacturing liquids dispersions
Dip coating	10^{1}–10^{2}	Confectionery
Extrusion	10^{0}–10^{3}	Polymers, snacks, pasta, cereals, pet foods
Sheeting	10^{1}–10^{2}	Dough sheeting
Pipe flow	10^{0}–10^{3}	Pumping liquids
Spraying	10^{3}–10^{4}	Spray drying
Chewing and swallowing	10^{1}–10^{2}	Foods
Rubbing	10^{3}–10^{5}	Application of creams and lotions to the skin
Pouring from a bottle	10^{1}–10^{2}	Foods, cosmetics
Draining under gravity	10^{-1}–10^{1}	Liquid layer brought on product

Sources: Adapted from various sources.

denotes the time derivative of the shear). Both shear rate and rate of shear are abbreviations for "shear strain rate." In view of Equation 3.7,

$$\dot{\gamma} = \mathrm{d}y/\mathrm{d}t = \mathrm{d}\Delta x/\mathrm{d}\Delta y)/\mathrm{d}t = \mathrm{d}v_x/\mathrm{d}y = v_x/h \tag{3.9}$$

To obtain some idea about the practical range of $\dot{\gamma}$ during several familiar processes, see **Table 3.1**.

For a homogeneous ideally viscous fluid material, behavior is characterized by

$$\sigma = \eta\dot{\gamma}, \tag{3.10}$$

where η is the (dynamic) *shear viscosity* (N m^{-2} s), mostly shortened to *viscosity*. For Newtonian liquids, η is a constant independent of $\dot{\gamma}$.

3.2 Relationship between rheological quantities

For a homogeneous material, relationships can be deduced between the rheological quantities K, G, and E. A complicating factor is that during elongation or uniaxial compression, one normal stress on the material becomes lower or higher than the other two. This may give rise to a volume change in the material. The numerical values in the relations between K, G, and E depend on this volume change. A measure of the change in volume upon uniaxial elongation or compression is the *Poisson ratio* (μ), which for elongation in the x-direction reads

$$\mu = \frac{\varepsilon_y}{\varepsilon_x} = \frac{\varepsilon_z}{\varepsilon_x} = \frac{1}{2}\left(1 - \frac{\partial V}{V\partial\varepsilon_x}\right) \tag{3.11}$$

In a uniaxial tensile experiment (**Figure 3.1**), a stress σ_e will result in a tensile strain ε. The lateral strain in the direction perpendicular to the direction in which σ_e acts will be equal to $(-\mu\varepsilon)$. Under conditions where the strains produced by the stresses are additive,* an isotropic tensile stress, σ_e, will give strains $(\varepsilon - 2\mu\varepsilon)$ in each coordinate direction. The fractional increase in volume, $\partial V/V$, will then be $[1 + \varepsilon(1 - 2\mu)]^3 - 1 \approx 3\varepsilon(1 - 2\mu)$. The bulk modulus K can then be written as

$$K = \frac{\sigma_{ii}}{\varepsilon_v} = \frac{\sigma_{ii}}{3\varepsilon(1-2\mu)} = \frac{E}{3(1-2\mu)} \tag{3.12}$$

For soft solids and liquid materials that change their shape already under the action of stresses much smaller than those needed to change their volumes significantly, $E \ll K$ and $\mu \approx 0.5$.

To derive the relationship between E and G, consider in a thought experiment the deformation of a cube of an isotropic elastic material by simultaneous application of a tensile stress, σ_{xx}, in the x direction and an equal compressive stress σ_{yy} in the y direction (**Figure 3.4**). The cube deforms into a rectangular block in such a way that there is no change in dimension in the z direction, and the longitudinal strains in the x and y directions are of equal magnitude, but of opposite sign. The square, $ABCD$, will be deformed into a rhombus, $A'B'C'D'$, without any rotation of AC and BD. The elongational strain ε_x in the x direction will be

$$\varepsilon_x = \frac{\sigma_{xx}}{E} + \mu\frac{\sigma_{yy}}{E} \tag{3.13}$$

The prism APD will be in equilibrium if a shear stress, σ, acts on AD toward A. The x component of the force will be proportional to $\sigma \cdot PD$ to the right, and $\sigma \cdot AD/\sqrt{2} = \sigma \cdot PD$ to the left. Such shear stresses act on all faces of the square, $ABCD$. The linear strains, $\varepsilon_x = DD'/OD = -AA'/OA$, correspond to a shear strain equal to $2\varepsilon_x$ (Whorlow 1992; see also the discussion on shear flow in Section 3.3), so we obtain $2\varepsilon_x = \sigma/G$. In combination with Equation 3.13, this gives

$$E = 2G(1 + \mu) \tag{3.14}$$

For an incompressible material, $\partial V/V = 0$, which requires $K = \infty$. Then one obtains $\mu = 0.5$ and $E = 3G$. This relation holds by good approximation as long as $K \gg E$ (or G).

* Additivity here means that if several stresses are applied consecutively, the resulting total deformation at time t equals the sums of the individual deformations, had they been applied independently (*Boltzmann superposition principle*).

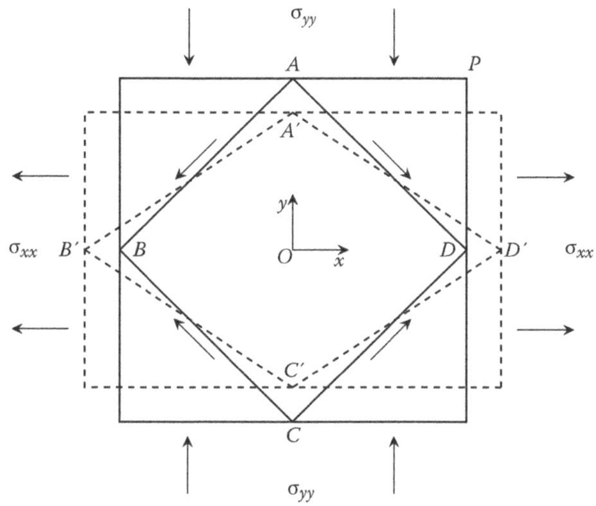

Figure 3.4 *Cross section of a cube before and after the application of equal tensile and compressive stresses, σ_e, in perpendicular directions (for further explanation, see text). (Reprinted with small adaptations from Whorlow, R.W.,* Rheological Techniques, *Ellis Horwood, Chichester, 1992. With permission.)*

The relationship $E = 3G$ may be extended to a Newtonian liquid, or fluid in general, by interpreting E and G as being the ratio of a stress to the corresponding strain at time t. In shear,

$$G = \frac{\sigma}{\dot{\gamma} t} = \frac{\eta}{t} \tag{3.15}$$

and in tension

$$E = \frac{\sigma}{\dot{\varepsilon} t} = \frac{\eta_e}{t} \tag{3.16}$$

It follows that

$$\eta_e = 3\eta \tag{3.17}$$

The ratio of the elongational viscosity over the shear viscosity is known as the *Trouton ratio.* For a Newtonian fluid, Equation 3.17 is not limited to small strains. On deformation, such a fluid does not "remember" its previous shape, and Equation 3.17 will apply for each successive small strain. For non-Newtonian fluids, such as polymer solutions, the Trouton ratio may become very large (up to 10^4) and will depend on the strain and the strain rate applied. For other flow types, the Trouton ratio may be different; for example, for Newtonian liquids in biaxial extension (see Section 3.3), it will be 6.

3.3 Types of flow

During manufacturing and consumption, foods are often subjected to flow. For liquid food products containing dispersed particles or macromolecules, not only the effect of the particles on the flow behavior of the system is of interest, but also the effect of the flow on the particles. Flow may cause, for instance, the breakup and the aggregation of particles, the extent depending on the flow conditions and the properties of the particles.

The most important types of flow are laminar and turbulent flow. *Laminar flow* is characterized by the presence of one velocity gradient dv/dx, and the streamlines, that is, the trajectories of small volume elements, exhibit a smooth and regular pattern. Different types of laminar flow can be distinguished. Some examples are shown in **Figure 3.5**. Depicted flows are two-dimensional, implying the absence of a flow component in the z direction perpendicular to the plane of the figure. *Turbulent flow* is characterized by unordered velocity fluctuations in the direction of the main flow component

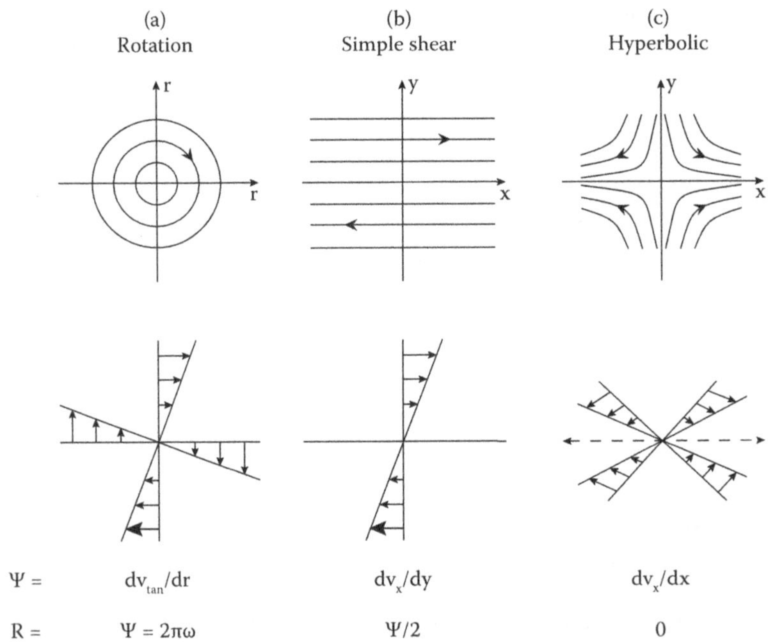

***Figure 3.5** Cross sections through three types of laminar flow. The upper row gives the streamlines, the second row the velocity profiles. ψ, velocity gradient (rad s^{-1}); v = linear flow velocity (m s^{-1}); R, rotation rate; ω, revolution rate (s^{-1}). For (simple) shear flow, ψ is equal to $\dot{\gamma}$ and in the center of hyperbolic flow, is equal to $\dot{\varepsilon}$. (Reprinted from Walstra, P.,* Physical Chemistry of Foods, *Marcel Dekker, New York, 2003. With permission.)*

as well as in both directions perpendicular to it. Flow is chaotic. Laminar flow exert viscous (or friction) forces on dispersed particles, whereas in turbulent flow inertia forces are dominant.

Figure 3.5a shows pure *rotational flow*, characterized by circular streamlines. A volume element in the center will only rotate and will not be displaced. Those outside the center will only rotate around the center, all at the same speed. These volume elements are displaced, but there will be no change in the distances in and between the volume elements. The volume elements are not deformed. No work of deformation is applied to the volume elements. Rotational flow does not lead to energy dissipation as a result of friction between fluid components because the whole system rotates as a block. The velocity gradient ψ equals the rotation rate R since

$$\psi = \frac{\mathrm{d}v_{\tan}}{\mathrm{d}r} = \frac{\mathrm{d}2\pi\omega r}{\mathrm{d}r} = 2\pi\omega \tag{3.18}$$

Figure 3.5c shows *hyperbolic flow*. A characteristic of this type of flow is that in the center, flow is elongational (also called extensional flow). A volume element in the center will be elongated, as depicted in **Figure 3.1**, at a rate increasing exponentially with the distance from the center. It will not rotate. Similarly, two fluid particles on the x-axis will move apart in time also at an exponentially increasing rate. The velocity gradient $\mathrm{d}v_x/\mathrm{d}x$ is in the direction of the flow and is equal to $\dot{\varepsilon}$. Finally, it should be noted that a volume element in the center of the two-dimensional hyperbolic flow situation will not only be extended in the x direction, but will also be compressed in the y direction at a strain rate $-\dot{\varepsilon}$.

Figure 3.5b shows what is called *simple shear flow* although, as will be shown below, it is not such a simple flow type. It is characterized by straight streamlines. Simple shear as well as simple shear flow contains both a rotational and an elongation component as is illustrated in **Figure 3.6**. For simple shear flow, the contribution of the rotational component and the elongational component is equal. This is illustrated in **Figure 3.6** for infinitesimal deformations.

A volume element subjected to shear will be rotated as well as elongated. Elongation takes place in the direction of the diagonal *OB* and compression in the direction *AC*. The rotation and elongational component together result in a situation where a point on the outside of a volume element subjected to shear flow is successively submitted to elongational and compression forces at a speed equal to half the rotation rate. The same holds for liquid droplets and solid particles or their aggregates dispersed in the fluid as a result of the friction or viscous stresses ($\eta\dot{\gamma}$) exerted on them. This makes shear flow much less efficient than elongational flow for deforming dispersed liquid droplets or aggregates, whereas in rotational flow they will not deform at all. A deformable droplet subjected to shear flow will take an ellipsoidal shape with an orientation of about 45° to the direction of flow. The fluid inside the droplet will rotate. Much the same holds for a (random) polymer coil (see Section 11.1).

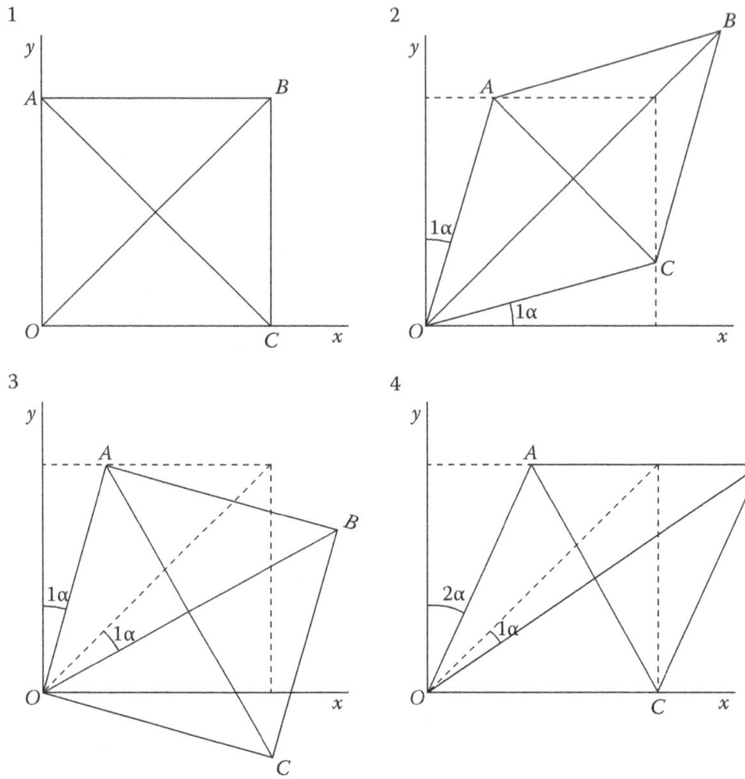

Figure 3.6 *Deformation of a volume element subjected to simple shear. Application of simple shear deformation over an angle α in x direction and in positive y direction results in elongation of diagonals (situation 2), whereas application of simple shear in x direction and in negative y direction results in rotation of volume element over an angle α (situation 3). Adding up deformations shown for situations 2 and 3 gives a volume element subjected to simple shear over an angle 2α. Note that this figure only holds for infinitesimal deformations.*

It appears that all flows can be decomposed in an elongational and a rotational component except pure rotational and elongational flow (e.g., Lodge 1979; Darby 1976). The contribution of both flow types to the overall flow varies depending on the precise flow conditions. Another factor is that most flows are not two-dimensional. This can best be illustrated by considering the hyperbolic flow depicted in **Figure 3.5**. Rotation of the flow pattern around the x or y axis results in the so-called *axisymmetric flow*. Rotation around the x-axis results in *uniaxial elongational flow* in the x direction. A practical example is flow through a constriction in a tube. Rotation around the y-axis results in *biaxial elongational flow* in the x,z plane. Practical examples of this type of flow are the deformation of material around a gas cell growing in size as occurs during proofing and baking of bread and lubricated squeezing flow between two approaching plates (**Figure 3.7**). For the latter case, there will be a shear component in the flow if

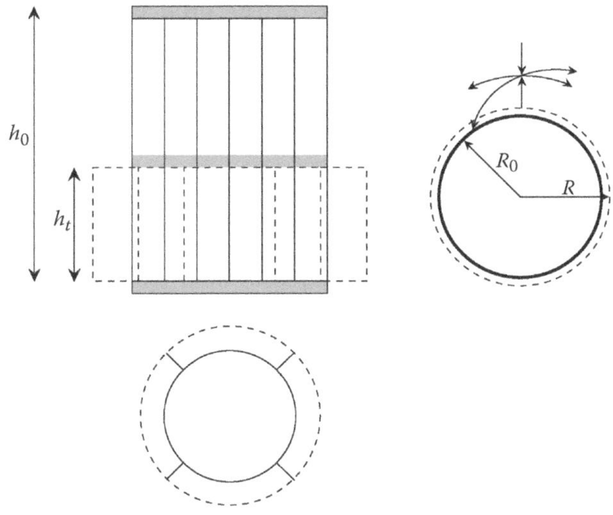

Figure 3.7 *Illustration of occurrence of biaxial elongational flow as a result of uniaxial compression between two lubricated plates (left panel) or due to growth of gas cells (right panel).*

there is friction between the liquid and the plates. The flow around a growing gas cell will also contain a shear component that will be larger for smaller gas cell sizes. In general, there will always be an elongational flow component when liquid flow is accelerated or decelerated in the direction of the flow.

Turbulent flow will develop at higher flow rates, with the exact transition flow rate depending on the flow conditions. At low flow rates, a local instability as a local acceleration or deceleration of a fluid element does not grow because the accompanying energy fluctuations are damped by viscous dissipation. At higher flow rates, inertia forces are no longer outweighed by viscous dissipation, causing—with increasing flow rates—the streamlines to become more and more wavy and eddies to develop. The flow becomes more chaotic, the extent increasing with flow rate. Any volume element is subjected to rapid fluctuations both in velocity and direction of flow. A flow profile will give the time average of the flow velocities of the volume elements as a function of space.

During turbulence, flow eddies (vortices) will develop with different sizes. The size of the largest eddies with relatively low flow rates will be of the order of the smallest dimension of the vessels containing the fluid (e.g., the pipe diameter or the distance between the plates in **Figure 3.3**). These large eddies transfer their kinetic energy to smaller eddies, which transfer it to still smaller eddies, and so on. With decreasing eddy size, the local velocity will be lower, but less than proportional, which implies that the local velocity gradient ψ will be higher. In general, energy dissipation as a result of liquid flow is given by the force times the distance the liquid element involved moves per second. Replacing the force by stress and the distance traveled per second by the velocity divided by the thickness of the fluid layer, one obtains for the energy dissipation for any type of flow (in J m^{-3} s^{-1})

$$\text{Energy dissipation} = \sigma\psi = \eta\psi^2 \tag{3.19a}$$

and for shear flow

$$\text{Energy dissipation} = \sigma\dot{\gamma} = \eta\dot{\gamma}^2 \tag{3.19b}$$

The smallest eddies have such a high ψ that their excess kinetic energy is dissipated as heat. This implies that, in turbulent flow, more energy is dissipated as heat than in laminar flow and the other way around that more energy (higher stress) is required for turbulent flow.

An important point is at which flow rates the transition occurs from laminar to turbulent flow. This depends on when inertial stresses become dominant over frictional or viscous stresses. The former are proportional to ρv^2, and the latter are equal to $\eta\dot{\gamma}$ for laminar flow, where ρ is the mass density of the liquid. For shear flow, $\dot{\gamma}$ is proportional to v/h, where h is a characteristic length perpendicular to the direction of flow (the distance between the plates in **Figure 3.3** or the pipe diameter for flow through a cylindrical pipe). The ratio between inertial and viscous stresses is proportional to a dimensionless quantity, the *Reynolds number* (*Re*), which is given by

$$Re \equiv \frac{h\bar{v}\rho}{\eta}, \tag{3.20}$$

where $\bar{v}$ is the average flow velocity, that is, the volume flow rate divided by, for example, the cross section of the flow channel. When *Re* surpasses a critical value, turbulence starts to develop. Details of Formula 3.20 depend on the geometry of flow. **Table 3.2** gives the Reynolds number for some flow geometries as well as critical values of *Re*.

Table 3.2 Reynolds Number and Critical Values of *Re* for Transition of Laminar Flow to Start of Development of Turbulence for Various Flow Geometries

Flow Geometry	*Re*	Critical Value
Cylindrical pipe with diameter *d*	$d\bar{v}\rho/\eta$	2000
Flow in the gap between concentric cylinders with radii *r* and *Kr*, *K* < 1, as a result of rotation of the outer one[a]	$r(1-K)\bar{v}\rho/\eta$	Depends strongly on (1 – *K*)[b]
Between two flat plates at distance *h* apart	$h\bar{v}\rho/\eta$	~2000
Film of thickness *h* flowing along an inclined flat surface	$4h\bar{v}\rho/\eta$	4–25
Flow around a sphere of diameter *d*[c]	$d\bar{v}_\infty\rho/\eta$	~1

[a] Rotation of the inner cylinder will lead to Taylor vortices at much lower *Re* (Section 8.2.1.1).

[b] For (1 – K) = 0.01, 0.02, 0.05, and 0.1, *Re* is ~2000, 2200, 2500, and 7000, respectively; for (1 – K) > 0.06, *Re* increase more rapidly with (1 – K).

[c] Here, $\bar{v}_\infty$ is the velocity of the sphere relative to the fluid at a "large" distance.

In the case of turbulence, flow is not only in the main direction, but also perpendicular to it. This side flow will be less close to a wall and disappear very close to it, resulting in laminar flow in a small boundary layer near a wall. The velocity gradient in this boundary layer will be much higher than the overall velocity gradient perpendicular to the direction of flow in the rest of the liquid where flow is turbulent.

The explanation of the value of the viscosity of liquids is outside the scope of this book. It is related to how easily molecules can move with respect to each other, which depends strongly on the free volume present in the liquid and the interaction forces between the molecules. In the presence of a velocity gradient, it becomes more difficult for molecules to move along the other ones. A temperature increase leads to an increase in the free volume (lower density) and with that, to a lower viscosity.

3.4 Definitions of stress and strain at large deformations

In the preceding discussion, stress and strain in uniaxial extension (or compression) were defined as the force per area and the strain ε as the ratio between the change in length (height) and the original length of the material. This definition holds well for small deformations. However, for large deformations, two complications arise. First, upon deforming a material in uniaxial or biaxial extension, the material is not only extended but also compressed in the direction(s) perpendicular to the tensile strain(s), and second, the definition of the strain becomes less straightforward.

For the calculation of the stress, the force should not be divided by the initial surface area A_0, but by the actual surface area A. For cylindrical test pieces with a Poisson ratio of 0.5, this gives for the *true* tensile or compressive *stress* σ_e:

$$\sigma_e = \frac{f}{A_t} = \frac{f}{A_0(l_0/l_t)} \tag{3.21}$$

where the subscript 0 indicates the original cross section and length of the test piece and t the actual one.

In Section 3.1.2, the strain ε was defined as dl/l_0, where dl is the change in length of the test piece during the compression test. This strain measure is usually called *Cauchy* or *engineering strain*. However, this strain measure leads to problems when large deformations are considered. For instance, when a test piece of 10 cm height is compressed over a distance of 4 cm, the strain will be 0.4. On the other hand, when it is first compressed over a distance of 2 cm and later again over a distance of 2 cm, the calculated strain is 0.2 + 2/8 = 0.45. For elongation, the reverse will be the case. A good measure of the strain should automatically correct for the change in the actual height of the test piece. A common measure of the strain is the so-called *Hencky* or *true* or *natural* or *log strain*, which reads as

$$\varepsilon_H = \int_{l_0}^{l_t} \frac{dl}{l_t} = \ln\frac{l_t}{l_0} \tag{3.22}$$

Because the term "Hencky strain" is used most frequently in food science, this term will be used in the following discussion. The relation between the Hencky strain and the Cauchy strain, ε_C, is given by

$$\varepsilon_H = \ln\frac{l_t}{l_0} = \ln\frac{l_0 + dl}{l_0} = \ln(1+\varepsilon_C) \tag{3.23}$$

To show the importance of the use of the right mathematical expression for the strain measure, force versus Cauchy strain and true stress versus Cauchy and Hencky strain curves are shown in **Figure 3.8** based on data of a compression experiment on old Gouda cheese.

The expression of the strain rate in terms of the Hencky strain (*Hencky strain rate*) reads as

$$\dot{\varepsilon}_H = \frac{dl}{l_t dt} = \frac{v}{l_t}, \tag{3.24}$$

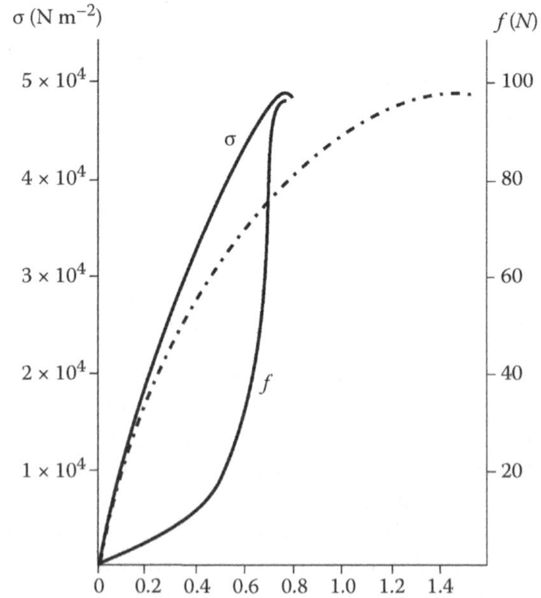

Figure 3.8 *Force F (right axis) versus Cauchy strain and true stress σ (left axis) versus Cauchy (full curve) and Hencky strain (dashed curve). The slope at start of stress versus strain curves gives Young's modulus. (After van Vliet, T., Luyten, H., in* New Physico-Chemical Techniques for the Characterization of Complex Food Systems, *ed. Dickinson, E., Blackie Academic & Professional, London, 1995.)*

where v is the speed at which the test piece is extended or compressed. In many apparatuses, the applied v is constant, implying that—according to Equation 3.24—$\dot{\varepsilon}_H$ increases or decreases during uniaxial compression or extension experiments, respectively.

As discussed in Section 3.1.2 and illustrated in **Figure 3.7** on uniaxial compression or extension, the material will also be deformed in the direction perpendicular to the direction in which it is compressed or extended. If the volume of the material stays constant, the strain in those directions is equal to half the uniaxial compression or tensile strain. This implies that for the uniaxial compression of a cylindrical test piece, its biaxial extension strain $\varepsilon_{b,H}$ in the direction normal to the compressive strain should be half the uniaxial compression strain, so

$$\varepsilon_{b,H} = -0.5 \ln \frac{l_t}{l_0}. \tag{3.25}$$

References

Darby, R. 1976. *Viscoelastic Fluids: An Introduction to Their Properties and Behavior*. New York: Marcel Dekker.

Lodge, A.C. 1979. *Body Tensor Fields in Continuum Mechanics with Applications to Polymer Rheology*. London: Academic Press.

van Vliet, T., and H. Luyten. 1995. Fracture mechanics of solid foods. In *New Physico-chemical Techniques for the Characterization of Complex Food Systems*, ed. E. Dickinson. London: Blackie Academic & Professional.

Walstra, P. 2003. *Physical Chemistry of Foods*. New York: Marcel Dekker.

Whorlow, R.W. 1992. *Rheological Techniques*. Chichester: Ellis Horwood.

4

Descriptive Rheology

4.1 Classification of materials according to their rheological behavior

Phenomenologically, it is possible to classify materials according to their rheological behavior. As a starting point, we will first consider ideally viscous and ideally elastic behavior. These two groups of materials differ with respect to the fate of the energy supplied to the material by an external force. This energy is the product of the stress (N m^{-2}) and the strain (relative displacement (–)) and measured in J m^{-3}. For materials showing *ideally elastic behavior*, this energy is stored, and upon terminating the stress it is fully and immediately released. For those showing *ideally viscous behavior*, all the energy supplied is immediately and completely dissipated as heat and is no longer available for allowing the material to regain its original shape after terminating the stress.

A first complication is that material properties often depend on strain or strain rate (*nonlinear behavior*). Moreover, the properties of a material may depend on the deformation time, resulting in *nonequilibrium* behavior. Finally, for many materials, the reaction on a stress or strain generally consists partly of a viscous contribution and partly of an elastic one, that is, they behave *viscoelastically*. The ratio between these two contributions mostly depends on the speed of deformation or, to be more precise, on the time scale over which forces are applied on a small volume element of the material. Below, first equilibrium behavior will be discussed, followed by nonequilibrium behavior and rate effects. The classification of materials presented below is independent of the type of strain or flow involved. To keep the overview convenient, the discussion will be limited to shear deformations, except where mentioned specifically.

4.1.1 Equilibrium behavior

In **Figure 4.1**, a graphical overview is given of the simple basic relationships between stress and strain rate for fluids and fluid-like materials subjected

Observed behavior		Examples
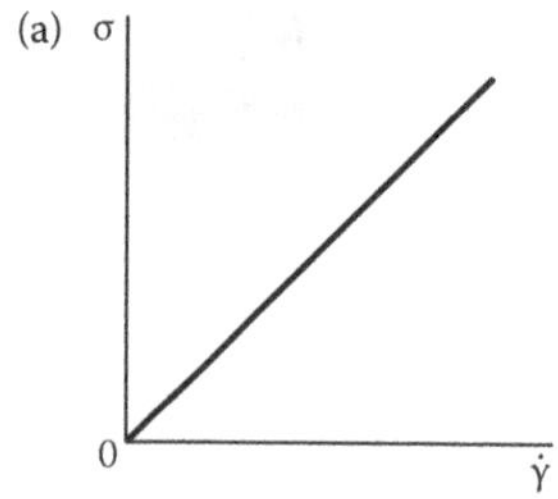	Newtonian liquid Linear viscous $\sigma = \eta\dot{\gamma}$ η = a constant	Low molecular liquids. Dilute dispersions of spherical hard particles (e.g., dilute emulsions). Water, honey, sugar syrup, water alcohol mixtures, pasteurized milk.
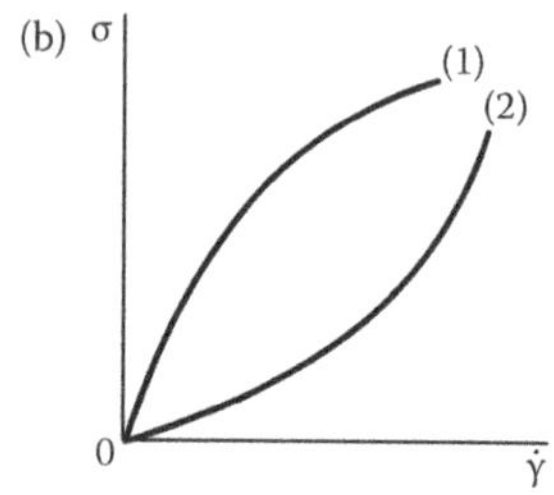	Non-Newtonian liquid $\sigma = \eta_{app}\dot{\gamma}$ $\eta_{app} = f(\dot{\gamma})$ (1) shear thinning (pseudo-plastic) (2) shear thickening (dilatants) η_{app} = apparent viscosity	(1) Solutions of macromolecules, concentrated milk, drinkable yogurts, cheese spread, fruit juices and some dispersions with a high volume fraction particles as homogenized peanut butter. (2) Concentrated suspension of native starch granules.
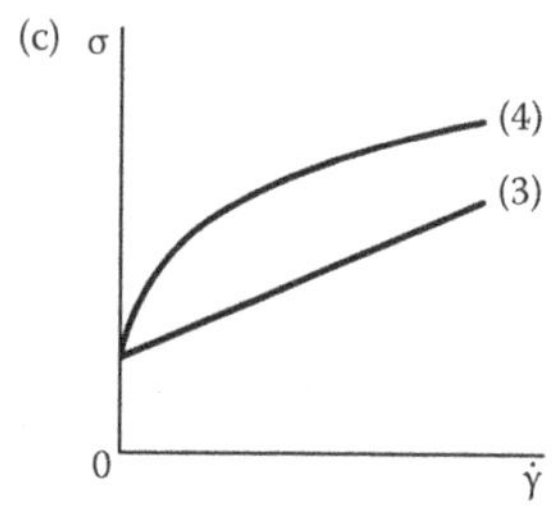	(3) Bingham liquid $\sigma - \sigma_y = \eta_B\dot{\gamma}$ (4) plastic liquid $\sigma - \sigma_y = \eta_{app}\dot{\gamma}$ η_y = yield stress η_B = Bingham viscosity	(3) Bingham model is a so-called ideal model. (4) Margarine, butter, mayonnaise, tomato ketchup, whipped cream, apple sauce.

***Figure 4.1** Classification of materials with a liquid or liquid-like character according to their rheological behavior. Left column: shear stress versus shear rate diagrams characterizing rheological behavior. Middle column: name of indicated behavior and equations describing shear stress shear rate diagram. Right column: examples of materials showing this behavior over the shear rate range considered (for further discussion, see text).*

to a shearing deformation. In addition, examples are given of materials that behave according to the given basic relationship. One should be aware that, in practice, often a combination of these simple relationships can be observed, particularly when the mechanical behavior is studied over a large range of shear rates or stresses. For instance, solutions of macromolecules behave as a Newtonian liquid (**Figure 4.1a**) at (very) low and high shear rates, whereas in between they exhibit shear thinning behavior (**Figure 4.1b**).

Figure 4.1a illustrates the relationship for a linear viscous liquid (Equation 3.10). Liquids obeying such behavior are called Newtonian liquids. Only one material parameter, the viscosity η, suffices to fully define their rheological behavior under shear. The viscosity is given by the slope of the shear stress versus shear rate line, and is independent of shear rate (**Figure 4.2**, line 1) and shearing time.

For non-Newtonian liquids, the relationship between σ and $\dot{\gamma}$ is not linear (**Figure 4.1b**). A shear rate-dependent viscosity is obtained, called *apparent viscosity* (η_{app}), where $\eta_{app} = f(\dot{\gamma})$. For non-Newtonian liquids, Equation 3.10 reads as

$$\sigma = \eta(\dot{\gamma})\dot{\gamma} = \eta_{app}\dot{\gamma} \qquad (4.1)$$

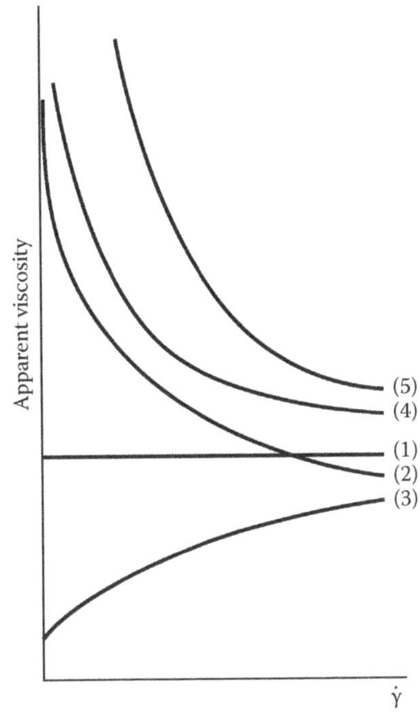

Figure 4.2 *(Apparent) viscosity as a function of shear rate for the liquid and liquid-like materials shown in* ***Figure 4.1****. (1) Newtonian behavior; (2) shear thinning; (3) shear thickening; (4) Bingham liquid; (5) plastic flow behavior.*

The apparent viscosity is given by the slope of the line connecting the origin of the stress versus strain rate curve and the point considered on that curve. It decreases or increases with shear rate (**Figure 4.2**, curves 2 and 3, respectively). If η_{app} decreases with increasing $\dot{\gamma}$, one speaks of *shear thinning* (or *pseudoplastic*) behavior, and when an increase is observed, of *shear thickening* behavior. If the latter is accompanied by a volume increase of the material, the term *dilatant behavior* is often used. In daily practice, the above terms are often used in a rather imprecise way. According to the given definitions, it would be more precise to speak of shear rate-thinning/thickening, respectively or, if the phenomenon is observed for other flow types, of strain rate-thinning/thickening. The use of these terms avoids confusion for a number of products, including bread dough (and some polymer melts), for which the resistance against deformation increases with increasing strain, the so-called *strain hardening*, but decreases with increasing strain rate (strain rate-thinning).

There are many liquid food products that exhibit shear thinning behavior over at least a part of the full stress versus shear rate curve. Clear examples are products thickened by polysaccharides such as xanthan gum and starch, or by a concentrated suspension of deformable particles as gelatinized, but still intact starch granules, or fruit particles. Shear thickening behavior is much less common. A concentrated suspension of starch granules may show shear thickening behavior.

Shear thinning behavior also complicates the relationship between extensional and shear viscosities as discussed in Section 3.2. Equation 3.17 only holds for Newtonian liquids. Since both η and η_e depend on strain rate, the extensional strain rate should be recalculated into an equivalent shear rate or vice versa, and next the elongational and shear viscosity should be compared at these equivalent rates. Based on consideration of viscoelastic and inelastic flow behavior, it has been argued by Jones et al. (1987) that, for uniaxial flow, the equivalent shear rate is $\sqrt{3}\dot{\varepsilon}_H$, which gives for the Trouton ratio (*Tr*)

$$Tr_{\text{uniaxial}} = \frac{\eta_e(\dot{\varepsilon}_H)}{\eta(\dot{\gamma} = \dot{\varepsilon}_H\sqrt{3})} \tag{4.2}$$

Using similar considerations, Huang and Kokini (1993) argued that *Tr* for biaxial flow should be calculated as

$$Tr_{\text{biaxial}} = \frac{\eta_B(\dot{\varepsilon}_B)}{\eta(\dot{\gamma} = \dot{\varepsilon}_B\sqrt{12})} \tag{4.3}$$

For Newtonian liquids, the first and second normal stress difference (Equations 2.2a and b) is zero. However, for non-Newtonian liquids, this will no longer be the case when their reaction on a stress contains an important elastic component (viscoelastic liquids). For these liquids, the first normal stress difference N_1, in particular, may be substantial compared to the shear stress, leading to such phenomena as die swell and rod climbing. In the latter case, the viscoelastic liquids climbs in the impeller shaft during stirring as can be noticed, for example, during mixing of cake batter. Die swell may occur during extrusion of viscoelastic liquids, resulting in an increase of the diameter of the liquid jet to a diameter substantially larger than the die diameter. These phenomena will be discussed further in Section 11.2.

Many food materials such as margarine, tomato ketchup, buttermilk, whipped cream and common materials (e.g., numerous paints), and clay suspensions behave like solids under small stresses, and are liquid-like under large stresses. Clear flow is only noticed above a certain threshold value, the so-called ***yield stress***, σ_y (**Figure 4.1c**). For such systems, one speaks of ***plastic flow*** behavior. Above the yield stress, they exhibit a shear thinning character. The viscosity is given by the slope of the line connecting the origin of the stress and strain rate axes and the point considered on that curve. Below the yield stress, $\eta_{app} = \infty$, and above it, η_{app} is given by Equation 4.1. The apparent viscosity decreases with increasing shear rate (**Figure 4.2**, curve 5). A Bingham fluid is an ideal model fluid that in practice does not exist. It is characterized by a linear relation between the stress corrected for the yield stress and the strain rate. For many fluids exhibiting plastic flow behavior, this model is good enough to allow engineering calculations. Note, however, that although the so-called Bingham viscosity is assumed to be a constant, η_{app} is not. It decreases with increasing shear rate (**Figure 4.2**, curve 4).

Establishing σ_y is often problematic for two reasons: (1) the slope of the stress versus strain rate curve is often very steep near the origin, and (2) its value often depends on the measuring time (see below). The latter implies that, depending on the purpose for which one wants to know σ_y, one may have to accept different values for different measuring times. However, this does not imply that σ_y cannot be a useful material characteristic.

Mathematical models that can be used to describe the different types of flow behavior are discussed in Section 4.3.1.

A corresponding overview to that given in **Figure 4.1** for predominantly fluid materials is presented in **Figure 4.3** for predominantly solid materials. The main difference is that in **Figure 4.3**, stress is plotted as a function of strain

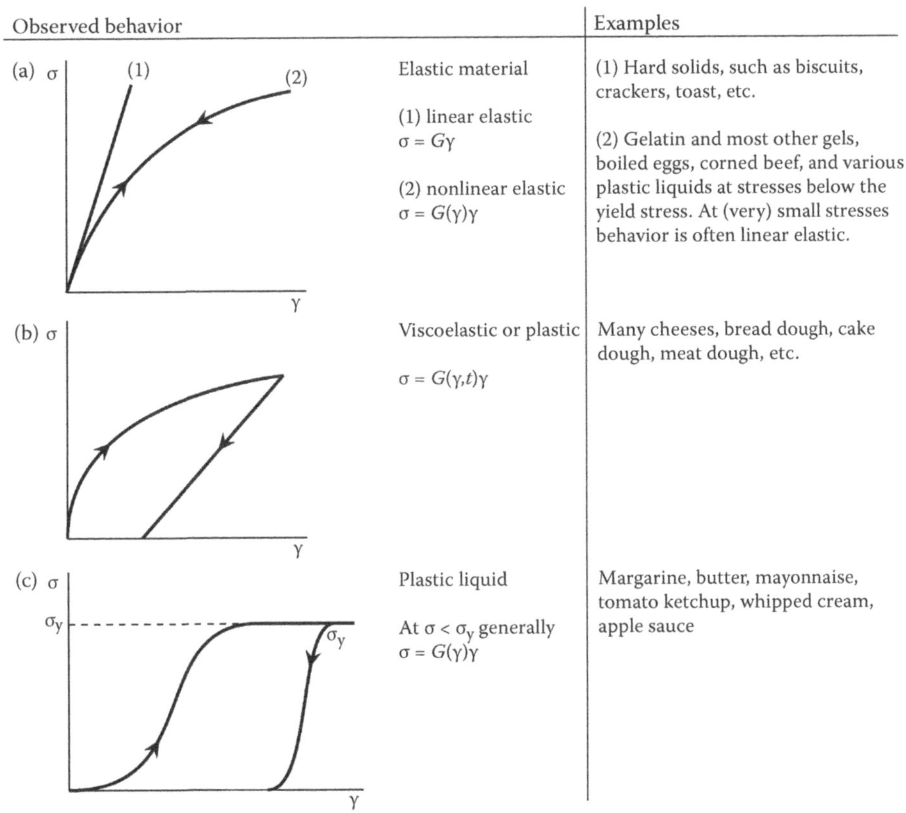

***Figure 4.3** Classification of solid and solid-like materials according to their rheological behavior. Shear stress σ, as a function of the shear strain γ, for various materials with solid-like character. Left column: shear stress versus strain diagrams characterizing rheological behavior. Middle column: name of indicated behavior and equations describing shear stress shear strain diagram. Right column: examples of materials showing this behavior over the strain range considered (for further discussion, see text).*

instead of strain rate. Various simple relationships between stress and strain can be observed. However, as in the case for fluids, in practice usually a combination of these simple relationships is found when mechanical behavior is studied over a large range of strains and/or time.

For an ideally elastic material (Hooke solid), the relationship between σ and γ is linear (**Figure 4.3a**, curve 1) (Equations 3.5 and 3.8 for elongation and shear deformation, respectively). Such behavior is observed for all solid materials at low γ, although the strain range over which the relationship is linear varies considerably, for example, 2–3 for rubbers, 0.2–1 for most polymer gels, ~1 for gelatin gels, ≈0.003–0.03 for many particle gels (e.g., yogurt), and ~0.0002 for bread dough and margarine. Only brittle materials such as potato crisps, various types of chips and hard biscuits, and engineering materials (e.g., cast iron) and ceramic products are linearly elastic up to the point where they fracture. For most other materials, the behavior becomes nonlinear at larger deformations (**Figure 4.3a**, curve 2).

For nonlinear materials, *G* and *E* are functions of strain, but the material behaves reversibly with respect to deformation. Equations 3.5 and 3.8 should be replaced respectively by

$$\sigma_e = E(\varepsilon)\varepsilon \tag{4.4}$$

$$\sigma = G(\gamma)\gamma, \tag{4.5}$$

where Young's modulus *E* and the shear modulus *G* are a function of the strain.

The reversibility with respect to deformation (stress) is lost for so-called plastic materials, such as cheese and dough for bread and many other bakery products that behave according to **Figure 4.3b**. During deformation, their behavior is apparently nonlinear elastic, but in addition the material does not return to its original shape after the stress is removed. Part of the deformation is permanent, indicating a viscous contribution; the material behaves in a viscoelastic manner. The relative contribution to the total deformation of the elastic and the viscous component depends on the value of the applied stress and the time that it has been applied. This is further discussed in Section 4.2.

As mentioned already above, many solid-like products start to flow at stresses above the yield stress (**Figure 4.3c**). They exhibit plastic flow behavior as discussed above in the part on predominantly fluid materials. Below the yield stress, these materials show nonlinear elastic (e.g., margarine) or plastic behavior (e.g., cheese spread).

Plastic and many plastic fluid materials behave viscoelastically under low stresses. One speaks of *linear viscoelastic* behavior if the elastic component can be described by Equations 3.5 and 3.8 for an elongational and shear deformation, respectively, and the viscous component by Equations 3.6 and 3.10. The so-called *linear region* is that part of the strain in which the rheological behavior of a material is linear viscoelastic.

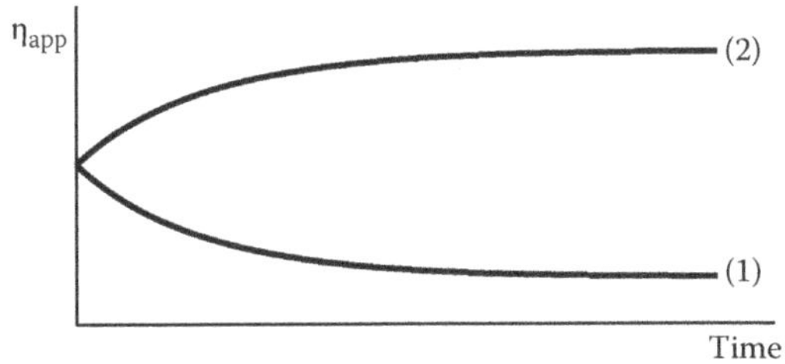

Figure 4.4 *Apparent viscosity, η_{app} as a function of flow time during shear at constant σ or $\dot{\gamma}$; curve 1, thixotropic behavior; curve 2, antithixotropic behavior.*

4.1.2 Nonequilibrium behavior

For many non-Newtonian fluids and for products with a plastic flow character, the relationship between σ and $\dot{\gamma}$ depends on the flow time (duration of flow) and flow history. Two main types can be distinguished: *thixotropy* and *antithixotropy* (sometimes referred to as *rheopexy*) (**Figure 4.4**).

For thixotropic materials, η_{app} decreases with increasing time of flow (**Figure 4.4**, curve 1), whereas in the second case it increases (curve 2). Both thixotropy and antithixotropy are reversible phenomena. The apparent viscosity has to return to its original value during rest or during shearing at a lower $\dot{\gamma}$.

The σ versus $\dot{\gamma}$ diagram of thixotropic materials is usually equal to that of materials with a shear thinning or plastic flow character, especially if one has waited at every applied σ or $\dot{\gamma}$ until η_{app} has become constant (solid curve in **Figure 4.5**). If after reaching an equilibrium state (state 1 in **Figure 4.5**) $\dot{\gamma}$ is increased instantaneously from $\dot{\gamma}_1$ to $\dot{\gamma}_2$, σ will initially increase to state 2, which is characterized by the same η_{app} as in state 1. Next, η_{app} will decrease in time to a new equilibrium situation (state 3). If next the shear rate is decreased instantaneously from $\dot{\gamma}_2$ to $\dot{\gamma}_1$, σ will decrease to state 4 and subsequently increase to the equilibrium value 1. If one does not wait until equilibrium at every $\dot{\gamma}$, intermediate values of σ will be measured, for example, those characterized by state 2′ and 1′ in **Figure 4.5**. Note that equilibrium value implies that the same value for σ at set $\dot{\gamma}$ is obtained coming from a lower or a higher value of $\dot{\gamma}$.

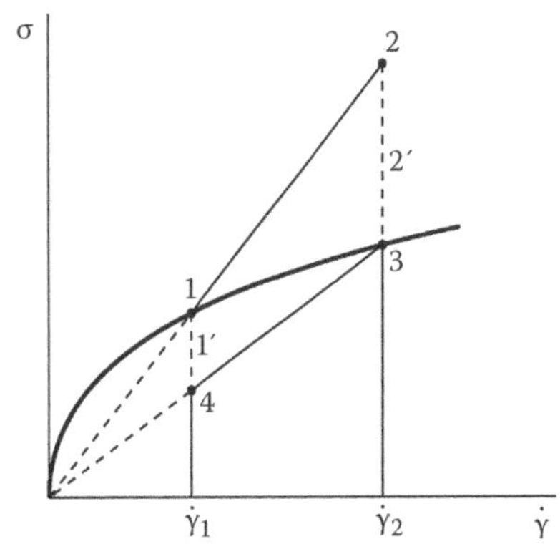

Figure 4.5 *Relation between σ and $\dot{\gamma}$ for a thixotropic material when $\dot{\gamma}$ is increased to a higher $\dot{\gamma}$ and subsequently decreased to lower $\dot{\gamma}$. For further explanation, see text.*

In general, thixotropy and antithixotropy will be recognized as such when the time scale of the structural changes leading to this behavior is longer than $1/\dot{\gamma}$, and the response time of the apparatus is shorter than about 10^3–10^4 s, depending on the patience of the person carrying out the

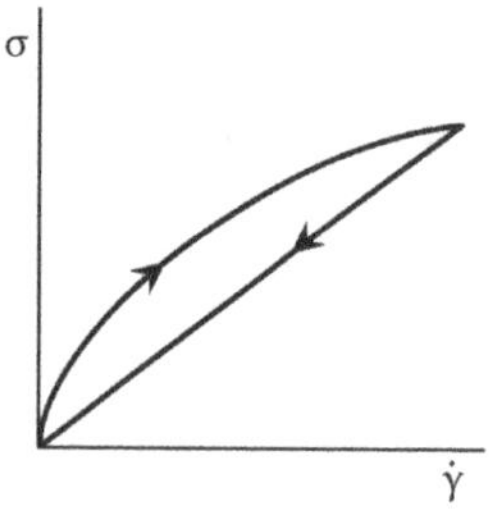

***Figure 4.6** Hysteresis in a thixotropic material.*

experiment. These two phenomena may not be confused with situations in which η_{app} changes irreversibly owing to irreversible changes in structure during shearing. A direct consequence for these types of behavior is that the measured η_{app} is affected by all the handling involving flow of the material, such as filling the measuring body of a rheometer. Thixotropy is a common phenomenon not only for many food materials such as tomato ketchup and mayonnaise, personal care products, and pharmaceutical products, but also for many concentrated suspensions (e.g., paints) and polymer solutions.

Thixotropic behavior is often studied by subjecting the material to increasing $\dot{\gamma}$ or σ followed by a decrease at a known rate. If the time during which the material is sheared at each $\dot{\gamma}$ or σ is shorter than that required to reach the equilibrium value of η_{app}, the relationship obtained between σ and $\dot{\gamma}$ shows a so-called *hysteresis loop* (**Figure 4.6**). Its size will depend on the rate at which $\dot{\gamma}$ or σ was increased and reduced and on their maximum value and, hence, depends on measuring conditions. Note that the presence of a hysteresis loop does not automatically imply thixotropic behavior of the liquid. It may also be due to the irreversible breakdown of the structure of the product during shearing.

Antithixotropic behavior is not common for food materials. Sometimes it can be observed, for example, in some ropy yogurts, although in these cases it is observed at shear rates outside the range applied in the mouth during consumption. However, during processing of (concentrated) polymer (thickener) solutions, one should be aware of the possibility of antithixotropic phenomena.

4.2 Dynamics: The role of time scale

As mentioned in the Introduction, dynamic features play an important role in the mechanical behavior of many materials. The consequences of the duration of deforming a material under constant conditions of shear rate or stress have been discussed in Section 4.1.2 and were shown to be related to thixotropic and antithixotropic behavior. In this section, the effect will be discussed in terms of time scale, that is, the time that a stress of a certain magnitude and direction acts on a material. The time scale of an experiment is important for the reaction of a material being more elastic or viscous and for shear rate thinning behavior. The following discussion will be primarily focused on the first aspect.

For a steady-state experiment, in which a certain compressive or tensile stress is applied to an ideally elastic material, the applied time scale and the duration of the experiment coincide. However, this is no longer the case if the material is subjected to a varying stress (e.g., to a sinusoidally oscillating stress). The time scale of the experiment is then roughly equal to the reciprocal of the oscillation frequency in rad s^{-1}. Upon shear flow, the time scale and the duration of shear stress or shear rate application are (usually) also completely different. Owing to the rotational component in a shear flow (Section 3.3), dispersed (aggregates of) particles or macromolecules will start to rotate. This implies that, owing to the elongational component of the flow, a certain point on the surface of the particle will be subjected successively to a tensile stress and a compressive stress equal to the shear stress. The time scales of these stresses depend on the rate of rotation and, hence, on $\dot{\gamma}$. The time scale is roughly a quarter of the rotation time of the particle = $\pi/4\dot{\gamma} \approx \dot{\gamma}^{-1}$.

An example from daily life, showing the importance of time scale for rheological behavior, is that the inner part of a mature soft cheese, such as Camembert, flows if one stores it for some time after slicing the cheese. However, the inner part of the same cheese can also be cut with a knife, and one has to chew on it during mastication, implying elastic behavior over the time scale of cutting and chewing. In general, one sees for viscoelastic systems that they behave relatively more elastically over short time scales and are more viscous over longer time scales.

The origin of time scale-dependent behavior is in the structure of a material. The characteristics of ideally elastic materials are, first, that on application of a stress the resulting strain is independent of the duration over which the stress is applied and, second, that they recover their original shape after the release of the applied stress. This implies that all energy supplied during the deformation was stored in the material, for example, in deformed bonds, and that this stored energy is used to enable the material to return to its original shape. Hence, the bonds between the molecules or structural elements* forming the material must be permanent over the time scale of deformation. The characteristics of liquids (viscous materials) are as follows: first, they start to flow on application of a stress and, second, they keep the shape they have at the moment the stress is released (neglecting inertia effects). To allow flow, bonds have to break and re-form, stress-free, during the time scale of the experiment. All supplied energy is dissipated as heat and therefore not available to allow the material to recover its original shape to any extent. This shows clearly that the bonds between the molecules constituting the liquid cannot be permanent. Still, there must be bonds between the molecules, at least for low molecular weight liquids; otherwise, they would not form a condensed phase. The essential difference with elastic materials is that the bonds are not permanent over the time scale of the deformation.

---

* The physical building blocks of a liquid-like or solid structure may be called structural elements, that is, regions bounded by a close surface, where at least some of their properties differ from those in the rest of the system. Structural elements can be particle cells, air bubbles, emulsion droplets, crystals, starch granules, etc.

In viscoelastic materials, the process of spontaneous disruption and stress-free reformation of the bonds proceeds over time scales between seconds and days. It can easily be followed by first deforming the material to a certain extent and then measuring the stress required to keep this deformation constant. Except for ideally elastic materials, stress will decay at a certain rate—it relaxes. The time required for the stress to decrease to $1/e$ (i.e., 36.8%) of its value at the moment the deformation was stopped is called the *relaxation time* (**Figure 4.7**).

Relaxation processes can be visualized by considering a *Maxwell element* consisting of an ideal spring in series with an ideal dashpot, representing elastic and viscous deformation, respectively (**Figure 4.8**). If such an element is stressed over a short time scale, the resulting deformation will be almost exclusively attributable to deformation of the spring, whereas for long-lasting stresses, deformation will be caused by displacement of the piston in the dashpot. For intermediate time scales, both contribute to the reaction of the element. The ratio of both contributions depends on the "Maxwell" modulus, G_M of the spring, and the effective resistance/friction (which can be modeled as a viscosity η_M) experienced by the piston in the dashpot. If a Maxwell element is subjected to a stress deformation, the rate of deformation of the spring and dashpot are additive, in shearing terms,

$$\frac{d\gamma}{dt} = \frac{1}{G_M}\frac{d\sigma}{dt} + \frac{1}{\eta_M}\sigma \tag{4.6}$$

For constant deformation $d\gamma/dt = 0$, one obtains

$$\frac{d\sigma}{\sigma} = -\frac{G_M}{\eta_M}dt \tag{4.7}$$

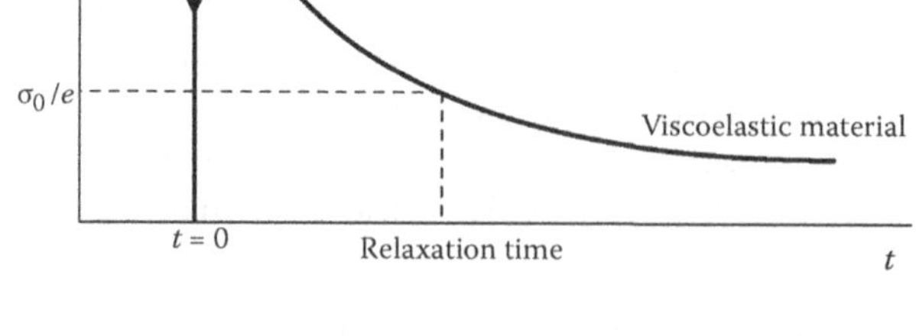

Figure 4.7 *Stress relaxation at fixed strain of an ideally elastic material, a liquid, and a viscoelastic material. At time t = 0, a strain γ or ε is applied instantaneously, leading to a stress σ_0.*

This gives for the decrease in stress as a function of time

$$\sigma(t) = \sigma_0 e^{-(G_M/\eta_M)t} = \sigma_0 e^{-t/\tau_{rel,M}} \qquad (4.8)$$

where $\tau_{rel,M}$ (= η_M/G_M) is called the relaxation time of the system. For an instantaneous deformation, the stress, σ_0, obtained directly after the deformation is $G_M \times \gamma_0$, where γ_0 is the deformation at $t = 0$. From Equation 4.8, it follows immediately that, for $t \ll \tau_{rel,M}$, the reaction of the element is primarily elastic, whereas it is primarily viscous for $t \gg \tau_{rel,M}$.

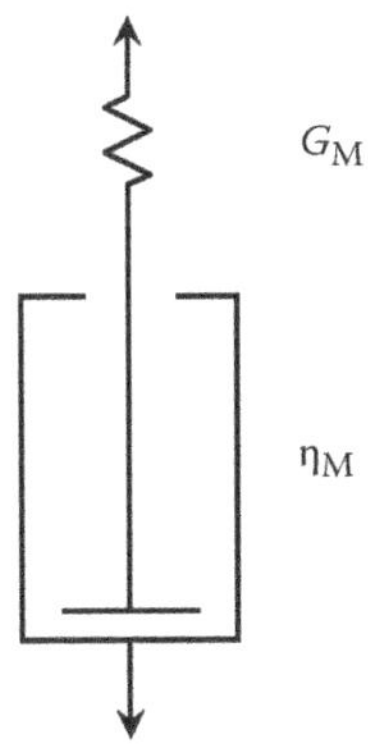

***Figure 4.8** Maxwell element consisting of an ideal spring in series with an ideal dashpot. G_M is elastic modulus of the spring of Maxwell element, η_M a parameter characterizing viscous resistance of dashpot, and $\tau_{rel,M}$ is relaxation time of Maxwell element.*

For liquids, observed relaxation times are very short (e.g., about 10^{-13} s for water and 10^{-8} s for oil), whereas for solids, τ_{rel} is very long (e.g., for concrete and steel, $\gg 10^{10}$ s). For viscoelastic materials, τ_{rel} is in between and corresponds roughly with the human time scale (e.g., for polymer melts 10^{-3}–10 s, strongly depending on the characteristics of the polymer such as molecular weight). This indicates that the ratio between the relaxation time and the characteristic time scale of observation, t_{obs}, is important for the observed rheological behavior of a material. The ratio is expressed in the *Deborah number* (*De*), defined through

$$De = \tau_{rel}/t_{obs} \qquad (4.9)$$

The rheological behavior of materials with one single relaxation time can be classified according to their Deborah numbers as follows: elastic or solid behavior when $De \gg 1$, viscous or liquid behavior when $De \ll 1$, and viscoelastic behavior when *De* is of the order of 1. In daily life, we take implicitly for t_{obs} the time scale over which we experienced the mechanical behavior of materials (about 10^{-1} to 10^6 s).

The concept involved in the Deborah number was proposed by Reiner (1964), after the Old Testament, Judges 5.5, where in the original version Deborah sings "the mountains flow before the Lord." The mountains do not flow for man because his lifetime is too short, but they do for God whose observation time is (assumed to be) infinite. The important conclusion from this discussion is that the distinction between solid and fluid/liquid behavior not only depends on an intrinsic property of the material, but also on the duration of observation. What we consider as solid may behave as a liquid over centuries. Over very long observation times, the motto "panta rhei" (everything flows) of rheology as science is more of a reality than one would anticipate when considering only the human time scale ranging from, say, 0.1 s to 100 years ($\sim 3.2 \times 10^9$ s).

The relevance of the role of the time of observation for the mechanical behavior of food products can be further illustrated by considering the time scale of processes involved in the production and handling of solid-like cheeses such as Gouda and Edam cheese. These cheeses should contain some holes, the so-called round eyes, whereas cracks should be absent. During their production, first a milk gel is formed by enzyme action that is cut into pieces. Therefore, the gel has to behave in an elastic manner (time scale of cutting process <0.1 s). After these pieces have lost most of their water, they are brought together first by molding and pressing (>1 h) and afterward by resting. This results in the fusion of particles over a period of about 3–6 days (time scale of fusion, $\sim 5 \times 10^5$ s). During ripening, some holes may be formed at strain rates of about 10^{-6} to 3×10^{-5} s^{-1} as a result of gas production by bacteria (therefore, the cheese must flow). To prevent sagging of the cheeses during ripening due to flow under their own weight, they have to be turned around roughly every 2 weeks, implying observation times of about 10^6 s. On the other hand, these cheeses behave as a solid during eating and cutting—they fracture under the exerted stresses. Time scales involved in eating and cutting are about 10^{-2}–1 s. It should be mentioned that in this paragraph, we neglect the effects due to changing pH and water content, which will affect the reaction of the material. However, as long as pH and water content are in their normal range (4.9–5.7 and 40%–46%, respectively), the varying rheological behavior is to a large extent due to time scale effects.

As mentioned above, the time scales about which stresses are exerted on suspending molecules or particles is about $1/\dot{\gamma}$ and therefore varies, for instance, between 10^2 and 10^{-3} s for shear rates increasing from 10^{-2} to 10^3 s^{-1}. For molecules with a relaxation time in this observation time range (or for aggregates of particles characterized by spontaneous breakup and reaggregation processes in this time range), their behavior will be different at low and high shear rates. This may lead to shear (rate) thinning behavior. This will be discussed further in Section 10.3.1.

In practice, most materials contain structural elements connected by various bonds, each with different τ_{rel} values. Such materials cannot be characterized by a single τ_{rel}, but often a broad spectrum of relaxation times is needed to describe their mechanical behavior. Similarly, many viscoelastic materials exhibit a predominantly viscous behavior over longer time scales; in practice, however, the relationship between viscous and elastic behavior as a function of time scale is more complicated than when all bonds have the same relaxation time. A spectrum of relaxation times can be modeled as a series of Maxwell elements in parallel. Such a spectrum can be calculated from, for instance, stress relaxation measurements and oscillatory measurements (Sections 7.1 and 7.4).

Relaxation of the bonds between the structural elements in a material should not be confused with fracture of the bonds as a result of a force exerted on them. The latter may result in fracture or yielding of the material, which will be discussed in **Chapter 5**.

4.3 Descriptive modeling of rheological behavior

There is an urgent need for the ability to characterize the relation between shear rate (or strain rate in general) and the accompanying stresses during liquid flow using a few parameters that allow handlers to calculate the liquid flow through a geometry, the ensuing pressures, the dimensions of the required apparatus, etc., that is, for process engineering calculations. Furthermore, for quality control, it can be very convenient if the flow behavior of a product can be characterized with some simple parameters that can be compared with standard values. Similarly, there is a need for characterizing the relation between stress and strain for solid materials. The significant danger of modeling the rheological behavior of materials is that obtained characteristics are unconditionally translated into physical characteristics of the studied material. In the literature, yield stress data obtained by extrapolating shear stress versus shear rate data to zero shear rate by using the Bingham or Herschel–Bulkley model (see below) are often considered real yield stresses of the undisturbed material, although this is nearly never the case. In this context, it can be useful to distinguish between the so-called static yield stresses (i.e., the yield stress of the undisturbed material) and dynamic yield stresses (i.e., the yield stress of sheared material). However, in the latter case, the determined "yield stress" will generally strongly depend on the shearing history.

The discussion in this section will be limited to relatively simple phenomenological models with a restricted number of adjustable parameters.

4.3.1 Modeling of liquid flow behavior in shear flow

The simplest models are those for materials showing linear viscous behavior, Equation 3.10 ($\sigma = \eta\dot{\gamma}$). All the other models used are empirical or, at best, semiempirical ones. In view of this empirical character, it is not possible to extrapolate experimentally determined model parameters to much higher or lower σ or $\dot{\gamma}$. Below, only the most popular models will be discussed. For a more extensive overview, readers are referred to Steffe (1996).

A very popular model for liquids is the so-called *power-law model* (an alternative name is *Ostwald–de Waele* model). As noted in Section 4.1, many products exhibit shear thinning and/or shear thickening over a limited range of $\dot{\gamma}$. For these products, the relationship between σ and $\dot{\gamma}$ can often be modeled by

$$\sigma = k\dot{\gamma}^n, \tag{4.10}$$

where k (the so-called *consistency index*, expressed in units of N m^{-2} s^n), and n (the *power law index*) are constants. The apparent viscosity is given by

$$\eta_{app} = k\dot{\gamma}^{n-1} \tag{4.11}$$

The advantage of the power law equation is that it is relatively easy to use in either analytical or numerical calculations for complex flow situations. Its main drawbacks are that, for constant values of k and n, it predicts an unlimited increase or decrease of η_{app} with increasing $\dot{\gamma}$, and either zero or infinite η_{app} values for $\dot{\gamma} \to 0$ for n greater or less than 1, respectively. In practice, k and n are only constant over a limited range of $\dot{\gamma}$, mostly over 1 order of magnitude or even less. The consistency index, k, is in fact a measure of η_{app} at $\dot{\gamma} = 1$ s^{-1}. The power law index is $n = 1$ for a Newtonian fluid, $n < 1$ for shear rate-thinning, and $n > 1$ for shear thickening behavior. To avoid the dimensional problem in which the units of k depend on n, one can rewrite Equation 4.10 as

$$\sigma = k_p \left| \frac{\dot{\gamma}}{\dot{\gamma}_p} \right|^{n-1} \dot{\gamma} \tag{4.12}$$

where $\dot{\gamma}_p$ is a reference shear rate, usually taken to be 1 s^{-1}. Thus, k_p always has the dimension of viscosity.

Often, the flow behavior of materials exhibiting plastic flow behavior is modeled as

$$\sigma = \sigma_y + \eta_B\dot{\gamma} \tag{4.13}$$

where η_B is a constant, the so-called *Bingham viscosity* (**Figure 4.1c**). In practice, Bingham flow behavior is never observed; but often, for stresses clearly above σ_y, Equation 4.13 describes the flow behavior well enough to be used for engineering calculations. In fact, σ_y and η_B are fitting constants, and η_B is not even a viscosity because, for an ideal Bingham liquid, η_{app} decreases with increasing $\dot{\gamma}$ (**Figure 4.2**, curve 4). For $\sigma > \sigma_y, \eta_{app} = \eta_B + (\sigma_y/\dot{\gamma})$ and for $\sigma < \sigma_y$, $\eta_{app} = \infty$.

A commonly used alternative to Equation 4.10 for modeling plastic flow behavior is the Herschel–Bulkley model (Herschel and Bulkley 1926a, 1926b), which reads as

$$\sigma = \sigma_y + k\dot{\gamma}^n \tag{4.14}$$

Above the yield stress, the flow curve is not straight as for the Bingham model. The apparent viscosity is given by

$$\eta_{app} = \frac{\sigma_y}{\dot{\gamma}} + k\dot{\gamma}^{n-1} \tag{4.15}$$

For $\sigma < \sigma_y$, $\eta_{app} = \infty$.

The effect of temperature on the viscosity of liquids is usually modeled by an Arrhenius-type equation, which reads for a Newtonian liquid as

$$\eta = A\mathrm{e}^{-E_a/R_g T}, \tag{4.16}$$

where A is a preexponential constant (here, in Pa s), E_a is the (Arrhenius) activation energy for flow (J), R_g (J K^{-1} mol^{-1}) is the universal gas constant, and T (K) is the absolute temperature. In the derivation of Equation 4.16, A and E_a are (mostly implicitly) assumed to be independent of T. A higher E_a implies a stronger dependence of the viscosity on the temperature. Taken literally, the definition "activation energy for flow" for E_a is wrong. A liquid flows already if only the slightest force is applied, implying that there is no activation energy for flow as there is for chemical reactions. In addition, R is a well-defined quantity for simple liquids, but is no longer so for more complex liquids, for example, concentrated solutions/dispersions of various macromolecules and cell particles such as fruit juices. Thus, it is difficult to imagine what exactly a mole of the dispersion is and with that, what the physical meaning of E_a may be. For these reasons, it would be better to model the temperature dependence of the viscosity of liquid foods using the following equation (van Boekel 2009)

$$\eta = A\mathrm{e}^{-B/T}, \tag{4.17}$$

where A and B are fit parameters without any (suggested) physical meaning.

Modeling of thixotropic behavior is usually done by introducing a structure breakdown and reformation function, with its equilibrium depending on the shear rate. For an overview of these models, we refer to the literature (Steffe 1996; Holdsworth 1993).

4.3.2 Modeling stress versus strain curves of solids

Phenomenological models used to describe the deformation behavior of solids are less popular than those used for liquids. A likely reason for this trend is the availability of models with at least a partly theoretical basis.

An important characteristic of the large deformation behavior of solid materials is the manner in which the stress increases with increasing strain. In the case of the so-called linear or Hookean behavior, the stress increases linearly with increasing strain according to Equation 3.8 at a constant strain rate. For most food products, the increase in stress with strain will be stronger or less than linear at large deformations (**Figure 4.9**). The products exhibit strain hardening or softening (weakening). Strain hardening and softening are defined as the phenomenon that the stress required to deform a material increases or decreases, respectively, more than proportional to the strain (at constant strain rate and increasing strain).

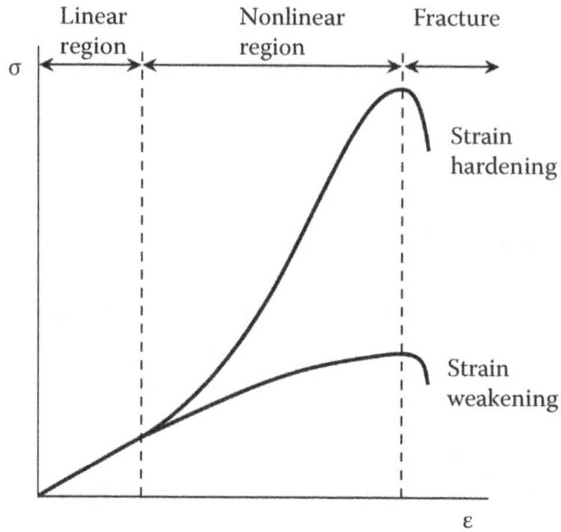

Figure 4.9 *Illustration of strain hardening and strain weakening of materials on large deformation. (Redrawn after Foegeding, E.A. et al.,* J Texture Stud.*, 42, 118, 2011.)*

An equation commonly used to fit nonlinear stress strain relations is the so-called Hollomon equation that reads as

$$\sigma = K_H \varepsilon^n, \tag{4.18}$$

where σ and ε are the true stress and strain (Hencky strain for uniaxial compression experiments) above the yield point (here defined as the point where the relation becomes nonlinear), respectively, K_H is the strength coefficient (equal to σ at γ or ε = 1), and n is the strain hardening exponent (also called the strain hardening index). Note that Equation 4.18 has a similar form to the so-called power law equation for liquids (Equation 4.10). It is an empirical equation and often fits stress–strain curves for various materials to a reasonable degree over a limited range of strain values. Deviations from this relationship are often observed at low and large strains. Most authors who use this equation assume implicitly that the yield point is at zero stress and strain, which may lead to mistakes, similar to the case of fitting plastic flow behavior by Equation 4.10 instead of Equation 4.14.

References

Boekel, M.A.J.S. van. 2009. *Kinetic Modeling of Reactions in Food.* Sections 5.4 and 5.5. Boca Raton: CRC Press.

Foegeding, E.A., C.R. Daubert, M.A. Drake, G. Essink, M. Trulson, C.J. Vinyard, and F. van de Velde. 2011. A comprehensive approach to understanding textural properties of semi- and soft-solid foods. *J. Texture Stud.* 42: 103–136.

Herschel, W.H., and R. Bulkley. 1926a. Konsistenzenmessungen von Gummi-Benzollösungen. *Kolloid Zeitschrift* 39: 291–300.

Herschel, W.H., and R. Bulkley. 1926b. Measurement of consistency as applied to rubber–benzene solutions. *Proc. Am. Soc. Testing Mater.* 26: 621–633.

Holdsworth, S.D. 1993. Rheological models used for the prediction of the flow properties of food products: a literature review. *Trans. Inst. Chem. Eng.* 71 **(Part III)** 139–179.

Huang, H., and J.L. Kokini. 1993. Measurement of biaxial extensional viscosity of wheat flour doughs. *J. Rheology* 37: 879–891.

Jones, D.M., K. Walters, and P.R. Williams. 1987. On the extensional viscosity of mobile polymer solutions. *Rheologica Acta* 26: 20–30.

Reiner, M. 1964. The Deborah number. *Physics Today* 17: 62.

Steffe, J.P. 1996. *Rheological Methods in Food Processing Engineering.* East Lansing, MI: Freeman Press.

5

Fracture and Yielding Behavior

The fracture and yielding behavior of foods is an important quality mark, affecting aspects such as (1) eating quality; (2) usage properties, for example, ease of cutting, grating, and spreading; (3) handling properties during storage and/or further processing in connection with, for example, shape retention and robustness during transport. These aspects pertain to the fracture and yielding behavior of foods under widely varying conditions and time scales. During eating, for example, food is quickly deformed until it fractures; the time scale of this deformation is less than a second. On the other hand, the time scale of shape retention during storage may be of the order of days or (much) longer.

Fracture and yielding are properties that are concerned with large deformations. However, in spite of their importance, at present no conclusive descriptions of fracture and yielding exist that hold for all food products. In fact, in several cases there is no clear distinction between fracture and yielding. Both cause a sudden and significant change in the mechanical properties of the material placed under stress. This will also happen during the breaking of, for instance, the stem of a flower, causing both parts to become separated. Buckling of thin constructions is another example. Buckling of cell walls or long relatively thin beams may happen if cellular material or materials with an open sponge structure are subjected to a compressive stress. A food example is compression of a slice of bread. The discussion in this chapter will mainly focus on the fracture and yielding behavior of food materials. First, some basic notions will be discussed. Next, various aspects of fracture mechanics will be presented. No extended overview of the theoretical aspects will be given. Focus will be placed on aspects that are relevant for manufacturing, handling, usage, and consumption of food products (Section 5.2).

5.1 Basic notions

General characteristics of macroscopic fracture are: (1) simultaneous breaking of the bonds between the structural elements forming the network (e.g., atoms, molecules, particles) in one or more macroscopic planes; (2) breakdown of the structure of the material over length-scales clearly larger than the structural elements, resulting in the formation of cracks; and (3) ultimately, the falling apart of the material into pieces. The first characteristic applies also to viscous flow; bonds will break and reform within the time scale of the experiment. Yielding does not include the third characteristic; it results in a material that flows, but also includes the second characteristic. The initial processes leading to fracture and yielding are the same. Sometimes, yielding precedes fracture—as may be the case, for example, for ductile steel and young Gouda cheese (especially at slow deformation rates) (**Figure 5.1**). In this example, local yielding of the cheese mass takes place at a strain of ~0.1, resulting in a much lower increase in stress with increasing strain than at lower strains. In other cases, it may be unclear whether one has to deal with fracture or yielding. For instance, this becomes evident during strongly deforming or even cutting of rennet-induced milk gels, an important step in cheese making. Then the protein network clearly fractures along visible macroscopic planes, resulting in curd particles, but the whole remains continuous because of the abundance of the solvent, which stems from the fast expulsion of it (syneresis) by the curd particles formed. When syneresis is nearly absent, such as occurs during cutting of milk gels with a pH of about 4.6, the whole system tends to

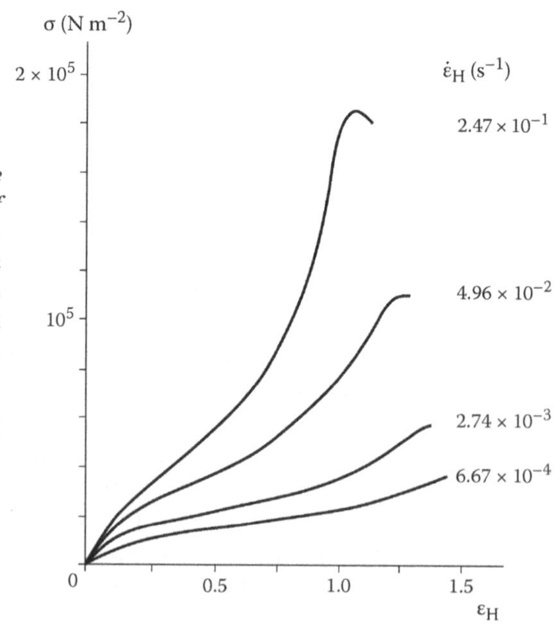

***Figure 5.1** Stress–strain curve in uniaxial compression of 2-week-old Gouda cheese, pH 5.2. (Reprinted from van Vliet, T. et al. in* Food Colloids and Polymers: Stability and Mechanical Properties, *ed. Dickinson, E., Walstra, P., Royal Society of Chemistry, Cambridge, 1993. With permission.)*

fall apart. Therefore, we will speak of fracture in such cases. Visible yielding clearly occurs during spreading of butter or margarine, although at the scale of the bonds between the crystals, fracture will occur.

For the example shown in **Figure 5.1**, fracture has occurred at the end of the curves. It is common practice to define *fracture stress*, σ_{fr} (Pa or N m^{-2}), and *fracture strain*, ε_{fr} (–), as the stress and strain, respectively, at the maximum in the curve (**Figure 5.2**). The area below a stress versus strain curve up to the fracture strain gives the (total) energy required for fracture W (J m^{-3}) of the specimen or the toughness (note that the term "toughness" used in fracture mechanics does not have to be equivalent to the same term used in sensorial description of foods. Moreover, another definition of fracture toughness will be given in Section 5.2.1. The various definitions of toughness will be compared more extensively in Section 5.2.5). It should also be noted that although these definitions of σ_{fr} and ε_{fr} are straightforward, the determination of the fracture stress and strain of a material is less so. For instance, both may depend on such conditions as deformation speed and the way in which the material has been deformed. Furthermore, (large) cracks may already develop in a material before the maximum in the stress is measured, whereas visible fracture may only occur after the maximum. In fracture mechanics, the resistance to deformation of a material is usually called *stiffness* and is quantitatively expressed as the ratio between the stress and strain (therefore, it is expressed in Pa). Often, the slope of the linear portion of the stress–strain curve is taken (**Figure 5.2**), and then the stiffness is equal to Young's or shear modulus in rheology.

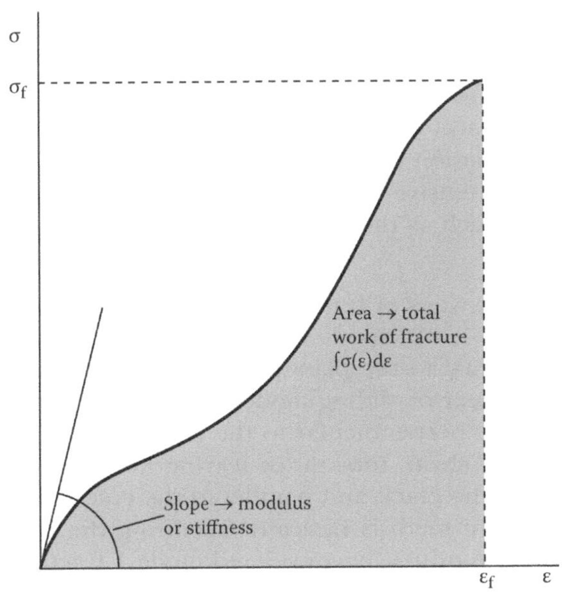

***Figure 5.2** Example of a stress (σ) versus strain (ε) curve for a nonlinear deforming material. Maximum stress and accompanying strain are indicated.*

In the next section, we will first discuss the theoretical aspects of fracture behavior. A distinction will be made between linear elastic, elastic plastic and time-dependent fractures. Linear elastic fracture mechanics has been studied most extensively in view of its importance for mechanical engineers involved in construction work. These aspects are studied in the research area *fracture mechanics* (see below). On the other hand, yielding of materials is studied in rheology. A complete characterization of the mechanical behavior of food materials, including properties related to breakdown during handling, usage, and eating of solid and semisolid food products, requires the combined approaches of rheology and fracture mechanics. We will discuss this more explicitly in Section 5.2.6.

5.2 Fracture mechanics

Fracture mechanics was originally developed for homogeneous brittle, linearly elastic materials by engineers involved in the constructions of buildings, bridges, etc. The basic assumption is that in all materials, small (or larger) defects/cracks are present, which weaken its *strength* (fracture stress). Fracture mechanics deals with the behavior of such cracked bodies subjected to stresses and strains that may arise from applied loads or from self-equilibrated stress fields, as may develop, for example, during drying or cooling of materials. Its power stems from the fact that crack tip phenomena can, to a first order, be characterized by relatively easily measured global parameters, such as crack length and global stress (stress calculated in the absence of the defect). Concepts developed later were extended to include (some) plastic deformation of the material near the crack tip. For many foods, plastic deformation is the largest part of the total deformation before fracture takes place, making fracture behavior time-dependent (or, in other words, deformation rate-dependent). For the latter type of materials, fracture mechanics has been developed to a lesser extent, but these concepts can still be used to understand the fracture behavior of these materials. The discussion will be limited to some basic concepts relevant for food. For a more extensive discussion, we refer the readers to textbooks on fracture mechanics such as those written by Atkins and May (1985) and Anderson (1991).

In fracture mechanics, three different ways of fracturing a material are commonly distinguished. These are mode 1, which refers to fracture in tension or by crack opening (tension stress normal to the plane of the crack); mode 2, which deals with fracture in plane shear or sliding mode (shear stress acting parallel to the plane of the crack and perpendicular to the crack front); and mode 3, which refers to out-of-plane shear, torsion, or tearing mode (shear stress acting parallel to the plane of the crack and parallel to the crack front (**Figure 5.3**). Biting by using the front teeth is basically fracturing the food by wedge penetration and is therefore primarily considered mode 1 fracture. Chewing food between the molars primarily involves mode 3 fracture. In the

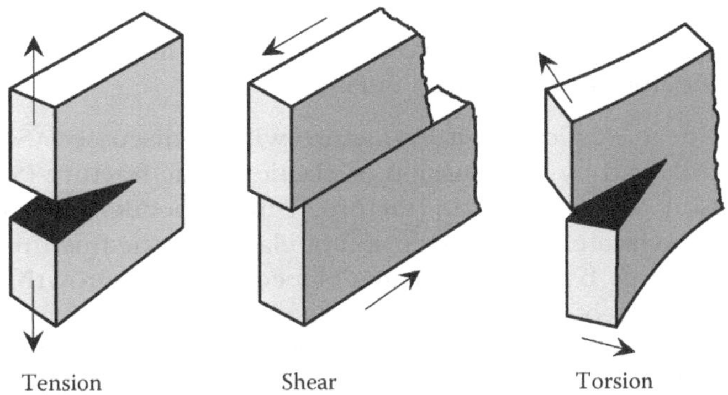

Figure 5.3 *Three modes of fracture: (1) tension, (2) in-plane shear, and (3) out-of-plane shear or torsion. (Adapted from Vincent, J.F.V., in* Biomechanics—Materials: A Practical Approach, *ed. Vincent, J.F.V., Oxford University Press, Oxford, 1992.)*

following discussion, we will primarily consider mode 1 fracture except when indicated otherwise.

In uniaxial compression or tensile testing, the fracture mode will not only depend on the way the material is deformed, but also on material properties. When a cylindrical piece of material is compressed uniaxially, it is deformed both in shear and in tension. During the compression, the circumference of the cylinder will increase, giving tensile stresses along its circumference (**Figure 5.4**, upper panel). In addition, as schematically illustrated in **Figure 5.4** (lower panel), the upper and lower parts of the specimen are compressed in the direction of the applied force, whereas the outside parts are forced sideways, giving rise to shear forces along the diagonals. For the first case, start of

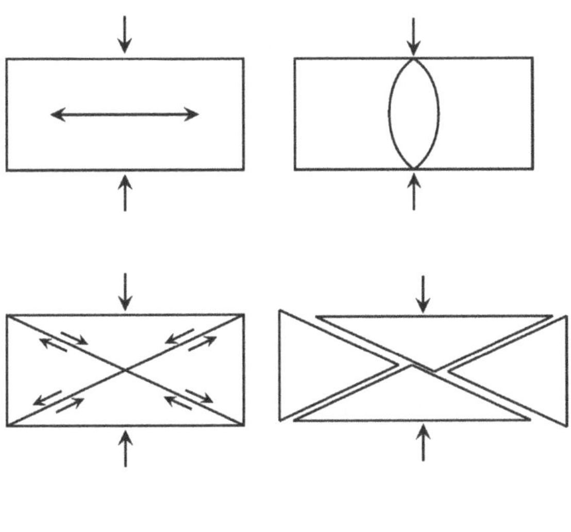

Figure 5.4 *Features of fracture in tension (upper panel) and shear mode (lower panel) during uniaxial compression of a test piece.*

fracture can mostly be noticed with relative ease, but when fracture is due to the formed shear stresses, extensive crack formation may already occur inside the test piece before it can be seen outside.

Below, first linear elastic or brittle fracture will be discussed (Section 5.2.1), and will be followed by a discussion of elastic plastic fracture (Section 5.2.2) as an extension of linear elastic fracture. Time-dependent fracture (Section 5.3.3) will not be treated as an extension of elastic plastic fracture, but will be based on the energy balance governing (speed of) crack growth. The discussion on the aspects of fracture mechanics will be finalized by brief discussions of halting crack propagation (Section 5.2.4); the parameters fracture stress, work of fracture, toughness, and fracture toughness (Section 5.2.5); and the difference between fracture and yielding (Section 5.2.6).

5.2.1 Linear elastic or brittle fracture

Materials showing *linear elastic* or *brittle fracture* are characterized by a linear relation between the stress and strain up to the fracture stress σ_{fr} and fracture strain ε_{fr} and ideally elastic behavior. The area below the stress versus strain curve is given by $(1/2)\sigma_{fr}\varepsilon_{fr}$. Linear elastic or brittle fracture mechanics generally holds for crystalline and glassy materials or their combinations. Food examples are sugar and salt crystals, potato crisps, toasts, and many dry powder particles. As noted above, the main assumption in fracture mechanics is that all materials are inhomogeneous, so that on deformation local stress concentrations arise. Below, first effects due to these stress concentrations will be discussed for isotropic materials and after that for materials with a clear anisotropic structure such as meat.

5.2.1.1 Effects due to stress concentration

The most important inherent properties of a material determining its fracture behavior are the mechanical properties of the structural elements, the strength and number of the bonds in between, and the inhomogeneity of the material. For homogeneous crystalline materials, the fracture stress can be predicted from the known strength and the geometry of the crystal structure. However, it is generally observed that the experimentally measured fracture stress is much (1–3 orders of magnitude) smaller than the theoretical value. For instance, for a homogeneous piece of material made of glass, the theoretical fracture stress is 10 GPa, whereas the actual fracture stress is about 0.1 GPa. This discrepancy is primarily attributed to the material being inhomogeneous due to, for example, imperfections in crystal structures or in gel network structure, or presence of gas cells, large pores, or even small cracks. In fact, all inhomogeneities present in a material can be considered tiny cracks. Stresses exerted on an isotropic material will be concentrated at the tip of these cracks, the extent of which will depend on the size and the shape of the "cracks." The stress σ at the tip of a through thickness elliptical crack in an ideal, isotropic, infinite plate of elastic material loaded in tension perpendicular to the direction of the long axis of the crack is given by

$$\sigma = \sigma_0(1 + 2l/b) = \sigma_0\left[1 + 2\sqrt{(l/r)}\right], \tag{5.1}$$

where l is half of the long axis of the ellipse, b is half of the short one, σ_0 is the average stress remote from the crack, and r is the radius of curvature at the crack tip (= b^2/l). As r decreases, the crack tip becomes sharper and the stress concentration factor [1 + $2\sqrt{(l/r)}$] increases. For sharp cracks, this factor may become very high (e.g., 200 for a crack 10 μm long and 1 nm wide as may be present in crystals). A well-known example of the effect of stress concentration is that caused by bringing about a linear notch (scratch) in a glass pane by a diamond knife. This enormously decreases the force needed to break the pane. For circular holes, $l = r$ and the stress concentration factor will be 3, independent of their size as long as this is substantially smaller than the size of the material.

A disadvantage of Equation 5.1 is that it gives σ only at one point of the sample and does not indicate how it falls off to the remote stress σ_0. In general, the stress σ_{yy} in an infinite plate loaded in tensile deformation (mode 1) in the y direction at a certain place near the crack tip is given by

$$\sigma_{yy,1} = \frac{K_1}{\sqrt{(2\pi R)}} f_{yy,1}(\theta), \tag{5.2}$$

where K_1 is the so-called *stress intensity factor* for tensile deformations (mode 1), R is the distance from the crack tip, and θ the angle with respect to the plane of the crack for the point where $\sigma = \sigma_{ij}$ (**Figure 5.5**). The subscripts yy denote the component of the stress tensor involved, that is, for tensile deformation xx, yy, or zz for loading in the x, y, or z direction, respectively, and xy, etc., for shear loading (**Chapter 2**). The parameter $f(\theta)$ is a dimensionless function of θ. K depends only on the deformation mode, whereas $f(\theta)$ depends both on the deformation mode (subscript 1) and the component of the stress tensor being considered (subscript yy), and with that, so does σ. Because $f_{yy}(\theta)$

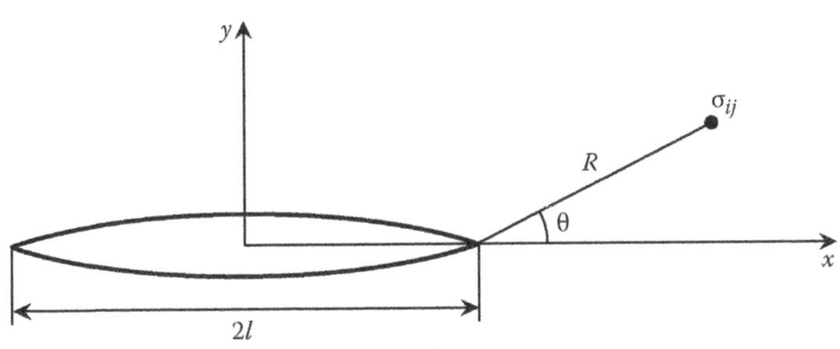

Figure 5.5 *Stress at a point in front of a crack tip. (Redrawn from Ewalds, H.L., Wanhill, R.J.H.,* Fracture Mechanics, *Delftse Uitgevers Maatschappij, Delft, 1985.)*

is dimensionless, K_1 is expressed in N $m^{-1.5}$ or Pa $m^{0.5}$. Similar equations, such as Equation 5.2, hold for the other two deformation nodes (for an overview of these equations and the equations defining various $f(\theta)$, see, e.g., Anderson, 1991, **Chapter 2**). The decrease in σ with $1/\sqrt{R}$ near the crack tip is independent of deformation mode and the component of the stress tensor considered. At larger distances from the crack tip, σ will no longer be proportional to $1/\sqrt{R}$ but levels off to the remote stress σ_∞. Finally, note that Equation 5.2 predicts that σ goes to infinity for $R \to 0$. We will return to this point in Section 5.2.2.

The stress intensity factor is used in fracture mechanics to more accurately predict the stress state (stress intensity) near the tip of a crack caused by a remote load or residual stress. It is useful for providing a failure criterion for brittle materials. The critical stress intensity factor K_C at which fracture of a material occurs is often called *fracture toughness.* Note that in the literature, (fracture) toughness may also be used to indicate the area below the stress versus strain curve up to fracture (Sections 5.1 and 5.2.5).

K gives the magnitude of the elastic stress field near the crack tip. It is given by

$$K = \sigma_0 \sqrt{(\pi l)}\, Y, \qquad (5.3)$$

where Y is a dimensionless parameter that depends on the geometries of the crack and the object and on the manner in which the crack is loaded. Thus, the relation between σ and l is as shown in Equation 5.1. For a throughout crack with a length $2l$ in an infinite plate under uniform tension, $Y = 1$ and 1.12 for a surface crack (notch) in a plate with a semi-infinite width. For an extensive discussion of the value of Y, readers are referred to the literature (e.g., Anderson 1991, **Chapter 2**).

5.2.1.2 Notch sensitivity in relation to material structure

In the preceding discussion on stress concentration, the starting point was an isotropic material; however, many foods are not isotropic. For instance, meat consists of fibers with relatively weak connections between them. It will be immediately clear that the strength of the meat and of the original muscle is much higher in the direction of the fibers than that perpendicular to it. Another difference with homogeneous materials is the much lower notch sensitivity, for example, the sensitivity to a small cut by a knife. A general definition of *notch sensitivity* is the extent to which the sensitivity of a material to fracture is increased by the presence of a surface inhomogeneity such as a notch or a crack. Notch sensitivity has been found to depend on the material structure and plasticity. In this section, only the first dependence will be discussed. The second one will be discussed briefly in Section 5.2.2.

Notch sensitivity can be illustrated using **Figure 5.6**, which shows a schematic layout of a material built from long strands of aggregated structural

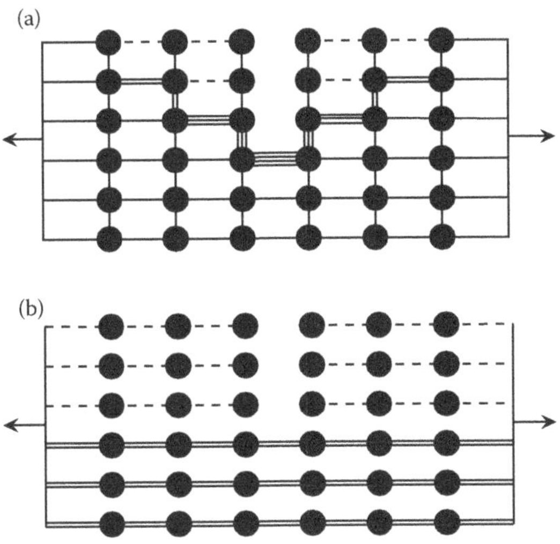

Figure 5.6 *Schematic representation of a material without shear connections between stress carrying trajectories (b) and with shear connections in between (a). Both materials are loaded in horizontal direction. Black circles denote structural elements in the material, and straight lines denote connections between structural elements. Full lines indicate connections carrying a load and dashed ones indicate stress-free connections. Number of full lines between structural elements is roughly proportional to stress exerted on the connection between the structural elements involved.*

elements with no connection between the fibers (**Figure 5.6b**) and a material with bonds between the structural elements perpendicular to the loading direction (the so-called shear connections between the "strands of aggregated structural elements") (**Figure 5.6a**). Examples of the first type of material/products are meat, puff pastry, and rope, and for the second type of materials, examples include food gels, cheese, and glass. In **Figure 5.6**, in both types of materials, a notch (crack starting at the surface of the material) is brought about with a depth of three strands. When material b is subjected to a uniform tensile stress, only the bonds between the structural elements in the fibers that are not cut will be loaded. The stress on these bonds will be

$$\sigma = \sigma_0 \frac{w}{w - l} \tag{5.4}$$

where w is the width of the specimen and l is the depth of the notch. For a linearly elastic material, Equation 5.4 results in curve 1 in **Figure 5.7** for the fracture stress as function of the notch length. Such a relation is typical for the so-called notch insensitive materials.

For the material illustrated schematically in **Figure 5.6a**, the stress exerted on the end points of the "strands" that are cut in two will be transferred to the adjacent "strands" (stress trajectories), resulting in stress concentration at the tip of the crack to an extent that is, in principle, given by Equation 5.1. Curve 2 in **Figure 5.7** gives the stress as a function of the notch length for such a material.

The discussion given above for isotropic materials implies that these will be sensitive for artificial notches applied to them by, for example, cutting with a knife or by biting with the front teeth since these are larger than half

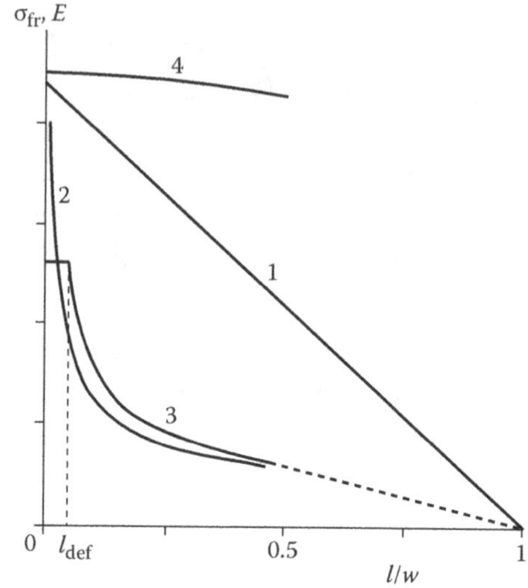

***Figure 5.7** Dependence of fracture stress σ_{fr} on length l of an applied notch for a notch insensitive material (curve 1) and for a notch sensitive material (curve 2) and curve 3 for a material with large inherent defects with length l_{def}. Curve 4 illustrate dependence of shear or Young's modulus on length l of a through notch for a test piece with a length four times its width; w is test piece width (Redrawn from Walstra, P.,* Physical Chemistry of Foods, *Marcel Dekker, New York, 2003.)*

the length of the inherent defects present in the material. No effect will be observed on the fracture stress as long as the applied notch is shorter than half of the inherent defects present in the material (**Figure 5.7**, curve 3). In this way, it is possible to determine the inherent defect lengths present in materials. Some examples are given in **Table 5.1**. The data for the potato starch gels agree well with the size of the partly swollen starch granules. Brittle materials such as potato crisps are extremely notch sensitive. In fact, most food products are notch sensitive, although its extent varies between products; examples are semihard and hard cheeses, fresh carrots, and apples. Products such as Gouda and Cheddar cheese become more notch sensitive upon aging, that is, they exhibit more brittle behavior. The inherent defect length of chocolate bars is quite high (1–2 mm), which causes them often to fracture next to the notch that can be seen on top of the bar.

Figure 5.7 illustrates the strong dependence of the fracture stress of a material on the applied notch length, and with that also on the size of the defects inherent to the specimen. This behavior is highly different from the dependence

Table 5.1 Inherent Defect Lengths Present in Some Food Materials (Approximate Numbers)

Food	Inherent Defect Length
Fresh cheese/pressed curd for Gouda cheese	2–3 mm
Gouda cheese	0.1–03 mm
Chocolate bars	1–2 mm
Potato crisps	10 μm
10% potato starch gels	0.2 mm
30% potato starch gels	0.1 mm

of the modulus of a material on the size of inherent defects present in the material. For notches smaller than half the width of a test piece, the modulus decreases only to a small extent with their increasing length, whereas for notch-insensitive materials σ_{fr} decreases linearly with the length of the notches or cracks present, and for notch-sensitive materials the decrease in σ_{fr} is much stronger still. The reason for this difference is that a modulus is given by the ratio of the stress over the strain. For a constant strain, the stress will be proportional to the number of bonds that are deformed in the material. This implies that the stress and with that the modulus will decrease only proportionally to the number of bonds that become elastically inactive (see Section 13.1 for further elucidation). The latter ensures that the decrease in modulus depend not only on l/w, but also on the length of the test piece. A notch with a length of 60% of the test piece width and 15% of its length leads to a decrease in modulus by only ca. 9%.

5.2.1.3 Crack propagation

Cracks and notches will start to grow (fracture is initiated) if the local stress at the crack tip exceeds the strength of the bonds between the structural elements giving the solid(-like) properties to the material. Below, the growth of a through-thickness crack in a plate is considered in a rectangular sheet of material subjected to a tensile stress in its longest direction and perpendicular to the crack (mode 1 deformation). During the growth of a crack, the strain energy in the material in a certain volume adjacent to the growing crack is released (**Figure 5.8a**, gray area). This energy can be used for further crack growth if it is transported to the tip of the crack. The strain energy density W_e (J m^{-3}) stored in a deformed ideally elastic material is

$$W_e = \int_{\varepsilon=0}^{\varepsilon_f} \sigma_0(\varepsilon)\, d\varepsilon \tag{5.5}$$

During crack growth, the amount of energy needed is proportional to the crack length, and the strain energy that is released will be proportional to the square of the crack length formed. Fracture will propagate spontaneously when the differential strain energy released during crack growth surpasses the differential energy required (**Figure 5.8b**),

$$W_e \pi d[(l + \delta l)^2 - l^2] \geq G_C d\delta l, \tag{5.6}$$

where d is the width of the crack, l is half its length (or the length of a notch), δl is the infinite small increase in crack length, and G_C is the *critical strain energy release rate* (often also called work of fracture) (J m^{-2}), that is, the energy needed to create one unit area of new crack area (i.e., two new surfaces). G_C is two times the *specific fracture energy*, R_s, which is often used in older literature on fracture mechanics. This quantity is expressed in energy per unit area surface formed.

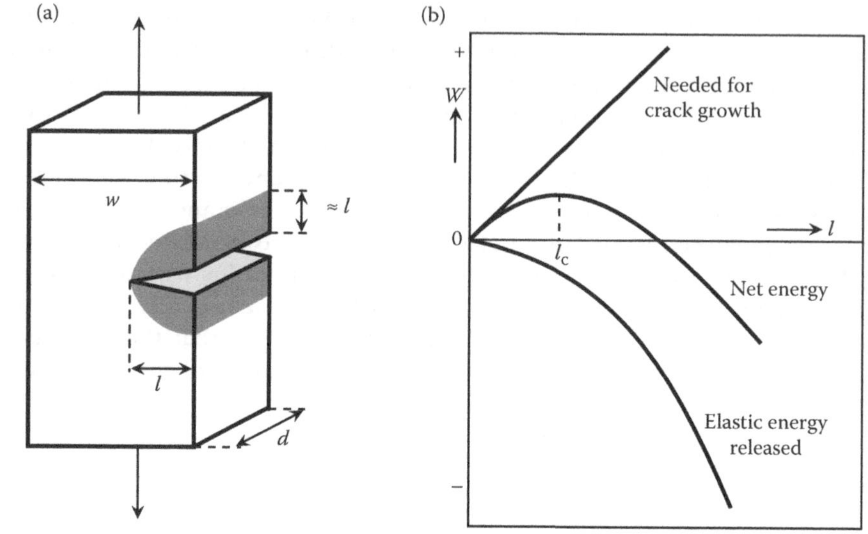

***Figure 5.8** Fracture in tension of a test piece with a width w and thickness d containing a notch of length l. (a) Geometry of the test piece. (b) Energies W involved as function of the growing notch length. l_c critical notch length for crack propagation. (Redrawn after Walstra, P.,* Physical Chemistry of Foods, *Marcel Dekker, New York, 2003.)*

During fracture propagation, crack growth speeds can become very high. Beyond a certain velocity, the crack growth in brittle materials becomes faster and unstable; the fracture surface is no longer smooth, but becomes increasingly complicated with side cracks and often splinter formation. Above this speed, sound is emitted (Fineberg and Marder 1999). This critical speed is related to the maximum speed of stress waves in a material, the so-called Rayleigh speed and is likely about one-fourth to one-half of this speed. The Rayleigh speed is equal to $\sqrt{(E/\rho)}$, where ρ is the density of the material and thus depends on material properties. For a porous material, one should consider the properties of the solid component. For most brittle food products, E is about 1–2 × 10^9 Pa and ρ is about 1–1.5 × 10^3 kg m^{-3}, implying a critical crack speed for sound emission of about 300–500 m s^{-1}.

Equation 5.6 gives as lower limit for spontaneous crack (fracture) propagation:

$$G_C = 2\pi l_c W_e, \qquad (5.7)$$

where l_c is the critical crack length for fracture propagation. This equation clearly shows the importance of the size of inhomogeneities present for the onset of crack propagation.

Basic assumptions in linear elastic fracture mechanics are that initially all deformation energy supplied to a material is stored and can be used for fracture propagation and that $E = \sigma/\varepsilon$ is constant independent of ε, and $W_e = (1/2)\sigma\varepsilon$. G_C is then given by

$$G_C = 2\pi l_c W_e = \frac{\pi l_c \sigma_{fr}^2}{E}, \quad (5.8a)$$

where σ_{fr} is the overall fracture stress at the onset of fracture propagation (for linear elastic or brittle fracture equal to the so-called fracture stress). Because G_C and E are material constants, σ_{fr} is proportional to $1/\sqrt{l}$. Hence, a curve that shows σ_{fr} at propagation as a function of the crack length is similar (but not identical) to the crack initiation line. Equation 5.8a can be rewritten to the following expression for the critical crack length at which fracture propagation will start

$$l_c = \frac{G_C}{2\pi W_e} = \frac{EG_C}{\pi\sigma_{fr}^2} \quad (5.8b)$$

Equation 5.8b can also be read in the sense that the critical crack length for a material to fracture depends strongly on the stress applied on it. At the moment that the applied stress becomes so high that the related l_c becomes shorter than cracks present in the material, fracture propagation will occur. This will also be the case for materials without added notches. The parameter l_c is than the characteristic length of the inherent defects, cracks, and weak points present in the material. This implies that the fracture stress/strength of a food product can be regulated by adapting the size of the inherent defects in the material, for example, by processing.

An alternative and at present often more favored way to the energy approach discussed above is the so-called critical stress intensity approach. The *critical stress intensity factor* K_{IC} is the minimum value of the stress intensity factor above which crack propagation occurs. The subscript I denotes fracture in tension (mode 1). Based on Equation 5.3, one may write for K_{IC}

$$K_{IC} = C_1 \sigma_{fr} \sqrt{l_c}, \quad (5.9)$$

where C_1 is a constant. For a through-thickness crack of length $2l_c$ in an infinite plane subjected to a tensile stress at a right angle to l_c (so-called plane stress state), $C_1 = \sqrt{\pi}$. Since, for a linearly elastic material, the strain (or deformation) energy W_e to be supplied up to fracture is $(1/2)\sigma_f\varepsilon_f = G_C/2\pi l_c$ and $E = \sigma_f/\varepsilon_f$, Equation 5.9 can also be written as

$$K_{IC} = C_2 \sqrt{EG_C} \quad (5.10)$$

This equation illustrates that it is primarily the energy balance that governs fracture behavior (especially crack propagation, and this is what is usually noticed as fracture) and not the stress or strain level. Although somewhat exaggerated, one can say that, in fact, the stress level is an essential boundary condition for fracture to take place. The combination of Equations 5.9 and 5.10 results in the same expression for the critical crack length l_c as Equation 5.8b for $C_2 = 1$. Equation 5.10 is often also written as $G_C = K_{IC}^2/E$.

K_{IC} is a fundamental material property (unit N $m^{-1.5}$ or Pa $m^{0.5}$), which depends on the structure of the material as a function of length scale, temperature, and—for many food materials—strain rate. For viscoelastic materials, this is, among others, due to the dependence of E and σ_{fr} and less often of ε_{fr} on strain rate. As has been shown by Vincent (2004), K_{IC} can be used to characterize the brittle fracture behavior of various types of fruit and vegetables (Section 18.3). A clear relation was observed between sensory hardness and K_{IC} for three different vegetables and three varieties of apples.

5.2.2 Elastic plastic fracture

Linear elastic fracture mechanics is only valid as long as nonlinear material deformation is confined to a small region around the crack tip. For many food products, this is not the case. Various food products show so-called *elastic plastic* or *ductile fracture.* These products are characterized by a higher energy (toughness) needed for fracture than products showing linear elastic or brittle fracture. Elastic plastic fracture mechanics applies for materials that exhibit nonlinear deformation rate-independent deformation behavior. Brittle products such as biscuit, wafers, and potato crisps will exhibit ductile fracture at somewhat higher water content than that found in the dry state, what can easily be recognized sensorially by much lower sound emission or its absence and increasing toughness. Moreover, an increase in temperature may lead to a transition from brittle linear elastic fracture to ductile fracture, as a result of its effect on the modulus as well as the yield stress of the food material.

As the name suggests, in linear elastic fracture mechanics, it is assumed that the increase in stress on deformation is a linear function of the strain. In elastic plastic fracture mechanics, this is no longer the case. To compensate for the nonlinear deformation behavior, a nonlinear energy release rate J is introduced that is equal to G_C for conditions where linear elastic fracture mechanics holds. For its calculation, the changes in the area under the stress versus strain curve during crack growth have to be considered (Anderson 1991). In view of the introductory character of this book and its focus on concepts instead of mathematical rigor, we will refrain from discussing this parameter further.

As discussed in Section 5.2.1.1, according to Equation 5.2 the stress σ will become infinite in the vicinity of a crack tip where R approaches zero, that is, there is a stress singularity at the crack tip. Many solid materials show yielding at a high stress, which implies that local yielding may occur near the crack tip. This causes the crack tip to become less sharp and limits the stresses to finite values (**Figure 5.9**). So, in reality, a plastic zone may be formed surrounding the crack tip, preventing the linear elastic fracture approach being applied unconditionally. Concepts developed in the scope of linear elastic fracture

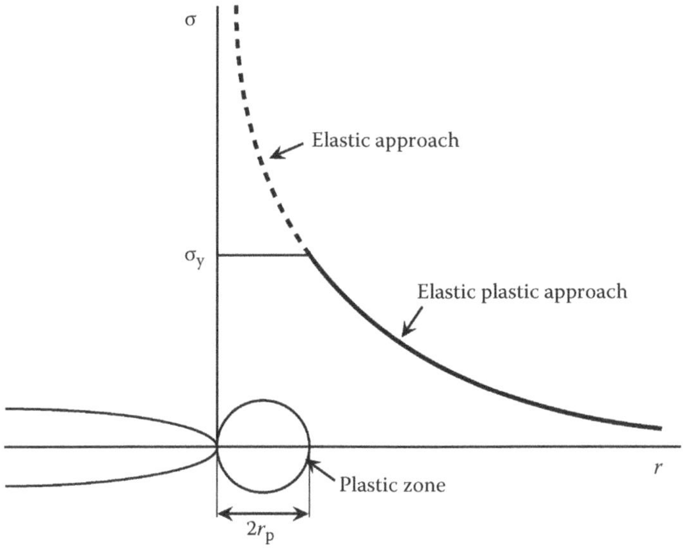

Figure 5.9 *Stress σ in front of a crack tip in absence and presence of yielding of material near the crack tip (schematic figure). r, distance from crack tip;* r_p*, radius plastic zone;* σ_y*, yield stress.*

mechanics can still be applied as long as the plastic zone is small compared to the crack length. For this condition, the effect of crack tip plasticity corresponds to an apparent increase in elastic crack length by an increment about equal to the radius of the plastic zone. Outside the plastic zone, the stresses and strains will be determined by K, and the stress intensity approach can still be used.

The consequences of extensive yielding of the material around the crack tip are as follows: (1) The tip becomes blunted, that is, a larger r in Equation 5.1, implying that the stress concentration becomes less. The stress at the crack tip will be equal to σ_y. Blunting of the crack tip has been observed visually during fracture experiments of Gouda cheese in tension. It causes the stress concentration at a certain defect length to become smaller (Equation 5.2). (2) Local plastic deformation occurs, which means that the pieces remaining after fracture no longer precisely fit each other as they will do after linear elastic fracture. (3) The work of fracture will be higher, because it now also includes the energy dissipated due to the local yielding. (4) The energy balance determining crack propagation is affected.

During crack propagation, the strain energy stored in the test piece is used to provide the extra energy necessary for the formation of two new surfaces (surface energy, G_s) and for the plastic deformation (G_p) near the crack tip (both expressed in J m^{-2}). As long as plastic deformation occurs only in a small zone around the crack tip compared to the test piece size, Equations 5.6

through 5.8 can still be used in a modified form taking into account that the critical strain energy release rate for elastic plastic fracture $G_{C,p}$ is given by

$$G_{C,p} = G_s + G_p \tag{5.11}$$

For a relatively small size of the plastic zone around the crack tip, G_p will already be much larger than G_s. It causes, for instance, the much higher work of fracture of mild or ductile steel (~10^5–10^6 J m^{-2}) than for pottery (1–10 J m^{-2}), whereas the fracture stresses are about the same (2–4 × 10^5 kN m^{-2}).

Another consequence of the occurrence of yielding of a material under high stresses is that brittle fracture will be inhibited if the size of the specimen becomes of the same order as the plastic zone formed at the tip of a crack. This has important consequences for size reduction processes such as during milling and grinding. The size l_p of the plastic zone around a crack tip can be estimated by substituting the yield stress σ_y in Equation 5.8b:

$$l_p = C\frac{EG_{C,p}}{\sigma_y^2}, \tag{5.12}$$

where C is a geometric factor depending on the size and shape of the specimen. If the diameter of a particle is smaller than l_p, ductile yielding will occur in the whole particle before it fractures and its size cannot be lowered any more by milling, etc. Expressed in another way, not enough elastic energy W_e can be stored in the material to provide the energy needed for fracture propagation. Since, for most materials, the yield stress is more strongly dependent on temperature than the elastic modulus, lowering the temperature may decrease l_p, thereby allowing grinding to smaller particles. Because E and σ_y depend on the deformation rate, the same will also apply to l_p.

By their nature, soft solids will not exhibit linear elastic fracture; some will show elastic plastic fracture, but many of them will, as a whole, behave viscoelastically and/or show extensive yielding over a major part of the deformed material. In such cases, the concepts developed for elastic plastic fracture mechanics cannot be applied unconditionally. Moreover, for these products, fracture behavior will depend on the speed of deformation.

5.2.3 Time-dependent fracture

As discussed above, fracture occurs when the rate at which strain energy is released per unit formed crack area exceeds the rate at which surface energy is adsorbed in creating new fracture surfaces. The energy analysis given in the part on crack propagation assumes that only the elastically stored energy and the critical strain energy release rate (or the critical stress intensity factor) play a role (Equations 5.7 and 5.9). In elastic plastic fracture, a relatively small contribution of energy dissipation is accounted for, due to the plastic deformation around the crack tip. However, for many products, energy dissipation into heat occurs in the main part (or whole) of the material during its deformation and during crack growth. This can be energy dissipation due to different mechanisms such as plastic (viscous) deformation of a viscoelastic material ($W_{d,v}$), friction between structural elements due to inhomogeneous deformation of the material ($W_{d,f}$), viscous flow of the liquid in the matrix $W_{d,l}$ (e.g., noted as serum release during compression of gels), or energy dissipation due to debonding of the interface between structural elements ($W_{d,b}$) (e.g., between fibrils in muscles). The extent of energy dissipation mostly depends on the deformation rate of the material causing the fracture behavior to become deformation rate-dependent. To emphasize this rate dependency, the term "time-dependent fracture" is used in this book—and not the alternative term "fully plastic fracture mechanics"—to indicate this type of fracture behavior.

For products exhibiting time-dependent fracture behavior, the quantitative relations between $G_c K_{1C}$ and strain energy developed within linear elastic and elastic plastic fracture mechanics are inadequate. However, this does not imply that concepts such as notch (defect) sensitivity due to stress concentration, difference between crack initiation and propagation, and the existence of a critical crack length for crack propagation, no longer hold. The main factor one has to be aware of is that the quantitative aspects of these concepts will depend on the speed of deformation of the material studied. In principle, this can be accounted for by making G_C (or J) a deformation rate-dependent parameter; however, this does not provide a direct understanding of the effect of the various mechanisms causing time-dependent fracture. Moreover, the mathematics involved is quite complex. Below, we will follow a qualitative approach directed on understanding the different factors determining time-dependent fracture behavior. An alternative may be based on using the critical strain energy release rate as a starting point, but this would allow a less straightforward illustration of the effect of different energy dissipation mechanisms.

For time-dependent fracture, the growth mechanism of cracks can be understood by starting with the following energy balance,

$$W = W_e + W_{d,v} + W_{d,f} + W_{d,l} + W_{d,b} \; (+ \; W_{fr}), \tag{5.13}$$

where W is the amount of supplied energy, W_e is the elastically stored energy, and W_{fr} is the fracture energy (all in J m^{-3}).

The term W_{fr} is put between parentheses since it is derived from W_e during the fracture process. Initially, during the deformation of the specimen, $W_{fr} = 0$. Fast growth of cracks occurs if W_e is so high that the energy released due to stress relation in the material in the vicinity of the formed crack exceeds W_{fr}.

In linear elastic fracture, the sum of W_e and W_{fr} is assumed to be constant as soon as fracture propagates spontaneously. In elastic plastic and much stronger time-dependent fracture, this is no longer the case, for several reasons: (1) part of the energy is dissipated because of the plastic deformation around the crack tip (similar to elastic plastic fracture) and, in addition, (2) energy dissipation takes place throughout the whole specimen. The latter will be the case both during deformation of the specimen before fracture occurs and during the fracturing process. Part of the elastically stored energy that comes free during stress relaxation around a growing crack is directly dissipated due to flow and/or inhomogeneous deformation of the material farther away from the crack tip.* Cause (2) will be dominant during time-dependent fracture.

As a result of the often extensive energy dissipation processes during crack (fracture) propagation, less energy is available for crack growth than there would be in their absence. The final result may be that crack growth speeds will be low, down to 1 cm/min or even less. This implies that, for larger test pieces, fracture propagations take some time to occur. During this time, deformation may continue, leading to an ongoing addition of W_e to the system. In addition, extensive plastic deformation in front of the crack tip leaves the tip strongly blunted. Therefore, a crack will be a much less effective stress concentrator than in a brittle material. These processes together cause fracture initiation (start of the growth) and fracture propagation (spontaneous crack growth) to become clearly separated, in contrast to what is the case for brittle fracture. An example of it is shown in **Figure 5.10**.

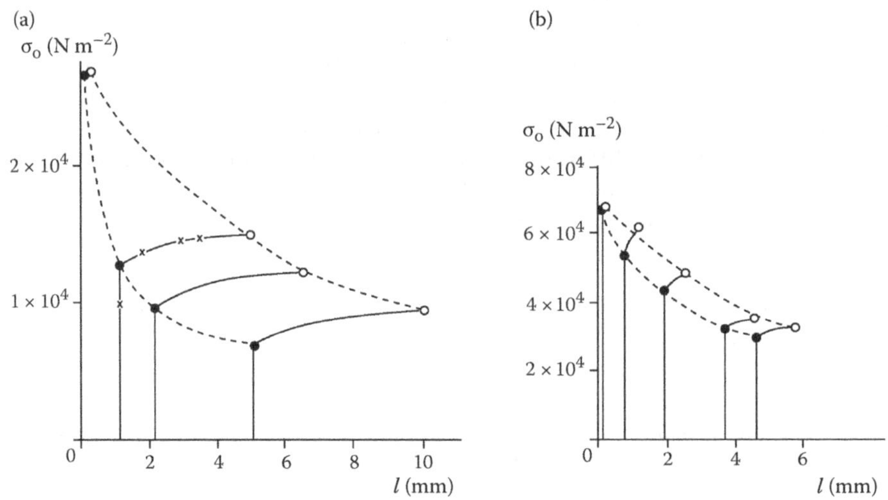

Figure 5.10 *Fracture initiation (●) (start of visible fracture) and propagation (○) (i.e., fast fracture) for young (a) and mature (b) Gouda cheese during tensile experiments. Width of test pieces, 20 mm. Measured average stress remote from crack tip* σ_0 *versus crack length* l *is shown for one example (×). (Reprinted from Luyten, H. et al.,* Neth. Milk Dairy J., *45, 55–80, 1991. With permission.)*

* Foods often consist of various structural elements having different moduli. As a result, when a stress is applied to the food, they may deform to varying degrees in such a way that at large applied deformations, they may become displaced with respect to each other at mutual interfaces, leading, for example, to energy losses due to friction or debonding processes.

The difference between fracture initiation and propagation can be noticed by biting into a cube of young (Gouda or similar) cheese (e.g., with ribbons of 1.5 cm). For a young cheese, it is necessary for the teeth to penetrate quite deeply into the cheese, which implies that a substantial amount of energy must be applied into the test piece before fracture can propagate. For a mature cheese, less penetration is required for a fracture to propagate (Luyten et al. 1991).

As mentioned above, energy dissipation can occur both during deformation of a specimen as well as during transport to the crack tip of the strain energy that becomes available because of stress relaxation during crack growth. The part of the energy that is dissipated during deformation can be estimated by cycle (or so-called hysteresis) experiments. In a hysteresis experiment a sample is loaded and unloaded, resulting in a stress–strain curve as shown in **Figure 5.11**. For an ideally elastic material, the loading and unloading curves will fall over each other, but for other materials they will not. Often, the part of the energy that is recovered depends on the applied strain and strain rate. For instance, for a 1-month-old Gouda cheese, the percentage of recovered energy decreased from about 90% to about 20% when the maximum Hencky strain was increased from 0.1 to 1.2 for a strain rate of 0.17 s^{-1}, and from 35% to 10% when the strain rate was 2.8×10^{-4} s^{-1}. The exact determination of recoverable energy for viscoelastic materials is difficult, because energy dissipation (viscous flow) may also occur during unloading, giving too low a value. Moreover, there may be some delayed elastic recovery after unloading has been completed.

Somewhat arbitrarily, in Equation 5.13, energy dissipation into heat has been split into four terms according to the mechanisms causing energy dissipation. Which term(s) are important will depend on the material (food) studied. Below, the effects of energy dissipation due to viscoelasticity of the material and as a result of friction will be discussed more extensively. Effects on fracture behavior

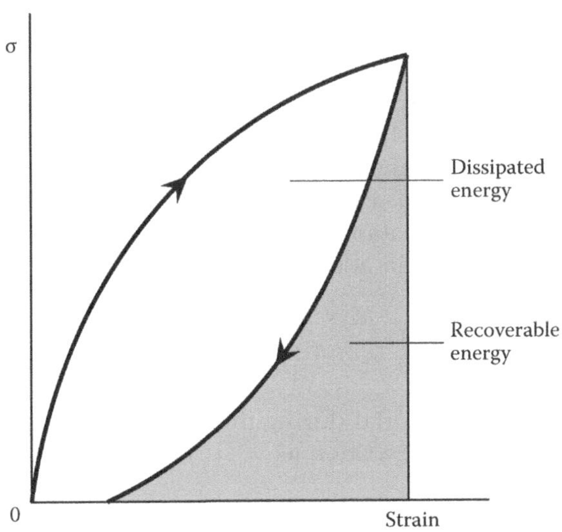

Figure 5.11 *Stress (σ) versus strain (ε_H) curve for an experiment in which the test piece is first loaded and next unloaded (cycle or hysteresis experiment).*

due to flow of the liquid in a gel are qualitatively very similar to effects due to viscoelasticity and therefore will not be discussed further. The effect of debonding of interfaces between structural elements will be discussed in Section 5.2.4.

5.2.3.1 Effects due to viscoelasticity

As discussed in Section 4.1, the viscoelastic behavior of a material implies that the reaction of such a material to a stress consists of a viscous component and an elastic component. In general, viscoelastic materials behave in a more viscous manner at low deformation speeds than at high deformation speeds. Assuming that this behavior is roughly independent of the strain, this implies that at low deformation speeds a relatively larger part of the supplied deformation energy W is dissipated as heat than at high deformation speeds. In terms of the Deborah number τ_{rel}/t_{obs} (Section 4.2), where t_{obs} is 1/strain rate $\dot{\varepsilon}$ or $\dot{\gamma}$, the ratio $W_e/W_{d,v}$ will be relatively high when De is high (high strain rate) and low when De is low. This will also be the case for the energy becoming available as a result of stress relaxation during crack growth. At low deformation speeds, the transport of this energy to the crack tip will be less efficient. Another effect is that at a low strain rate (deformation speed), there is enough time for stress relaxation, implying that the stress W involved in the deformation of the material stays low. This, the lower ratio of $W_e/W_{d,v}$, and the relatively less efficient transport of energy to the crack tip indicate that the material has to be deformed to a greater extent before W_e becomes high enough for crack propagation to occur, despite the fact that W_{fr} will also be lower.

There is no good and validated theory available that would allow the quantification of these different effects due to viscoelasticity. Only several qualitative remarks can be made. By subjecting a material to a sinusoidal oscillating stress or deformation, a measure can be determined of the energy stored in the material during a periodic (e.g., sinusoidal varying) application of a stress, the so-called storage modulus G', and of the energy dissipated in the material, the so-called loss modulus G'' (for an extensive discussion, see Section 7.4). The loss modulus can be recalculated in a dynamic viscosity η' by dividing G'' by the oscillation frequency (in rad s^{-1}) at which the stress was varied. If we assume, as a first-order approximation, that the ratio of the viscous and elastic contributions to the mechanical behavior of a material does not depend on the extent of deformation, the energy dissipation due to viscous flow may, in principle, be calculated from η'. Assuming, in addition, a homogeneous flow during deformation, one can write

$$W_{d,v} = \eta'\dot{\varepsilon}_v^2 t, \tag{5.14}$$

where $\dot{\varepsilon}_v$ is the viscous flow rate of the material during the deformation process. Using $t \approx \varepsilon_{fr}/\dot{\varepsilon}$, Equation 5.14 may be rewritten as

$$W_{d,v} = \eta'\dot{\varepsilon}_v^2 \varepsilon_{fr}/\dot{\varepsilon} \tag{5.15}$$

To solve Equation 5.15, the relation between $\dot{\varepsilon}_v$ and $\dot{\varepsilon}$ should be known. Moreover, η' will depend on both $\dot{\varepsilon}$ and ε, and ε_{fr} on $\dot{\varepsilon}$. These dependencies will be affected by the product involved. Because it is likely that $\dot{\varepsilon}_v$ will change in a roughly proportional manner to $\dot{\varepsilon}$, for viscoelastic materials Equation 5.15 implies that the absolute value of $W_{d,v}$ increases with the strain rate. And because $W_e/W_{d,v}$ also increases, this implies a higher W as a function of $\dot{\varepsilon}$ as well as a higher fracture stress when the strain rate is higher.

The following general statements can be made on the effect of a higher strain rate on the energy components involved in the fracture of a viscoelastic material:

- The energy dissipation due to viscous flow, $W_{d,v}$, will be relatively lower compared to W_e as long as the material behaves relatively more viscous at low strain rates. This will be the case if the relaxation times of (part of) the bonds is of the order of $1/\dot{\varepsilon}$.
- The elastically stored energy, W_e, will be higher because of the less extensive bond relaxation during the deformation process.
- The energy necessary for the formation of new surfaces, W_{fr}, will be somewhat higher as a result of the less extensive bond relaxation processes.

These trends have the following consequences:

1. The total amount of energy, W, required to deform a material to a certain strain increases with increasing $\dot{\varepsilon}$. This also implies that $\sigma(\varepsilon)$ increases with increasing $\dot{\varepsilon}$.
2. The material has to be deformed to a smaller extent to reach a value of W_e high enough for fracture propagation to occur. This implies that ε_{fr} decreases with increasing $\dot{\varepsilon}$. Another way of saying this is that during slow deformation, the material has more time to flow and therefore ε_{fr} is larger.
3. Because the effect due to higher W is generally more important than that due to lower ε_{fr}, σ_{fr} will increase with increasing $\dot{\varepsilon}$.
4. At very low $\dot{\varepsilon}$, W_e stays too low for fracture propagation to occur, and thus no fracture occurs. The material flows, or may first yield and then flow, as is for instance the case during hole formation in various types of cheeses, such as Gouda and Emmentaler, during their ripening process (see Section 4.2).

Another effect of deformation rate on the fracture behavior of viscoelastic materials is that at low $\dot{\varepsilon}$, yielding and flow around the crack tip will be more extensive, leading to less stress concentration. This will, in principle, lead to a higher σ_{fr}, but normally other effects of $\dot{\varepsilon}$ on σ_{fr} are more important.

Using the above reasoning, results can be explained that are obtained for the stress versus strain behavior as a function of strain rate for viscoelastic materials such as for a 2-week-old Gouda cheese (**Figure 5.1**).

5.2.3.2 Effects due to friction between structural elements as a result of inhomogeneous deformation

Most foods are composite materials that consist of various structural elements with different mechanical properties. This implies that these products will deform inhomogeneously, certainly at the larger deformations needed for fracture to occur. Such inhomogeneous deformations involve energy dissipation due to friction between the structural elements, for example, between the dispersed particles and the network in filled gels.

Energy dissipation due to friction $W_{d,f}$ will depend on certain factors such as the morphology of the structural elements, their mechanical and surface properties, and on the way the material is deformed. It also depends on the local rate of deformation and hence, on the crack speed. In general

$$W_{d,f} \propto v_{crack}^{a}, \tag{5.16}$$

where v_{crack} is the rate at which the crack advances, and parameter a is presumed to be constant over a certain range of v_{crack}. In general, it is observed that for spontaneously propagating cracks, v_{crack} increases proportionally to $\dot{\varepsilon}^{b}$, where b is about constant and smaller than 1. The important conclusion is that, in general, energy dissipation due to friction increases with the deformation rate of the material. For an elastic material, W_e will not be affected by the deformation rate, but the transport of energy that becomes available due to stress relaxation around the growing crack will proceed more inefficiently at higher $\dot{\varepsilon}$. In addition, the strong deformation of the material around the crack tip will result in strong local energy dissipation that will be higher for higher crack growth speeds. These effects will retard the speed of crack propagation, and in certain cases may lead to quite low crack growth speeds. The latter factor may lead to ε_{fr} increasing with increasing $\dot{\varepsilon}$ of the specimen when the crack growth speed becomes so slow that a measurable deformation of the whole test piece may occur during the time necessary for the crack to proceed through the test piece. So, the effect of friction between the structural elements forming the solid matrix of a product on ε_{fr} as a function of $\dot{\varepsilon}$ is the opposite to that of viscoelasticity.

The increase in energy dissipation with increasing $\dot{\varepsilon}$ implies that the total amount of energy, W, required to deform a material to a certain strain will also increase. This implies that just as for the effect of viscoelastic behavior, $\sigma(\varepsilon)$ and σ_{fr} increase with increasing $\dot{\varepsilon}$. Because W_e is about independent of $\dot{\varepsilon}$, the ratio $W_e/W_{d,f}$ decreases with increasing $\dot{\varepsilon}$, in contrast to the effect of viscoelasticity.

The relevance of the friction mechanism for giving time-dependent fracture properties has been shown (e.g., for various types of cheese and for starch gels). For 10% potato starch gels, W was found to be independent of $\dot{\varepsilon}$ at low $\dot{\varepsilon}$ and to increase with increasing $\dot{\varepsilon}$ at high $\dot{\varepsilon}$ in tensile experiments. Assuming that the increase in W was completely due to an increase in $W_{d,f}$, the constant a in Equation 5.16 was calculated to be about 0.3. Moreover, the time between the start of fracture and its completion was determined, allowing the calculation of the crack growth speed. It resulted in b, in the relation $v_{crack} \propto \dot{\varepsilon}^b$, being about 0.7. This value of b implies that v_{crack} increases less than linearly with $\dot{\varepsilon}$, and therefore that the test piece will be deformed more during crack growth at higher $\dot{\varepsilon}$ (van Vliet et al. 1993). This is illustrated in **Figure 5.12** for 10% potato starch gels. For potato starch gels, the energy dissipation was likely due to friction between the swollen starch granules at large deformations. During fracturing, several layers of starch granules around the crack tip were observed to move with respect to each other.

An increase in the strain at fracture ε_{fr} with increasing $\dot{\varepsilon}$ has, for instance, also been observed for Leicester (**Figure 5.13**), Cheddar, and Cheshire cheeses. These results could also be explained by assuming reasonable rates of crack growth of the order of 1–5 cm s^{-1} and that these increase less than proportionally with $\dot{\varepsilon}$ (van Vliet et al. 1993).

During deformation and fracture of various food products, different mechanisms causing energy dissipation may act at the same time. All mechanisms result in an increase of W, $\sigma(\varepsilon)$ as well as σ_{fr} with increasing $\dot{\varepsilon}$. Regarding the phenomenological aspects of fracture, the main difference between the various mechanisms is their effect on ε_{fr} being a decrease or an increase in ε_{fr} with increasing $\dot{\varepsilon}$ (**Figure 5.13**). When both energy dissipation due to viscoelasticity and energy dissipation due to friction between structural elements play a role, their effects on ε_{fr} may cancel each other out, as is likely the case during the fracturing of old Gouda cheese (**Figure 5.14**) and for young Gouda cheese with a (too) low pH (**Figure 5.13**).

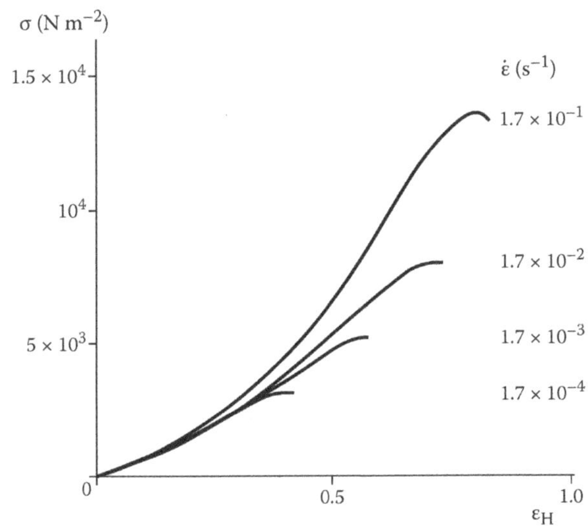

***Figure 5.12** Stress σ versus strain ε in uniaxial compression for 10% potato starch gels at various initial strain rates (indicated). (Reprinted from van Vliet, T. et al. in* Food Colloids and Polymers: Stability and Mechanical Properties, *ed. Dickinson, E., Walstra, P. Royal Society of Chemistry, Cambridge, UK, 1993. With permission.)*

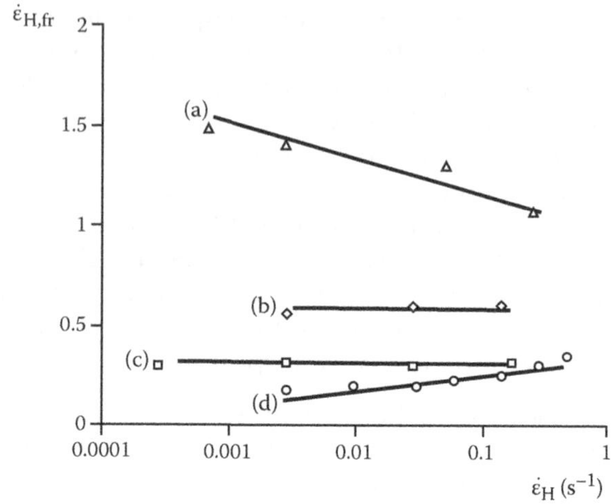

Figure 5.13 *Fracture strain $\varepsilon_{H,fr}$ in uniaxial compression as a function of strain rate $\dot{\varepsilon}_H$ for various cheeses. (a) Standard 2-week-old Gouda cheese, pH 5.24; (b) 1-week-old Gouda cheese with a low pH (4.94); (c) standard 9-month-old Gouda cheese; (d) Leicester cheese.*

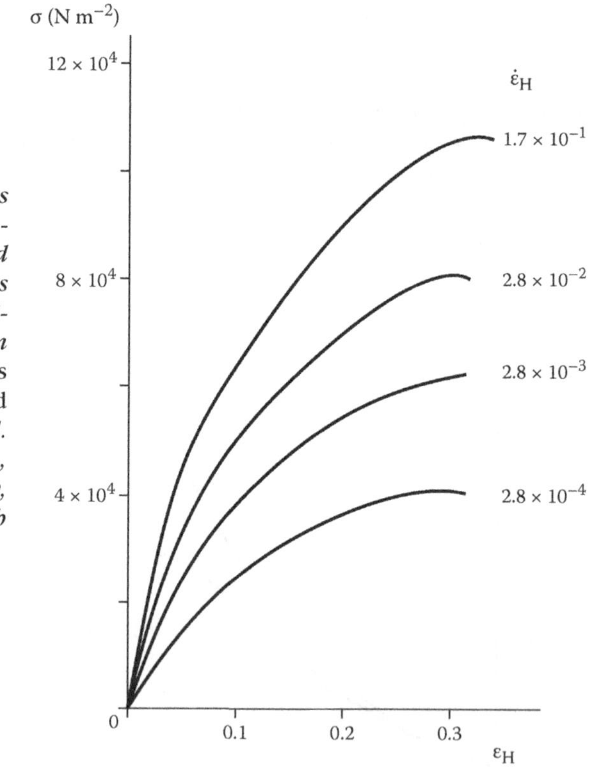

Figure 5.14 *Stress (σ) versus strain (ε_H) in uniaxial compression for 9-month-old Gouda cheese at various initial strain rates $\dot{\varepsilon}_H$ (indicated). (Reprinted from van Vliet, T. et al. in* Food Colloids and Polymers: Stability and Mechanical Properties, *ed. Dickinson, E., Walstra, P., Royal Society of Chemistry, Cambridge, UK, 1993. With permission.)*

5.2.4 Halting crack propagation

Crack propagation will stop when the differential energy released during crack growth becomes less than the differential energy required for further crack growth. This can occur when the crack come across an area with a much higher critical strain energy release rate (Equation 5.6) (or critical stress intensity factor) or as a result of extensive energy dissipation. In addition, crack growth may stop when the stored strain energy is released upon the removal of the external stress applied on the material before it has fractured completely. This can be the case for impulse loading, for example, when a product falls down on the ground. In this case, the results may be local damage only. It should be noted, however, that this mechanism will only be effective as unloading of the strain energy on removal of the applied stress proceeds faster than the crack growth. This implies that this mechanism cannot stop crack growth in very brittle materials such as pottery, potato crisps, and crispy snacks, where crack growth proceeds at very high speeds (about 300–500 m s^{-1}).

Composite materials may contain regions with a lower and higher critical strain energy release rate, G_C. For these products, the crack will normally go along the region with the higher G_C. This will normally be the case for hard particles in a soft continuous phase. Then crack growth does not really stop, but makes a detour around the part with high G_C.

In nature, a very important mechanism to control crack propagation is by opening a weak interface between fibers perpendicular to the direction of the crack growth. It plays a predominant role in the fracture behavior of muscles, tendons, wood, grass, etc. When one fiber is fractured, the crack will grow sideways along the weak interface, thereby dissipating strain energy (**Figure 5.15**). In addition, for ongoing fracture, a new defect has to grow out. This mechanism has a very positive impact for our health, in that it prevents the total rupture of our muscles and tendons. On the other hand, it makes cutting and biting of meat more difficult, especially, when it is not well prepared (i.e., cooked).

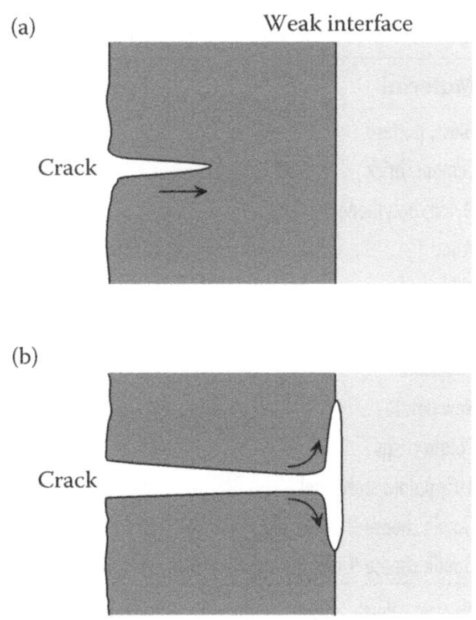

Figure 5.15 *Improving resistance to fracture due to opening up of a weak interface in path of a crack.*

5.2.5 Fracture stress, work of fracture, toughness, and fracture toughness

When people talk about comparing the strength of a material, they often implicitly refer to the fracture stress of these materials.* They are doing this, although when specifically asked, they know that a cup or a pan made of glass or pottery will fracture much faster than a cup or pan made of wood or mild steel—in spite of the fact that the fracture stress of these materials is roughly the same (**Table 5.2**). The answer is in the difference in work of fracture or, officially, the critical strain energy release rate between the different materials. This also implies that measuring the fracture stress or instrumentally measured hardness of a food is often not enough to characterize the sensory appreciation of the fracture behavior of a food. Therefore, concepts developed in fracture mechanics have to be taken into account. Several illustrations of this are given in **Chapter 18**. Below, first the difference between fracture stress and work of fracture (critical strain energy release rate) will be discussed further than has been implicitly done above. Second, the difference between the definitions of toughness used in fracture mechanics and some common ones in sensory science will be evaluated to avoid misunderstanding.

Table 5.2 Fracture Stress σ_{fr} and Work of Fracture (Critical Strain Energy Release Rate) G_C of Various Materials (Approximate Numbers)

Material	Fracture Stress, σ_{fr} (kN m^{-2})	Work of Fracture G_C (J m^{-2})
Glass, pottery	1.7×10^5	1–10
Cement, brick	4×10^3	3–40
Nylon, polythene	$(1.5–6) \times 10^5$	10^3
Wood	10^5	10^4
Mild steel	4×10^5	$10^5–10^6$
Bones, teeth	2×10^5	10^3
Chocolate	300	~90
Raw potato	100	300–1000
Potato crisps	$2–4 \times 10^3$	30–50
10% potato starch gel	4–15	1–5
Gouda cheese 2 weeks old	20–180	4–10
Gouda cheese 9 months old	40–110	≤ 1

Source: Partly after Gordon, J.E., *Structures or Why Things Don't Fall Down*, Penguin Books, Harmondsworth, UK, 1983; potato crisps data after Rojo, F.J., Vincent, J.F.V., *Eng. Failure Anal.*, 16, 2698–2704, 2009. With permission.

Note: Values for the potato starch gel and for Gouda cheese depend on the strain rate applied ($3 \times 10^{-4}–2 \times 10^{-1}$).

* In the literature, the term "gel strength" is often used in connection with the modulus of a gel. This use is inappropriate, because the modulus is a measure of the stiffness (resistance to deformation) of the gel and does not have to be related to the fracture stress of the gel.

When a test piece is fractured in tension, at least one crack must spread right across the material to divide it into two pieces, implying that two new surfaces must have been created that did not exist before. Therefore, all chemical or physical bonds that hold the two surfaces together have to be broken.

The energy needed to break these bonds can be estimated in two ways. First, one can take as a starting point an average value for the cross section of a bond. This gives the number of bonds per cross section, for instance, 4×10^{18} m^{-2} for an average cross section of the bonds of 0.25 nm^2. The energy content of a covalent bonds is about 200–500 kJ mol^{-1}. This gives for the work of fracture (critical strain energy release rate, G_C) 1–3 J m^{-2}, which is in reasonable agreement with observed values for glass and pottery, but far below that for wood, mild steel, bones, and teeth (**Table 5.2**). Second, one can take as starting point the excess surface energy (surface tension) of the two new surfaces. For foods, the surface tension will rarely be above that of water (70 mN m^{-2}), which would imply that $G_C \approx 0.1–0.15$ J m^{-2}. However, values observed for foods are generally between 1 and 10 J m^{-2}, 1 to 2 orders of magnitude higher.

Possible reasons for the discrepancy between these values and the observed work of fracture are as follows:

(1) The fracture surface is uneven, causing the surface area of the crack to be larger than it appears on a macroscopic scale.
(2) Crack formation is often accompanied by the formation of (small) side cracks or of small cracks about parallel to that of the main one, which also increases the effective surface area.
(3) Local yielding causes energy to be dissipated during fracture, and the increase of the work of fracture will be about proportional to the value of l_p in Equation 5.12. This effect will be much more enhanced if energy dissipation takes place over the whole volume of the studied material. For nonbrittle food products, this is usually the main factor determining the work of fracture.
(4) In many food products, the strength of the material is the result of a network of macromolecules (biopolymers). On extending such a network, all the bonds between the atoms forming the macromolecular chain will be extended. When one of these bonds or cross-links between two macromolecules fractures, energy in all these bonds will relax and will be dissipated as heat. For long chain lengths between the cross-links, this may in principle give substantial extra energy dissipation, provided that the concentration of the network-forming component is high enough. On the other hand, the work of fracture will be lower, because the volume fraction of macromolecules is generally far below 1.

The effect of reasons (1) and (2) can be dealt with by using the concepts developed in Section 5.2.1 on linear elastic or brittle fracture, whereas those of points (3) and (4) can be dealt with using concepts discussed in the sections on elastic plastic fracture and time-dependent fracture, respectively. The quantitative prediction of the magnitude of these effects is very difficult, because the parameters needed to do so are mostly not known. Therefore, G_C, σ_{fr}, as well as W_{fr} have to be determined experimentally. The appropriate methods/apparatuses will be discussed in **Chapters 7 and 8**.

Table 5.2 provides an overview of the measured values for the fracture stress and work of fracture of a series of materials. Glass and pottery are hard brittle materials. Linear elastic fracture is adequate to describe their fracture behavior. This is also the case for potato crisps. Its fracture usually involves the formation of (many) side cracks. Construction materials, including bones and teeth that must be able to resist tensile and bending stresses, exhibit a much higher work of fracture, whereas σ_{fr} is of the same order of magnitude as that for glass and pottery. For these materials, elastic plastic fracture and time-dependent fracture mechanics have been developed. When potatoes are cooked, the work of fracture decreases from 0.3 to 1 kJ m^{-2} to about 15–75 J m^{-2}, which significantly affects their fracture behavior. They become more brittle. Cooked potatoes fall apart when they are pricked with a fork, whereas uncooked ones do not. One-year-old Gouda cheese is somewhat harder than a 2-week-old cheese, but the main difference is that the old cheese exhibits brittle fracture and the young one shows a much more ductile behavior as is manifested in a higher G_C. These examples clearly illustrate that, in most cases, the work of fracture (critical strain energy release rate) is the critical material property that governs fracture. This does not only apply during handling of solid food products, but also during eating. Work of fracture determines, to a large extent, the sensory perception of food products, for example, crispy products should have a low work of fracture. Their fracture behavior can be described using concepts developed for linear elastic fracture.

The difference in sensory perception and in handling properties between 2-week-old and 1-year-old Gouda cheese is directly related to the large difference between the stress versus strain curves of both cheeses (**Figure 5.16**). The young cheese exhibits yielding at low strain followed by a region in which the increase in σ with increasing ε is much less. Only at ε close to ε_{fr}

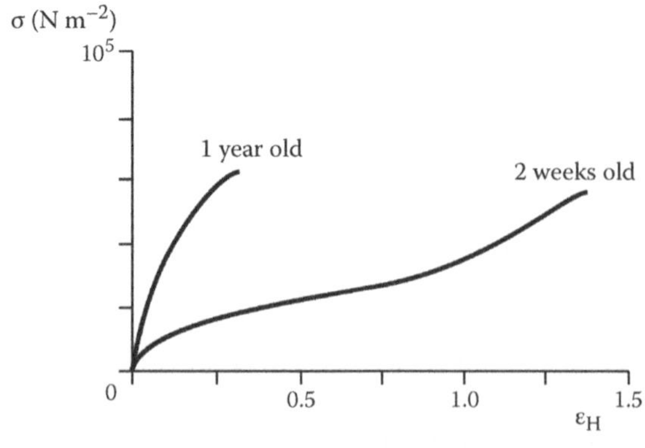

Figure 5.16 *Stress σ versus strain ε_H curves for 2-week-old and 1-year-old Gouda cheese. Strain rate 2.7×10^{-3} s^{-1}.*

is a relatively steep increase in σ observed. For the old cheese, σ increases rather steeply with ε until the curve levels off relatively close to macroscopic fracture. In the latter material, much more energy is required in order to deform the material to the same extent of strain. In this case, this implies that the stored energy increased much more steeply with ε—energy that can be used for crack propagation. The final result is a lower work of fracture and, in addition, fracture occurs at a lower strain.

As discussed above, *fracture toughness* (in Pa $m^{0.5}$) is the critical stress intensity factor required to cause a material to fracture. It describes the ability of a material containing a crack to resist fracture and is directly related to the critical strain energy release rate (or work of fracture) for brittle materials (Equation 5.10). Brittle fracture is very characteristic for materials with low fracture toughness. For unnotched material, a common definition of *toughness* (in J or J m^{-3}) is the energy required to propagate fracture by a given crack area. In this case, toughness is generally derived from the area under a force deformation (or stress–strain) curve. Another definition of toughness used in material science and sometimes also in fracture mechanics is the ability to absorb mechanical (or kinetic) energy up to failure. Its value will depend on the (size and) shape of the test piece. Therefore, on reading the literature, one should always check the definition of toughness used by the authors. An overview is given in **Table 5.3**.

For those involved in food science, it may become even more confusing, because in sensory science one can also distinguish various definitions of toughness. For comparison with those commonly used in fracture mechanics, several definitions of toughness used in sensory/food science are included in **Table 5.3**.

Table 5.3 Various Definitions of Toughness Used in Fracture Mechanics and in Sensory (Food) Science

In fracture mechanics

- Toughness of unnotched samples. The energy required for fracture of a material specimen (unit J) or the energy required to propagate fracture by a given crack area (unit J m^{-3}). In both cases, toughness is generally derived from the area under a force deformation (or stress–strain) curve.
- Toughness of notched samples (usually denoted as fracture toughness). The ability of a material containing a crack to resist fracture as given by the critical stress intensity factor K_{IC} (unit Pa $m^{0.5}$).
- Toughness is the ability to absorb mechanical (or kinetic) energy up to failure.

In sensory science (often related to the product group considered)

- Hard to cut or chew (http://www.thefreedictionary.com, meat).
- Sensory properties relating to the extent to which a product such as meat is hard to chew or cut because of its innate resistance, hardness, and leathery structure (International Food Information Service, 2006).
- An overall impression of the breakdown of the product, which includes factors such as how many chews does it take to prepare the sample for chewing (i.e., how does it break down on chewing), softness, and how much juice is expressed as you bite down on the sample when it is placed in the mouth (after Peachy et al., 2002).
- Combination of bite force and duration of chewing. Chewing efficiency is low (Dijksterhuis et al., 2007).

Note: A distinction is made between those used in fracture mechanics or more generally in material science and some ones customary in sensory science.

5.2.6 Fracture or yielding

As indicated in Section 5.1, the main difference between fracture and yielding is that in the latter case, the material does not fall apart into separate pieces but stays in a coherent mass. In both cases, there is a breakdown of the structure of the material over length scales clearly larger than the structural elements. In fracture mechanics terms, in both cases, inhomogeneities/defects must start to grow, leading to structure breakdown; however, in the case of yielding, no fracture propagation—resulting in the material falling apart into pieces—takes place. In general, yielding behavior will be favored if (1) the material contains many, more-or-less equal, weak points, so that fracture is initiated at many points at roughly the same time, and (2) strong energy dissipation occurs after fracture initiation so that not enough energy is available at the crack tip to allow fracture propagation. Presumably, the latter condition is essential and the former favors yielding. Energy dissipation may be due to friction between the different components as a result of inhomogeneous deformation and/or due to viscoelastic behavior (plastic flow). Because energy dissipation processes play an important role in the yielding process of materials and because these are time-dependent, in general, the yield stress σ_y will depend on the strain rate, being lower at lower deformation rates. In addition, the strain at yielding will also depend on the strain rate. An example of the effect of deformation rate on yielding is shown in **Figure 5.17**. Regarding strain rate dependency, yielding is very similar to the time-dependent fracture discussed in Section 5.2.3. The difference is mainly phenomenological;

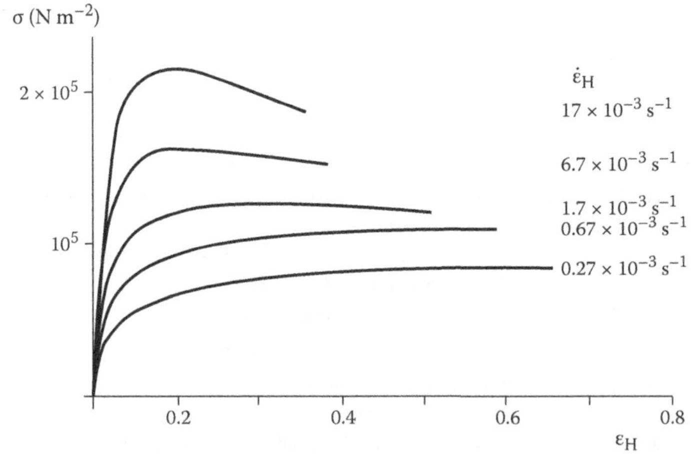

Figure 5.17 *Stress (σ) versus strain (ε_H) curves in uniaxial compression for 12 wt.% glycerol lacto-palmitate gels in a 2 wt.% sodium caseinate solution. Strain rate $\dot{\varepsilon}_H$ is indicated. Extensive yielding occurred at a strain of about 0.05. (Reprinted from van Vliet, T., in* Food Macromolecules and Colloids, *ed. Dickinson, E., Lorient, D., Royal Society of Chemistry, Cambridge, 1995. With permission.)*

during fracture, the material falls apart into pieces and after yielding it stays in a coherent mass.

Yielding does not have to take place throughout the whole volume of a piece of material. Very often, yielding is localized in certain shear planes (slip planes) that are separated by relatively undisturbed layers. The resulting deformation of the material is then very inhomogeneous at mesoscopic length scales. This is, for instance, mostly the case during spreading of butter or margarine. It depends on the scale of the inhomogeneities if it is noticed (or not) by the consumer spreading the product.

Various materials may exhibit yielding at relatively low strains and fracture at much larger strains. An example of such a material is young Gouda cheese (and other similar types of cheeses), which yields a strain of about 0.1–0.2 and fractures at much larger strains (1–1.5) (**Figure 5.1**), whereby both the yield and fracture strain decreases with increasing strain rate. A young Gouda cheese at pH > 5.2 only yields and does not fracture at strain rates below about 10^{-5}. In such a cheese, big holes can be formed because of gas production by bacteria (e.g., common in Emmentaler type of cheeses).

5.3 Strain hardening and stability against fracture in extensional deformation

During drawing thin long threads from a thick cylindrical shaped material, the material will be extended in a uniaxial direction and compressed in the directions perpendicular to the tensile strain. Similarly, a film between two growing gas cells in polyhedral foam will be extended in two directions (biaxial extension) and compressed in the third direction. This is, for instance, the case during proofing and the oven rise of bread dough.

Both during thread extension and biaxial extension of films, it will always happen that accidentally they become thinner locally. A thin spot in a film or a local neck in a thread will form frequently by chance, and this may lead to breaking of the film or thread. Consider a straight thread subjected to a force f acting on it at both ends in the direction of the thread. Then the tensile force in the thread will be constant over its whole length. This implies that the tensile stress σ ($= f/A$) will be higher in places where the thread is locally thinner for the case where Young's modulus E is constant or decreases with increasing ε. This indicates that the thread will be extended faster at the thin spot and will soon fracture in that place.

A thread or film can only be stable against extension and finally rupture if the extension of the thinner part is less than that of the thicker part when the tensile force increases. This will be the case if the force required for further thinning of the thin part $[A + dA] \cdot \sigma(A + dA)$ of the thread or film is higher than the force required for further thinning of the thicker part

$A \cdot \sigma(A)$ of the thread or film, respectively. For uniaxial extension of a thread, this will be the case if

$$[A + dA] \cdot \sigma(A + dA) > A \cdot \sigma(A), \tag{5.17}$$

where A is the cross section of the thread and $\sigma(A)$ is the accompanying stress. Equation 5.17 can be transformed to

$$\sigma dA + A d\sigma > 0 \tag{5.18}$$

Dividing by σA gives

$$\frac{dA}{A} + \frac{d\sigma}{\sigma} > 0 \tag{5.19}$$

For an incompressible material, $A \cdot l$ will be constant, so dA/A in Equation 5.19 can be replaced by dl/l, where l is the length of the test piece. Next, in view of Equation 3.22, $\varepsilon_H = \int (dl/l)$, one may write as the criterion for the stability of the thread against rupture

$$d \ln \sigma_e / d\varepsilon > 1 \tag{5.20}$$

This criterion is known as the so-called Considère criterion (Considère 1885), which was developed for the stability against necking of cylindrical bars of metals in uniaxial extension. The Considère criterion implies that the stress should increase more than proportional with the strain. The phenomenon that the stress required to deform a material increases more than proportional to the (Hencky) strain (at constant strain rate and increasing strain) is called ***strain hardening***. It is the opposite of strain weakening (or softening) (**Figure 4.9**). Strain hardening is an important property of materials with respect to the occurrence of necking during drawing of threads from cylindrical pieces of materials and during spinning operations. Of course, the thread will eventually break, but at a far larger strain than in the absence of strain hardening.

Similar to the derivation of the criterion for the stability against necking in uniaxial extension, a criterion can be derived for the stability of the films between two gas cells in polyhedral foams. For films, local thinning will stop when

$$[h + dh] \cdot \sigma(h + dh) > h \cdot \sigma(h) \tag{5.21}$$

The biaxial strain $\varepsilon_{b,H}$ in the direction perpendicular to the compressive stress is $-(1/2)\ln(h_t/h_0)$ (Equation 3.25). Because $d\varepsilon_{b,H}$ is positive, the obtained stability criterion reads as

$$d \ln \sigma / d\varepsilon_{b,H} > 2 \tag{5.22}$$

Because $\varepsilon_{b,H}$ is half the uniaxial strain ε_H, both criteria are, in essence, the same as long as the strain involved is defined in the right way, that is, the uniaxial strain $\varepsilon_H = -\ln(h_t/h_0)$ for thinning of a thread and the biaxial strain $\varepsilon_{b,H} = -(1/2)\ln(h_t/h_0)$ for thinning of a film (van Vliet 2008).

5.4 Concluding remarks

A thorough characterization of large deformation and fracture behavior of foods cannot be carried out by determining only the fracture stress and strain. Concepts developed in fracture mechanics have to be considered to gain a better understanding of the fracture behavior of foods. What consumers perceive as fracture during handling and eating of food is primarily not determined by the stress at which the bonds between the structural elements rupture, but by the energy balance that governs crack propagation and, in this way, the falling apart of the food into pieces. The local stress level at the tip of the crack is an essential boundary condition.

The role of energy balance in determining fracture behavior is especially clear for nonbrittle materials, such as gels, soft solids, and foams. The energy balance determines, among other things, if a material will yield or fracture on deformation. This energy balance is strongly affected by the shape of the stress versus strain curve, determining W_e as a function of ε, and by the amount of energy that dissipates during deformation and the fracturing process. Both energy storage and dissipation are considerably affected by the composition and structure of food and thereby its fracture behavior. This will be discussed further in **Chapters 13** through **17**. In **Chapter 18**, we discuss several examples of the use of concepts developed in fracture mechanics for establishing a relation between sensory perception and parameters that can be determined instrumentally.

References

Anderson, T.L. 1991. *Fracture Mechanics: Fundamentals and Applications*. Boca Raton, FL: CRC Press.

Atkins, A.G., and Y.M. May. 1985. *Elastic and Plastic Fracture*. Chichester, UK: Ellis Horwood.

Considère, M. 1885. Memoire sur l'émploi du fer et de lacier dans les constructions. In: *Annales des Ponts et Chausées*, vol. 9, 574–605. Paris: CH. Dunod.

Dijksterhuis, G., Luyten, H., De Wijk, R., and Mojet, J. 2007. A new sensory vocabulary for crisp and crunchy dry model foods. *Food Qual. Preference* 18: 37–50.

Ewalds, H.L., and R.J.H. Wanhill. 1985. *Fracture Mechanics*. Delft: Delftse Uitgevers Maatschappij.

Fineberg, J., and M. Marder. 1999. Instability in dynamic fracture. *Phys Rep.* 313: 1–108.

Gordon, J.E. 1983. *Structures or Why Things Don't Fall Down*. Harmondsworth, UK: Penguin Books.

International Food Information Service. 2006. *Dictionary of Food Science and Technology*. Oxford, UK: Blackwell Publishing.

Luyten, H., van Vliet, T., and P. Walstra. 1991. Characterization of the consistency of Gouda cheese: Fracture properties. *Neth. Milk Dairy J.* 45: 55–80.

Peachy, B.M., Purchas, R.W., and Duizer, L.M. 2002. Relationships between sensory and objective measures of meat tenderness of beef *M. longissimus thoracis* from bulls and steers. *Meat Sci.* 211–218.

Rojo, F.J., and J.F.V. Vincent. 2009. Objective and subjective measurement of the crispness of crisp from four potato varieties. *Eng. Failure Anal.* 16: 2698–2704.

van Vliet, T. 1995. Mechanical properties of concentrated food gels. In *Food macromolecules and Colloids*, ed. E. Dickinson and D. Lorient, 447–455. Cambridge: Royal Society of Chemistry.

van Vliet, T. 2008. Strain hardening as an indicator of bread-making performance: A review with discussion. *J. Cereal Sci.* 48: 1–9.

van Vliet, T., H. Luyten, and P. Walstra. 1993. Time dependent fracture behaviour of food. In *Food Colloids and Polymers: Stability and Mechanical Properties,* ed. E. Dickinson and P. Walstra, 175–190. Cambridge: Royal Society of Chemistry.

Vincent, J.F.V. 1992. Fracture. In *Biomechanics—Materials: A Practical Approach*, ed. J.F.V. Vincent. Oxford: Oxford University Press.

Vincent, J.F.V. 2004. Application of fracture mechanics to the texture of food. *Eng. Failure Anal.* 11: 695–704.

Walstra, P. 2003. *Physical Chemistry of Foods.* New York: Marcel Dekker.

III

Experimental Evaluation

6

Selection of Instrumental Method

The selection of a suitable instrumental measuring method to determine the rheological and fracture properties of foods is essential for obtaining the required information. It is just as essential as the selection of the preparation procedure of the test specimens and of the testing conditions (e.g., measuring temperature). This selection procedure must be the end point of a fixed procedure that involves the following questions:

1. What is the *purpose* of the measurement? Is it necessary to determine the fundamental rheological/fracture parameters, or will a global impression of the mechanical behavior of the food or raw material suffice in the context of a quality check? If one wants to be able to compare results from different tests using a different apparatus unequivocally and indifferently if the tests are executed at the same or at different places, the determination of fundamental rheological/fracture parameters is nearly always a prerequisite. This is also the case if the goal is to understand the relation between structure and the mechanical properties of a material or to instill quality control. This often requires the combination of different tests. If the aim is to perform a quality check, a well-standardized measurement, with a so-called empirical test method, may often suffice. For such a measurement, the main requirement is that the test must determine a relevant characteristic of the product under relevant conditions.
2. Which *parameters* have to be determined, and under what conditions (e.g., rate and time of deformation, type of deformation, stress level, temperature, pretreatment of the specimen)? Answering this question in the right way is especially essential for materials whose properties depend on the rate (shear thinning and viscoelastic materials) and

history of deformation (thixotropic materials and materials that may yield in an irreversible manner under a stress). On shaping brittle materials, small cracks will be formed easily in and at the surface of the material, which may significantly affect their fracture behavior.

3. Which *measuring method* can be used? A first subdivision of measuring methods can be made depending on whether the material is subjected to a stress (σ), a strain (γ or ε), or a strain rate ($\dot{\gamma}$ or $\dot{\varepsilon}$). A further distinction can be made between so-called *steady-state tests* and the so-called *dynamic tests*. During a steady-state test, the material is deformed by applying a steady σ, γ (or ε), or $\dot{\gamma}$ (or $\dot{\varepsilon}$), whereas during dynamic tests the applied variable varies in time (usually oscillates sinusoidally). The latter division is somewhat arbitrary. Experiments in which the reaction of a material on a sudden change in σ, γ (or ε), or $\dot{\gamma}$ (or $\dot{\varepsilon}$) is determined (the so-called *transient tests*) can be considered as either dynamic or steady-state tests. The same holds for tests in which the applied stress, strain, and strain rate are increased or decreased as a function of measuring time at a set rate. Altogether, this implies that a large set of measuring methods can be distinguished.
4. Which *measuring equipment* (apparatus) has to be selected and equipped with which measuring geometry? Based on the way the test piece is deformed, four main classes of instruments can be distinguished that are used commonly for studying food products. These are the tube viscometers, rotational rheometers, tension compression (or the so-called general material testing instruments), and empirical apparatuses. The choice between them has to depend on the type of material to be studied and the answers to questions (1) to (3). The first type of instruments is well suited for measuring the viscosity of liquids and liquid-like materials. Rotational instruments are well fitted for studying the rheological properties of non-Newtonian liquids and for studying the rheological and fracture behavior of materials that can solidify in the rheometer (e.g., gels). When equipped with a parallel plate geometry, they can also be used to study solid materials. Tension compression instruments are primarily used to determine the large deformation and fracture behavior of predominantly solid materials. Empirical test instruments are very popular in the food and other industries. They are not suited for determining fundamental rheological and fracture properties, but can be very useful for quality checking purposes. Other factors of importance during selection of measuring equipment are the scale of the inhomogeneities in the material to be studied, the possibility of the occurrence of slip between the test piece and the measuring body, and the requirements set for sample preparation.

Besides the types of instruments mentioned, other types can also be distinguished such as instruments based on wave propagation (e.g., by using ultrasound) or special instruments for determining extensional flow properties. However, it is beyond the scope of this book to discuss these more specialized instruments, which are less often used in the food industry.

A complication with regard to the determination of large deformation and fracture properties is that more parameters are required to give a full characterization of the material properties than is the case with small deformation properties. Moreover, these properties may be rate dependent even as small deformation properties are not (e.g., **Figure 5.12**). Fortunately, for many purposes, a few parameters can suffice to give an adequate characterization of the large deformation and fracture properties of a material; the specific parameters required will depend on the purpose of the experiments—for example, quality control regarding sensory or usage properties, stand up behavior, and determining the relation with product structure. For the latter case, in particular, an extra complication is that fracture properties depend on the size of defects present in the material to a much larger extent than small deformation properties do (Section 5.2.1). For large-scale inhomogeneous products (such as cookies, many snacks, and vegetables), this implies the need to use large test samples or to accept large variations in results. These large variations can as such be regarded as a product characteristic.

A mechanical test always involves a combination of measuring method, measuring equipment (instrument plus measuring geometry), test piece preparation, and test conditions. A complicating factor is that, with various instruments, a measurement can be executed according to different measuring methods and vice versa. In this book, no attempt will be made to give an overview of all possible combinations. The focus will be on generic principles.

In addition to the points discussed above, one should realize that the use of the best available measuring method and well-validated equipment is useless if the sample/test piece preparation and characterization are not done well. In particular, for inhomogeneous and thixotropic products, careful preparation of the test pieces is very important. Moreover, the preparation of test pieces from larger solid and semisolid products deserves special attention. Factors of importance are as follows:

- The sample should be representative of the product being studied in terms of composition and homogeneity.
- For thixotropic products or for products that are sensitive to irreversible structure breakdown, in general, the deformation history of the sample can have a large effect on the measured properties.
- Size of the test piece. As discussed in Section 5.2.1, fracture occurs because of the growth of defects already present in the material. In addition, crack propagation may stop at an inhomogeneity giving energy dissipation. Thus, the size of the test piece has to be large compared to the defects controlling the fracture. On the other hand, a test specimen often has to be homogeneous, which for most foods

sets an upper limit to the size. Another factor is that the relative inaccuracy in the determination of the stress and strain increases with the decreasing size of the test piece. This sets a minimum to the size of the test piece, depending on the material properties and the sensitivity of the apparatus.

- Shape of the test piece. The shape of the test piece must fit with the requirements set by the test. For instance, during uniaxial compression, the specimen height over diameter ratio should not be too high to avoid buckling of the specimen, and also not too low to avoid (excessive) friction between the plates and the test piece. Bending tests require test pieces that are long compared to their height (thickness or diameter) in the direction in which they are bent.
- Preparation of (deformable elastic) test piece. Care must be taken to ensure that test piece preparation does not irreversibly affect its mechanical properties. Test pieces can be taken from a food by, for example, using a special (cork) borer and cutting out with a mold or wire. First, it should be checked that this does not induce flaws at the surface of the test specimen, because these may act as starting points of crack growth (Section 5.2.1). Second, the resulting specimen shape and dimensions may be different from that of the borer or mold used to produce it (**Figure 6.1**). Third, for anisotropic materials (such as meat), the direction in which the test piece is taken from the product may greatly affect the measured mechanical properties.
- Environmental conditions during specimen preparation, storage, and testing. Temperature and temperature history are especially important for fat-containing products and for other products containing crystals, because for these materials the temperature history will affect the amount of solid material and the crystal modification. For instance, these factors have a strong influence on the mechanical properties of certain products such as cheese, shortcakes, butter, and margarine. For hygroscopic products (e.g., most dry products), the relative humidity during storage of the product and specimen can significantly affect their mechanical properties because of water uptake. For other products, drying out has to be prevented.

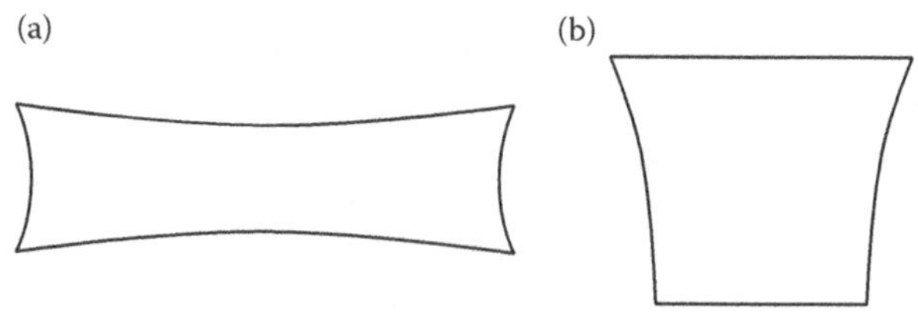

Figure 6.1 *Examples of shapes of a piece of young Gouda cheese (a) obtained with a cork borer and (b) by cutting with wires. (After van Vliet, T., Peleg, M.,* Bull. Int. Dairy Fed., *268, 5–15, 1991.)*

- Aspects such as mode and rate of deformation and the notion that the deformation of the measuring apparatus including attachments should be negligible under the required forces to deform the test piece should be part of the consideration given in questions (1) to (4).

Chapter 7 provides an overview of the main measuring methods, and **Chapter 8** gives an overview of the main measuring apparatus with an emphasis on those allowing the determination of fundamental rheological and fracture parameters. Tube viscometers, rotational rheometers, and the so-called tension compression apparatus (also called universal testing instruments) will be discussed. The latter types of apparatuses are primarily used for determining fracture parameters and the first two for measuring rheological parameters. Several empirical tests will be discussed in Section 8.4.

Summarizing, the selection procedure of a suitable instrumental measuring method has to involve a series of questions that should be addressed. Sufficient time should be set aside to tackle these issues. Selecting an appropriate method is the best way to avoid waste of resources (money and time) during the execution of tests and discussion of results.

Reference

van Vliet, T., and M. Peleg. 1991. Effects of sample size and preparation. In *Rheological and Fracture Properties of Cheese. Bull Int. Dairy Fed.* 268: 5–15.

7

Measuring Methods

The determination of the rheological and fracture properties of a material implies the measurement of the relation between its relative deformation (strain) and the stresses acting on it as a function of time. In this chapter, we shall discuss how stress, strain, or strain rate is applied as a function of time, and how the specimen can respond depending on its mechanical properties. The discussion will be limited to shear deformations. The same set of measuring methods discussed below can be distinguished for experiments involving compression or tensile deformations.

7.1 Tests at constant strain, stress relaxation

In these tests, a certain strain is applied instantly to the material, and σ required to maintain the deformation is measured as a function of time (the so-called *stress relaxation test*). Ideally, the material is deformed in a step function but, in practice, the deformation always takes some time (t). For a fluid (liquid), σ will be finite during deformation and zero for all positive t, whereas for an ideally elastic material, σ will increase linearly during deformation and stays constant in time afterward (**Figure 4.7**, Section 4.2). For other materials, σ will increase during deformation and after reaching constant strain initially decrease relatively rapidly and subsequently more slowly. Ultimately, σ may become zero or may approach some limiting constant value. During a stress relaxation test, the part before reaching the constant strain is mostly neglected in the following analysis and only the following part is analyzed. However, as will be discussed below, this is only justified if certain conditions are met.

From the stress response under steady strain, the *stress relaxation modulus* $G(t)$ can be calculated

$$G(t) = \sigma(t)/\gamma \qquad (7.1)$$

The decrease of stress in time is often characterized by the relaxation time τ_{rel} of the material, that is, the time required for the stress to decrease to $1/e$ (36.8%) of its value at the moment the deformation was stopped (Section 4.2). In practice, systems can rarely be characterized sufficiently just by one τ_{rel}. By doing so, one implicitly assumes that the stress decays exponentially to zero, although this is nearly never the case. For a full description, a spectrum of exponentials is often needed. In fact, we are interested in the spectrum of relaxation times governing the stress decay. To that end, it is useful to consider (just as in Section 4.2) a Maxwell element consisting of a single spring and dashpot (**Figure 4.8**). By applying this model, it is implicitly assumed that the relaxation behavior of the studied material is not affected by the application of the constant strain. This will only apply for strains that do not induce breaking of bonds between structural elements.

During deformation of a Maxwell element, the stress in the spring and the dashpot will be the same at any time; however, the strain redistributes itself over the spring and dashpot. Initially, it will reside completely in the spring (elastic deformation, $\gamma_{el} = \sigma/G_M$), but finally it may be entirely in the dashpot (viscous deformation, $\gamma_{visc} = (\sigma/\eta_M)t$), where G_M and η_M symbolize the viscosity and the elastic modulus of the Maxwell element, respectively. During deformation, $\gamma = \gamma_{el} + \gamma_{visc}$, which gives

$$\gamma = \frac{\sigma}{G_M} + \frac{\sigma}{\eta_M} t \tag{7.2}$$

During relaxation, the deformation will be constant, so $d\gamma/dt = 0$. Then, the time derivative of Equation 7.2 is

$$\frac{d\sigma}{dt}\frac{1}{G_M} + \frac{\sigma}{\eta_M} = 0 \tag{7.3}$$

On integration, one obtains for a single Maxwell element

$$\sigma(t) = G_M \gamma \, e^{-t/\tau_{rel,M}}, \tag{7.4}$$

where $\tau_{rel,M} = \eta_M/G_M$ is the relaxation time of the Maxwell element, as in Equation 4.8. Equation 7.4 describes the decrease in stress during a stress relaxation experiment for a system having the properties of a Maxwell element. Equations 7.4 and 4.8 are equivalent to each other with $\sigma_0 = G_M\gamma$.

For a set of parallel Maxwell elements and one element consisting of a single spring (**Figure 7.1**), one obtains for the total stress experienced by the material after the application of a sudden deformation γ

$$\sigma(t) = \gamma\left(G_e + \sum_{i=1}^{n} G_i e^{-t/\tau_{rel_i}}\right), \tag{7.5}$$

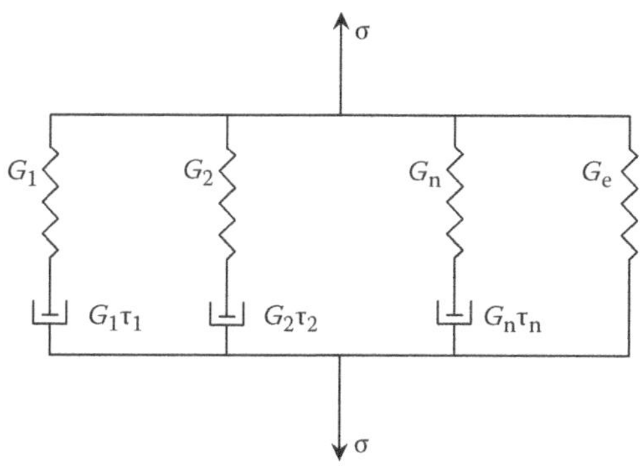

Figure 7.1 *Set of parallel Maxwell elements and one element consisting of a spring (for explanation of symbols, see text).*

where G_e is the modulus of the single spring element. For clarity, the subscripts M have been left out. For liquids, G_e will be zero as σ approaches zero for large t. Equation 7.5 gives, for the relaxation modulus $G(t)$,

$$G(t) = G_e + \sum_{i=1}^{n} G_i e^{-t/\tau_{\mathrm{rel}_i}} \tag{7.6}$$

Equation 7.6 is regularly used to fit measured stress relaxation data by a set of parallel Maxwell elements (often 2–3). The obtained values for the relaxation moduli and viscosities of the Maxwell elements cannot be related directly to material properties. They are, in essence, fitting parameters. Their values depend on measuring conditions as the deformation rate to obtain the set strain, time over which relaxation has been measured, and fitting conditions as the number of Maxwell elements used.

For a full description of the stress relaxation of a food, a spectrum of exponentials is often needed. For such a description, it is convenient to introduce a *modulus function* $H(\tau_{\mathrm{rel}})$ in such a way that $\sigma_0 H(\tau_{\mathrm{rel}})\mathrm{d}\ln\tau_{\mathrm{rel}}$ represents the contribution to the initial stress σ_0 of elements having logarithms of relaxation times between $\ln\tau_{\mathrm{rel}}$ and $\ln\tau_{\mathrm{rel}} + \mathrm{d}\ln\tau_{\mathrm{rel}}$. If the number of Maxwell elements becomes infinite, one obtains

$$G(t) = G_e + \int_{-\infty}^{\infty} H(\tau_{\mathrm{rel}})\, e^{-t/\tau_{\mathrm{rel}}} \mathrm{d}\ln\tau_{\mathrm{rel}} \tag{7.7}$$

Usually, $H(\tau_{\mathrm{rel}})$ is called the *relaxation spectrum* (spectrum of relaxation times), although strictly speaking, it is a *spectrum of modulus densities* (in N m^{-2} or Pa).

A complicating factor in relaxation tests is that stress relaxation already starts during the finite time required to attain the applied deformation. This relaxation process will be more extensive in the case of fast relaxing bonds and/or slow deformation rates. The result is a slower relaxation process directly after the intended deformation has been reached, and a longer apparent relaxation time (**Figure 7.2**). Moreover, this implies that relaxation time can

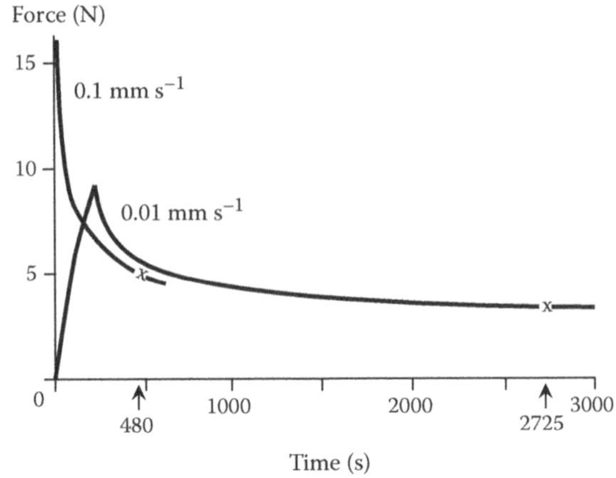

***Figure 7.2** Stress relaxation experiment on a 9-month-old Gouda cheese using two deformation speeds (indicated) for attaining applied deformation. x denotes calculated relaxation time using $\tau_{rel} = (1/e)\,\sigma_0$ for two measuring conditions.*

be calculated accurately only for $t \gg$ deformation time, roughly about 10 times as large. Another consequence is that experimentally determined stress relaxation times can only be compared if they are determined under identical experimental conditions in terms of the deformation history during the application of the constant strain. Evaluating the results of a stress relaxation experiment becomes even more complicated if the initial deformation is outside the region where the stress increases linearly with the strain or is the result of a shearing experiment at variable shear rates. Then, the relaxation behavior of the bonds between the structural elements will be affected by the value of the applied strain and/or the shear rate history. Results obtained can, at best, be used for qualitative comparison of data, but not for the calculation of, for example, a relaxation spectrum, because the basic assumption that the material structure is not affected by the applied deformation and with that its relaxation behavior, no longer holds.

In principle, it is possible to correct mathematically for stress relaxation during the application of the constant strain by introducing in the calculations a memory function that is a measure to which the strain at time t' affects the stress at the later time t. In general, its effect will decrease monotonically with increasing $t'-t$ (Whorlow 1992).

7.2 Tests at constant stress, creep test

In these tests, a constant stress is applied instantaneously to the material and the ensuing strain measured as a function of time. For an ideally viscous material the strain will increase linearly with time (**Figure 7.3**, left panel), and a viscosity can be calculated using Equation 3.6 or 3.10. For thixotropic or antithixotropic liquids, the relation between strain and time will be more complicated. The apparent viscosity is decreasing or increasing in

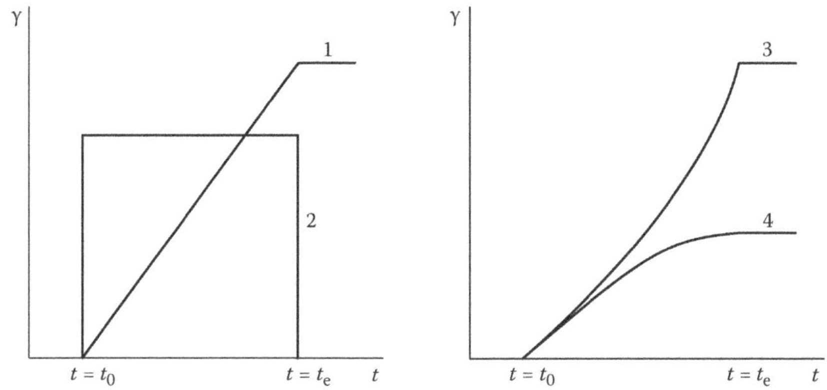

Figure 7.3 *Shear strain γ as a function of time t. At t = 0, a constant shear stress is (instantaneously) applied until $t = t_e$. At $t = t_e$, this stress is instantaneously removed. Left panel, behavior of a Newtonian liquid (curve 1) and an ideally elastic material (curve 2). Right panel, behavior of a thixotropic (curve 3) and an antithixotropic (curve 4) liquid.*

time, respectively, resulting in an increasing or decreasing shear rate, respectively. This results in an upward or downward curvature of the line representing the shear strain as function of time for thixotropic and antithixotropic liquids, respectively (**Figure 7.3**, right panel). For an ideally elastic material, the observed strain will be instantaneous and will remain constant in time (**Figure 7.3**, left panel). A Young's or shear modulus can be calculated using Equation 3.5 or 3.8, respectively. The material will regain its original shape after removal of the stress.

A particular type of test carried out by applying a constant stress is the so-called *creep test*. These tests are particularly useful for viscoelastic materials and allow the study of effects over long time scales (up to days as long as the material properties do not change). The applied stresses should be such that no fracture (structure breakdown) occurs.

For a viscoelastic material, three regions can be distinguished in terms of its reaction to an instantaneously applied stress, as illustrated in **Figure 7.4**. First an instantaneous increase, *AB*, in the strain is observed, the so-called *elastic response*, followed by a "delayed" *elastic response*, *BC*, and a *viscous response*, *CD*, which in the ideal situation shows a linear increase in γ in time. The term, "delayed elastic response" is used to indicate that it does not happen instantaneously after the sudden application of stress. When the stress is suddenly removed at $t = t_e$, the purely elastic part of the stored energy is immediately released, after which the delayed part follows more slowly. In the strain versus time diagram, these recoveries are reflected in the drops *DF* (≈*AB*) and *FG* (≈*BC*), respectively. The supplied energy giving the viscous response *CD* is

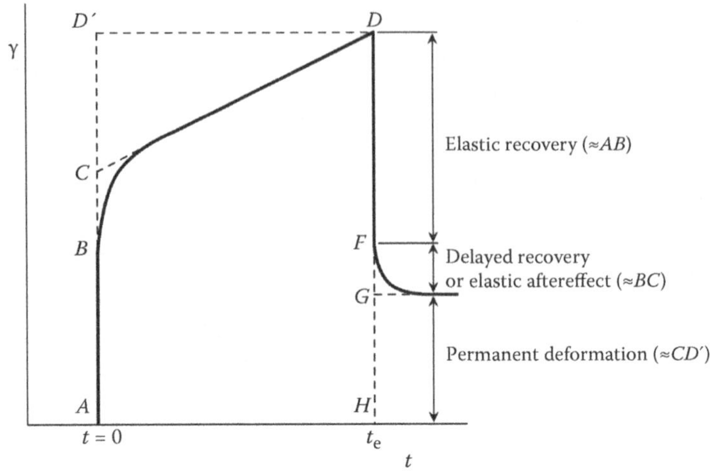

Figure 7.4 *Principle of a creep measurement for a viscoelastic material. At t = 0, a constant shear stress is (instantaneously) applied and maintained until $t = t_e$. Shear strain γ is followed as a function of time. At $t = t_e$, this stress is instantaneously removed. Idealized behavior. (After van Vliet, T., Lyklema, J., in Fundamentals of Interface and Colloid Science, vol. IV,* Particulate Colloids, *ed. Lyklema, J., Academic Press, Amsterdam, 2005.)*

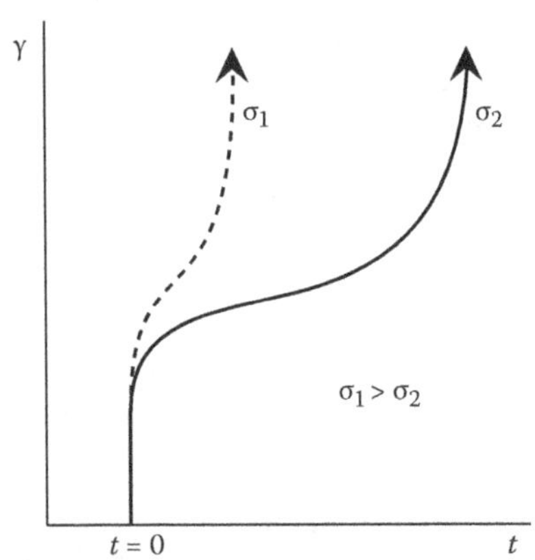

Figure 7.5 *Shear strain γ as a function of time for a viscoelastic material after application of such a shear stress σ that after some time, fracture occurs ($\sigma_1 > \sigma_2$). (Reprinted from van Vliet, T., in* Food Texture Measurement and Perception, *ed. Rosenthal, A.J., Aspen Publishers, Gaithersburg, MD, 1999a. With permission.)*

dissipated as heat. Therefore, it gives rise to the permanent deformation *GH* (≈*CD′*). From the slope of line *CD*, representing dγ/d*t*, an (apparent shear) viscosity can be calculated. For many materials, line *CD* will only become straight after very long creep times.

Figure 7.4 is idealized; in practice, the sections *AB* and *DF* and *BC* and *FG* are not always identical. For some systems, *DF* decreases with increasing t_e. It may indicate a slow breakdown of network structures (or imperfections in the experimental procedure; e.g., friction in the apparatus). An upward curvature in line *CD* is usually an indication of the slow structure breakdown of the material. If this structure breakdown becomes extensive, finally yielding or fracture of the specimen may occur. The loading time after which fracture occurs will depend on the applied stress. It will be shorter for larger stresses (**Figure 7.5**).

7.2.1 Analysis of creep curves in terms of retardation spectra

Usually, creep test results are expressed in terms of *(creep) compliance J*, which is defined as

$$J(t) \equiv \gamma(t)/\sigma \tag{7.8}$$

For an ideally elastic material, $J = 1/G$.

In older literature, in particular, creep curves are often described by a set of Kelvin (or Voigt) elements, consisting of a parallel spring and dashpot, in series with an additional Maxwell element (**Figure 7.6**). The spring is characterized by compliance J_K and the dashpot by viscosity η_K. Next, the material involved is characterized by fitting the curves with 1–3 Kelvin elements in combination with a Maxwell element, whereby the compliance of the springs and the viscosity of the dashpots are used as fitting parameters. However, the obtained value of these parameters cannot be translated directly in material properties. They are curve fitting parameters. Mostly their value depends strongly on the execution of the creep experiment, for example, time span of the application of the constant stress, and on the number of elements considered during the fitting procedure. In general, for a full description of a creep curve, a spectrum of exponentials is required just as in the case of stress relaxation.

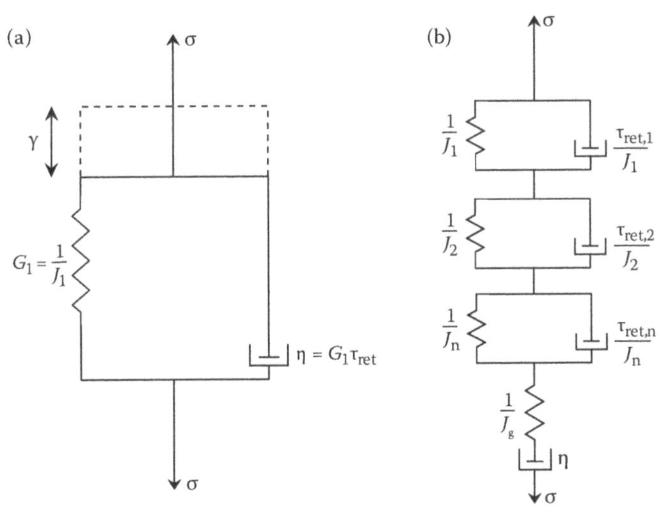

***Figure 7.6** Kelvin or Voigt model of viscoelastic behavior. (a) Single element, (b) set of elements in series with an additional Maxwell element. Springs are labeled with appropriate stress/strain ratio and dashpots with stress/strain rate ratios (for explanation of symbols, see text). (After van Vliet, T., Lyklema, J., in Fundamentals of Interface and Colloid Science, vol. IV,* Particulate Colloids, *ed. Lyklema, J., Academic Press, Amsterdam, 2005.)*

If a stress is suddenly applied to a Kelvin element, the spring cannot respond immediately because of the resistance caused by the viscous flow in the dashpot. It results in delayed elastic behavior. So, the increase in strain is retarded. After the stress is taken away, the energy stored in the spring relaxes, again with a rate determined by the viscosity of the fluid in the dashpot. This type of behavior is characteristic for *semisolids*. In the limit where the effective viscosity of the dashpot $\eta_K \rightarrow 0$, ideal elastic behavior is attained, characterized by the spring of the Kelvin element in **Figure 7.6**. Similarly, for the limit that the contribution of the spring can be neglected ($J \rightarrow \infty$), viscous behavior is obtained.

A retardation spectrum can be calculated from $J(t)$ in a manner analogous to the calculation of a relaxation spectrum from stress relaxation data. It should be noted that in rheology, it is customary to use the terms retardation and relaxation for creep and stress relaxation, respectively.

During deformation of a Kelvin element at any time, the strains of the spring and of the dashpot are the same. The stress applied on the element redistributes over the spring and the dashpot, representing elastic and viscous behavior, respectively. During deformation, $\sigma = \sigma_{el} + \sigma_{visc}$. This gives

$$\sigma = \frac{\gamma}{J_K} + \eta_K \frac{d\gamma}{dt} \tag{7.9}$$

Integration of Equation 7.9, with $\gamma = 0$ at $t = 0$, gives the total strain experienced by the Kelvin model after sudden application of a stress σ

$$\gamma(t) = \sigma J_K \left(1 - e^{-t/\tau_{ret,K}}\right), \tag{7.10}$$

where $\tau_{ret,K} = J_K \eta_K$ is the retardation time of the Kelvin element.

For a set of Kelvin elements and one Maxwell element in series, each characterized by a compliance J and a viscosity η, the total strain after a sudden application of a stress σ is given by

$$\gamma(t) = \sigma J_{M,g} + \frac{\sigma}{\eta_M} t + \sigma \sum_{i=1}^{n} J_i \left(1 - e^{-t/\tau_{ret,K}}\right), \tag{7.11}$$

where $J_{M,g}$ represents the compliance of the spring of the Maxwell element and η_M is the viscosity of the dashpot. The term $\sigma J_{M,g}$ represents the instantaneous elastic deformation of the material and σ/η_M denotes the linear increase in γ in time for long creep times. Just as for stress relaxation, for a full description of the creep behavior, a spectrum of exponentials is usually needed. Considering that for most materials, the range of retardation times to be taken into account is very extensive, it is more convenient to choose a logarithmic scale and to write

$$\gamma(\tau_{ret})\, d\tau_{ret} = \sigma L\, (\tau_{ret})\, d \ln \tau_{ret}, \tag{7.12}$$

where $L(\tau_{ret}) = \tau_{ret}\gamma(\tau_{ret})/\sigma$ is usually called the *retardation spectrum* (spectrum of retardation times) although, strictly speaking, it is a *spectrum of compliance densities*. It contains the same information as $\gamma(\tau_{ret})$. The creep compliance may then be written as

$$J(t) = J_g + \frac{t}{\eta} + \int_{-\infty}^{\infty} L(\tau_{ret})(1 - e^{-t/\tau_{ret}})\text{d}\ln\tau_{ret} \tag{7.13}$$

If the retardation spectrum is very low at long retardation times, say $\tau_{ret,x}$, and stays so for $t > \tau_{ret,x}$, the contributions of the term $L(\tau_{ret})e^{-t/\tau_{ret}}$ will be small. All significant contributions to the delayed strain have, by then, reached equilibrium, and the creep curve becomes linear. For a solid-like material the strain will be constant in time, but for materials with a liquid-like material strain, it will increase linearly with time. The *creep compliance* is then given by

$$J(t) = J_g + \frac{t}{\eta} + \int_{-\infty}^{\infty} L(\tau_{ret})\text{d}\ln\tau_{ret} \tag{7.14}$$

Equation 7.14 holds for $\exp(-t/\tau_{ret}) \ll 1$.

After unloading the sample, the curve for the strain versus time has to be the inverse of the deformation curve after loading, except for the deformation, equal to $(\sigma/\eta)t$, as a result of flow.

If either the relaxation spectrum $H(\tau_{rel})$ or the retardation spectrum $L(\tau_{ret})$ is known over the entire range of time scales, together with certain limiting values such as G_e, J_g, and η, the other of these two spectra can be calculated (Ferry 1980). Treatment of these interrelations is outside the scope of this book.

As mentioned above for an ideally elastic material, $J = 1/G$. Then, both parameters are independent of time. However, for viscoelastic materials they are not, and $J(t) \neq 1/G(t)$. This, although creep and stress relaxation are both manifestations of the same stress relaxation mechanism of the bonds between the structural elements at molecular and mesoscopic scales. It can be shown that, because of the different time pattern between creep and stress relaxation tests, the stress relaxation modulus moves more quickly to its final steady-state value compared with creep compliance (Ferry 1980). As a result, $J(t)G(t) \leq 1$.

7.3 Tests at constant strain rate

A set (shear) strain rate is applied, and the resulting stress σ is measured as a function of time. For a Newtonian liquid σ will be constant in time, whereas for non-Newtonian liquids σ may decrease (thixotropic liquids) or increase (antithixotropic or rheopectic liquids) in time. For systems showing a yield stress σ_y, $\sigma(t)$ may display *stress overshoot* (**Figure 7.7**). Such materials are first deformed "elastically" until σ exceeds σ_y, and the structure of the material responsible for the elastic properties is broken down. Then, the material yields. Upon continued deformation, the structure of the material is broken down further, resulting in a decrease in η_{app} with measuring time (thixotropic behavior). After a longer shearing time, a steady-state situation may be reached. If σ (steady state) is plotted as a function of the shear rate, a graph characteristic for shear thinning behavior is obtained. One should be aware that in reality structure breakdown starts already at stresses below σ_y, the extent of which depends on experimental conditions such as shear rate.

"Overshoot" measurements are useful for determining the large deformation and yielding (fracture) behavior of plastic fluids and for following the ongoing structure breakdown upon further deformation. The maximum stress during deformation is sometimes used as a measure of σ_y. The obtained σ_y usually depends on the applied shear rate or, in other words, on the time scale of the measurement. For many (viscoelastic) materials, this strain rate dependency is a real characteristic, just as is observed for the fracture stress of more solid-like viscoelastic materials (Section 5.2.3). Extrapolation of shear stress versus shear rate curves by, for example, the Bingham or the Herschel–Bulkley model, often leads to the false impression that σ_y is a fixed characteristic of a material, although this is mostly not the case. Moreover, the σ_y obtained often depends on the shear rate range involved in the extrapolation procedure.

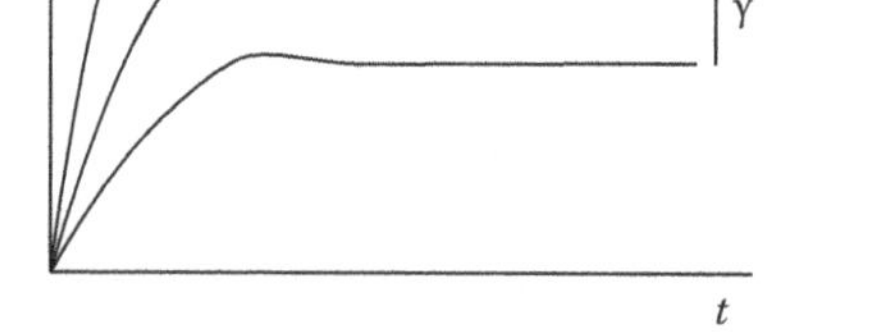

Figure 7.7 *Stress σ as a function of time t at three different shear rates $\dot{\gamma}$ for a material exhibiting stress overshoot. (Reprinted from van Vliet, T., in* Food Texture Measurement and Perception, *ed. Rosenthal, A.J., Aspen Publishers, Gaithersburg, MD, 1999a. With permission.)*

Fracture experiments on more solid-like semisolids are often performed by deforming them in uniaxial compression or tension at a set strain rate or for less sophisticated instruments at a set deformation rate. These experiments allow the determination of a fracture stress and strain, usually taken at the maximum in the stress versus strain curve. For a further discussion of these experimental methods, see Section 8.3. In addition, fracture tests of gels may also be performed by applying a torsion deformation (Section 8.2.4). To obtain a well-defined deformation, specially shaped specimens should be used.

Many new rheometers in which a constant strain rate can be applied are, in fact, constant stress rheometers. The constant strain rate is obtained by a back coupling procedure that adjusts the stress in relation to the measured strain or strain rate and their set values. For well-built instruments, this does not detract from their performance in constant strain tests.

7.4 Oscillatory tests

In an oscillatory test, a small periodic, mostly sinusoidally oscillating stress, strain, or strain rate is applied to the material at an angular frequency ω (rad s^{-1}), and the resulting strain or stress is measured. In these experiments, the time scale of the deformation (applied stress) is replaced as a variable by the frequency. Compared to the continuous ones discussed above, periodical measurements have an advantage in that relaxation phenomena can be studied at one well-defined time scale that can be varied easily by changing the frequency (time scale of experiment is roughly $1/\omega$). Therefore, they are very well suited for the characterization of undisturbed viscoelastic materials as a function of time scale. Both the elastic and viscous components can be obtained over a broad range of time scales. For viscoelastic systems, upon increasing ω, the response tends to become increasingly more "elastic." In addition, oscillatory tests are well suited for following gel formation and aging in time at time scales ($1/\omega$) that are short compared to those of the changes in the structure of the systems being studied.

For an experiment in which a harmonic (sinusoidal) oscillation of a shear strain γ is applied with a frequency ω, γ is given by

$$\gamma(t) = \gamma_0 \sin \omega t, \tag{7.15}$$

where γ_0 is the maximum shear strain (**Figure 7.8a**). Furthermore, it is assumed that the inertia of the studied material can be neglected and that the harmonic varying strain results in a harmonic varying stress proportional

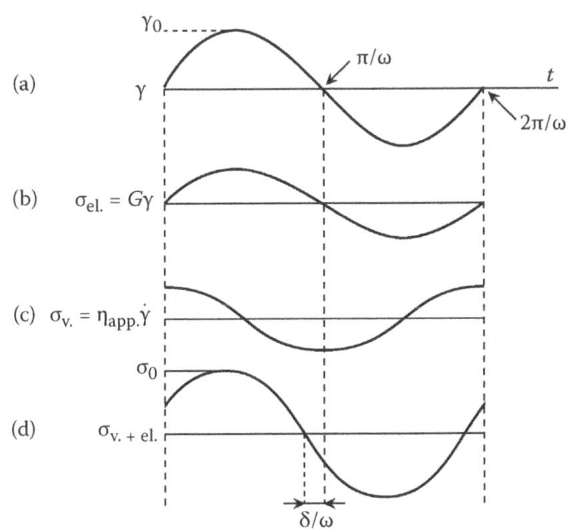

Figure 7.8 *Stress response of a (viscoelastic) material to a strain varying sinusoidally at a frequency ω (rad s^{-1}), δ phase difference between applied strain and stress response: (a) applied shear strain, (b) elastic, (c) viscous, and (d) total response of stress. Highly schematic.*

to the strain amplitude and with a phase lag δ relative to the strain, which is independent of these amplitudes (*linear viscoelastic behavior*). The magnitude of δ will depend on the type of material. Then, the stress σ is given by

$$\sigma(t) = \sigma_0 \sin(\omega t + \delta) = \sigma_0[(\cos\delta \sin \omega t) + (\sin\delta \cos \omega t)], \tag{7.16}$$

where σ_0 is the maximum shear stress. For an ideal elastic material, $\sigma_{el} = G\gamma$, and the stress is in phase with the strain (**Figure 7.8b**). For a viscous material, $\sigma_v = \eta\dot{\gamma} = \eta d\gamma/dt$. The shear rate, and with that the stress, is 90° or (1/2) π rad out of phase with the strain (**Figure 7.8c**). For viscoelastic materials, δ is between 0 and (1/2)π rad.

For a linear viscoelastic material, σ_0 is proportional to γ_0, and Equation 7.16 can be rewritten as

$$\sigma(t) = \gamma_0\left[\left(\frac{\sigma_0}{\gamma_0}\right)(\cos\delta \sin \omega t)\right] + \left[\left(\frac{\sigma_0}{\gamma_0}\right)(\sin\delta \cos \omega t)\right] \tag{7.17}$$

The first part of the right-hand side of Equation 7.17 is the part that is in phase ($\delta = 0$ and with that $\cos\delta = 1$ and $\sin\delta = 0$) with the strain. The second part is the out-of-phase part ($\delta = (1/2)\pi$, giving $\cos\delta = 0$ and $\sin\delta = 1$). It is convenient to define the quantities

$$G' = \frac{\sigma_0}{\gamma_0}\cos\delta \tag{7.18}$$

and

$$G'' = \frac{\sigma_0}{\gamma_0}\sin\delta \tag{7.19}$$

G' is called the (*shear*) *storage modulus* and G'' the (*shear*) *loss modulus*. G' is a measure of the amount of energy stored during a periodic application of a strain or stress, and the out-of-phase component G'' is a measure of the energy dissipated during the periodic application of a strain or stress. Inserting Equations 7.18 and 7.19 in Equation 7.17 gives:

$$\sigma(t) = \gamma_0[(G' \sin \omega t) + (G'' \cos \omega t)] = G'\gamma + \frac{G''}{\omega}\dot{\gamma} \tag{7.20}$$

If the strain varies as $\sin \omega t$, the rate of strain $\dot{\gamma}$ will vary as $\omega\cos \omega t$. The ratio of G''/G' gives the *loss tangent*, or

$$\tan\delta = \frac{\sin\delta}{\cos\delta} = \frac{G''}{G'}, \tag{7.21}$$

where δ is the loss angle. A higher $\tan \delta$ means that, during deformation of the test material, a relatively larger part of the supplied energy is dissipated as heat and less is elastically stored; the material behaves relatively in a more viscous and less elastic manner.

The ratio of σ_0/γ_0 gives the ***absolute shear modulus*** $|G^*|$. Summation of the squares of Equations 7.18 and 7.19 gives for $|G^*|$

$$|G^*| = \frac{\sigma_0}{\gamma_0} = (G'^2 + G''^2)^{1/2} \tag{7.22}$$

Remember that $(\cos\delta)^2 + (\sin\delta)^2 = 1$.

As $\sigma_v(t) = \eta'\dot{\gamma}$, where η' is the dynamic viscosity, one obtains the identity

$$G'' = \eta'/\omega \tag{7.23}$$

For a Newtonian fluid, $\eta' = \eta$.

In the above derivation, the moment that the shear strain was at maximum was arbitrarily taken as the origin for the time measurement. Alternatively, the same derivation can be given with the maximum of the stress as the origin of the time measurement. This results in the definition of a ***(shear) storage compliance*** J' and a ***(shear) loss compliance*** J'' given by (Whorlow 1992)

$$J' = \frac{\gamma_0}{\sigma_0}\cos\delta \tag{7.24}$$

$$J'' = \frac{\gamma_0}{\sigma_0}\sin\delta \tag{7.25}$$

The relation between the storage compliance and the storage modulus is given by

$$J' = \frac{G'}{G'^2 + G''^2} \tag{7.26a}$$

or

$$G' = \frac{J'}{J'^2 + J''^2} \tag{7.26b}$$

Those between the loss compliance and the loss modulus read as

$$J'' = \frac{G''}{G'^2 + G''^2} \tag{7.27a}$$

or

$$G'' = \frac{J''}{J'^2 + J''^2} \tag{7.27b}$$

The loss tangent is the same as those for the moduli

$$\tan\delta = \frac{J''}{J'} = \frac{G''}{G'} \tag{7.28}$$

Still another pair of quantities is obtained if the shear rate is taken as the leading parameter. This may be convenient if essential liquids are considered. Such a derivation results in the definition of a dynamic viscosity η' (the in-phase component) and η'' (the out-of-phase component) (Ferry 1980) where

$$\eta'' = G'/\omega \tag{7.29}$$

and $\eta' = G''/\omega$ (Equation 7.23). For an ideally viscous material, $\eta' = \eta$ and $\eta'' = 0$.

The main advantage of dynamic tests is that the contribution of both the elastic and the viscous components in the reaction of a material to an applied stress or strain can be determined over a large range of time scales (frequencies). An example of data that can be obtained is shown in **Figure 7.9** for skim milk gels formed by acidification or by enzyme (rennet) action. At a frequency of 1 rad s^{-1}, the ratio of $G''/G' = \tan\delta$ is the same for both types of skim milk gels, but at lower frequencies $\tan\delta$ is much higher for an enzyme (rennet)-induced skim milk gel, which implies that the gels exhibit clearly different properties. Rennet-induced skim milk gels exhibit a much stronger liquid-like behavior over long time scales.

For a relatively simple and unequivocal interpretation of the data, measurements should be performed in the linear region. It is generally assumed that over this range, measurements affect neither the structure of the material studied nor the structure formation and structure rearrangement processes. The moduli obtained have no predictive value for large deformation properties outside the linear region and for fracture properties (van Vliet and Walstra 1995). The strain at which the stress versus strain becomes nonlinear may provide information on the structure of the material involved, for instance, gels

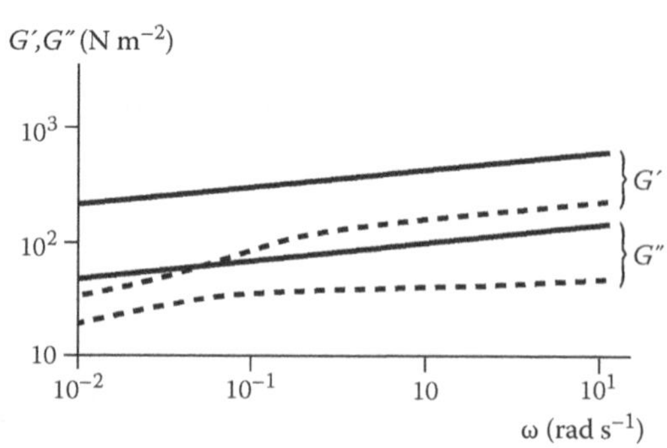

Figure 7.9 *Storage moduli G′ and loss moduli G″ as a function of frequency ω of skim milk gels formed by acidification (full curves) or by enzyme (rennet) action (broken curves).*

formed by flexible macromolecules (e.g., gelatin gels) will generally exhibit a much larger linear region than gels formed by hard particles (e.g., margarines containing a fat crystal network) (Section 13.3.2).

In recent years, techniques have been developed that allow the interpretation of measurements carried out somewhat beyond the linear region. A discussion of these techniques is considered to be outside the scope of this book.

7.4.1 Analysis oscillatory data in terms of relaxation spectra

As with the stress relaxation modulus, dynamic moduli can also be written as functions of the relaxation spectrum. As described above in Section 7.1, for a single Maxwell element the relationship $\gamma = \gamma_{el} + \gamma_{visc}$ holds. Differentiation with respect to time of Equation 7.2 gives

$$\dot{\gamma} = \frac{d\sigma/dt}{G_M} + \frac{\sigma}{\eta_M} \tag{7.30}$$

(the same as Equation 4.6) or

$$\sigma + \tau_{rel}\frac{d\sigma}{dt} = \eta_M \dot{\gamma} \tag{7.31}$$

Further elaboration of this equation gives for a material that can be described by a single Maxwell element, where $G' = G_M$ and $G'' = \omega\eta_M = \omega G_M\,\tau_{rel,M}$ for the time-dependent behavior of the storage and loss modulus (Ferry 1980; Whorlow 1992),

$$G'(\omega) = G\frac{\omega^2\tau_{rel}^2}{1+\omega^2\tau_{rel}^2} \tag{7.32}$$

and

$$G'' = G\frac{\omega\tau_{rel}}{1+\omega^2\tau_{rel}^2} \tag{7.33}$$

respectively (for clarity the subscripts M have been left out). Derivation of both equations is considered to be outside the scope of this book. It can be done in relatively simple manner by writing stress and strain as complex quantities (Appendix C). **Figure 7.10** shows a graphical representation of $G'(\omega)$ and $G''(\omega)$ versus ω for a material that can be described by a single Maxwell element. The loss tangent of such a material is given by

$$\tan\delta = \frac{G''}{G'} = \frac{1}{\omega\tau_{rel,M}} = \frac{t_{obs}}{\tau_{rel}} = \frac{1}{De} \tag{7.34}$$

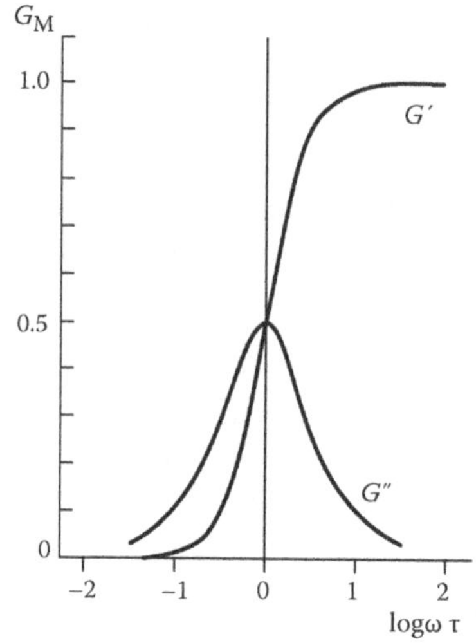

***Figure 7.10** Storage modulus G′ and loss modulus G″ as a function of frequency ω times relaxation time τ for a (hypothetical) material that is well described by one Maxwell element with a relaxation time τ. (After van Vliet, T., in* Food Emulsions and Foams: Interfaces, Interactions and Stability, *ed. Dickinson, E., Rodríquez Patino, J.M., Royal Society of Chemistry, Cambridge, 1999b.)*

For this case, tan δ is equal to the observation time (1/ω) divided by the relaxation time of the material or 1/*De* (Section 4.2). For more complex materials having a spectrum of relaxation times, the relationship between *De* and tan δ is much more complicated; rather, a spectrum of *De* numbers is required.

Equations 7.32 and 7.33 can be generalized by considering an infinite array of Maxwell elements. The results are (Ferry 1980; Whorlow 1992)

$$G'(\omega) = G_e + \int_{-\infty}^{\infty} \frac{\omega^2\tau_{rel}^2 H(\tau_{rel})}{1+\omega^2\tau_{rel}^2} \, d\ln\tau_{rel} \quad (7.35)$$

$$G''(\omega) = \int_{-\infty}^{\infty} \frac{\omega\tau_{rel} H(\tau_{rel})}{1+\omega^2\tau_{rel}^2} \, d\ln\tau_{rel} \quad (7.36)$$

The term $H(\tau_{rel})$ d ln τ represents the distribution function of shear moduli with relaxation times τ, of which the logarithms lie in the range between ln τ_{rel} and ln τ_{rel} + d ln τ_{rel}. As a first approximation, it can be derived that (Booij and Palmen 1982)

$$H(\tau_{rel}) \approx \frac{1}{\pi}[G'(\omega)\sin\delta]_{\omega=1/\tau_{rel}} \quad (7.37)$$

Comparison of Equation 7.37 with the definition of $H(\tau_{rel})$ shows that sin δ is a measure of the proportion of bonds with a relation time τ of about 1/ω, if δ is (about) independent of ω. For not too high values of δ (< π/4), the difference between sin δ and tan δ is only small. Then, for the case where δ is independent of ω, both may be used as a *qualitative* measure of the proportion of bonds with a relation time τ (van Vliet et al. 1991).

For an infinite array of Maxwell elements, *G*′(ω) is related to the number of elements that react elastically to a periodic applied stress or strain at angular frequency ω and *G*″(ω) to the number of elements that relax at frequencies roughly in between 0.1ω and 10ω. This implies that *G*″(ω) is related to the increase of *G*′(ω) with ω, and loss angle δ is related to the increase in the absolute modulus with ω. It can be shown that, by a first approximation, the following equations hold (Booij and Thoone 1982):

$$G''(\omega) = \frac{\pi}{2}\left[\frac{dG'\omega}{d\ \ln\omega}\right] \tag{7.38}$$

$$\delta(\omega) = \frac{\pi}{2}\frac{d\ \ln|G^*(\omega)|}{d\ \ln\omega} \tag{7.39}$$

For more exact relations, we refer to the literature (e.g., Ferry 1980).

References

Booij, H.C., and J.H.M. Palmen. 1982. Some aspects of linear and nonlinear behavior of polymer melts in shear. *Rheol. Acta* 21: 376–387.

Booij, H.C., and C.P.J.M. Thoone. 1982. Generalization of Kramers–Kronig transforms and some approximations of relations between viscoelastic quantities. *Rheol. Acta* 21: 15–24.

Ferry, J.D. 1980. *Viscoelastic Properties of Polymers*. 3rd ed, Chap. 3. New York: Wiley.

van Vliet, T., H.J.M. Van Dijk, P. Zoon, P. Walstra. 1991. Relation between syneresis and rheological properties of particle gels. *Colloid Polym. Sci.* 269: 620–627.

van Vliet, T., and P. Walstra. 1995. Large deformation and fracture behaviour of gels. *Faraday Discuss.* 101: 359–370.

van Vliet, T. 1999a. Rheological classification of foods and instrumental techniques for their study. In *Food Texture: Measurement and Perception*, ed. A.J. Rosenthal, 65–98. Gaithersburg, MD: Aspen Publishers.

van Vliet, T. 1999b. Factors determining small deformation behaviour of gels. In *Food Emulsions and Foams: Interfaces, Interactions and Stability*, ed. E. Dickinson and J.M. Rodríquez Patino, 307–317. Cambridge: Royal Society of Chemistry.

van Vliet, T., and J. Lyklema. 2005. Rheology. In *Fundamentals of Interface and Colloid Science*, Vol. IV. *Particulate Colloids*, ed. J. Lyklema, Chap. 6. Amsterdam: Academic Press.

Whorlow, R.W. 1992. *Rheological Techniques*. London: Ellis Horwood.

8

Measuring Apparatus

A large variety of instrumental techniques exists that allow the determination of rheological and fracture properties. In rheology, the two main classes are tube or capillary viscometers and rotational instruments. For determining fracture behavior, the so-called tension compression apparatuses are often used. Rotational instruments can be equipped with a range of different measuring geometries such as coaxial cylinders, cone and plate, parallel plates, or a spindle with thin vanes. Capillary instruments are well suited for determining accurately the viscosity of Newtonian liquids, and rotational instruments for studying rheological properties of non-Newtonian liquids and materials that can solidify in the rheometer (e.g., gels). Rotational instruments equipped with parallel plate geometry can also be used for studying solid materials. Tension compression apparatuses consist essentially of a fixed bottom plate and a moving bar, each of which has a part of the measuring geometry fixed to it. A load cell is placed in the moving bar or bottom plate, by which the forces, required to deform the test piece, can be registered. Because they are very popular in the food industry, in addition, several empirical test methods will be discussed in Section 8.4.

An excellent comprehensive overview of tube viscometers and rotational rheometers is found in Whorlow's (1992) book, *Rheological Techniques*. Below, the discussion of these types of instruments will be focused on those more regularly used for the study of food materials.

8.1 Tube viscometers

In a tube viscometer, the flow rate of a material caused by a pressure gradient ($\Delta p/l$) is determined, or conversely, the pressure gradient brought about by a known flow rate. For a low-viscous fluid, flow is mostly due to gravity. For liquids with a high viscosity, flow can be brought about by applying a gas or hydrostatic pressure to a reservoir at the entrance of a tube or by using a piston. Capillary viscometers are popular because they are (relatively) cheap and easy to use and to thermostat. Moreover, they give reproducible results. Below, both the terms fluid and liquid will be used. The term fluid encompasses liquids, gases, and plasmas. Within the context of the purpose of this book, the discussion will focus on liquids, although many of the equations that will be derived in this chapter hold for fluids in general.

8.1.1 Flow equations

The derivations of the equations that give the shear rate, shear stress, and viscosity, from a measured relationship between flow rate and pressure gradient are based on the assumptions that (Whorlow 1992)

(a) Flow is parallel to the tube axis.
(b) The velocity v of any fluid element is a function of the tube radius r only and independent of z (**Figure 8.1**).
(c) The velocity of the fluid at the wall is zero (no slip condition).
(d) The fluid is incompressible.
(e) Entry and exit effects can be neglected.
(f) A unique function $\dot{\gamma} = f(\sigma)$ relates the shear rate $\dot{\gamma}$ to the shear stress σ, implying, for example, no significant heat generation during flow.
(g) The normal stress is isotropic.

Because each fluid element inside a cylinder of length l is moving at constant velocity (assumption b), the net force on the cylinder must be zero; so,

$$\pi r^2 \Delta p = 2\pi r \sigma l, \tag{8.1}$$

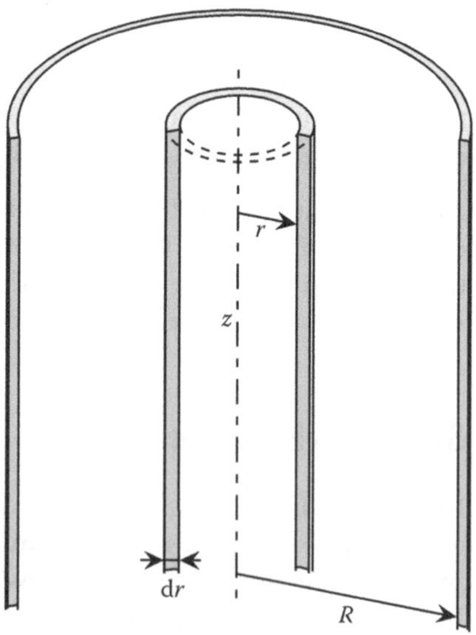

***Figure 8.1** Flow through a cylinder. Its length $l \gg R$ (for further explanation, see text). (After van Vliet, T., Lyklema, J., in* Fundamentals of Interface and Colloid Science, *Vol. IV, Particulate Colloids, Academic Press, Amsterdam, 2005.)*

where Δp is an abbreviation for $p_0 - p_1 + \rho g l$: $p_0 - p_1$ is the pressure drop over the cylinder from $z = 0$ to $z = l$, ρ is the density of the fluid, and g is the acceleration due to gravity. If the cylinder is inclined at an angle α to the vertical, the gravity contribution must be multiplied by $\cos\alpha$. The shear stress at radius r is then given by

$$\sigma = \frac{\Delta p}{2l} r \tag{8.2}$$

Hence, the shear stress at the wall σ_R equals

$$\sigma_R = \frac{\Delta p}{2l} R, \tag{8.3}$$

where R is the radius of the tube. The shear stress varies linearly from zero at the axis of the tube to a maximum value at the tube wall. The dependence of σ on r also causes $\dot{\gamma}$ to depend on r. The positive direction of σ has been taken opposite to the direction of flow and therefore for $\dot{\gamma}$, one gets

$$\dot{\gamma} = -\frac{\mathrm{d}v}{\mathrm{d}r} \tag{8.4}$$

This causes both σ and $\dot{\gamma}$ to be positive quantities. From Equation 8.2, it follows that for a Newtonian liquid with a viscosity η, the shear rate is given by

$$\dot{\gamma} = \frac{\sigma}{\eta} = \frac{\Delta p}{2\eta l} r \tag{8.5}$$

So, for a Newtonian liquid, the shear rate changes also linearly with r. Combination of Equations 8.4 and 8.5 gives for a Newtonian fluid the next expression for $\mathrm{d}v$

$$\mathrm{d}v = -\frac{\Delta p}{2\eta l} r\mathrm{d}r \tag{8.6}$$

Integration between $r = 0$ and $r = R$ gives the following parabolic velocity distribution (**Figure 8.2a**)

$$v = \int_{r=0}^{r=R} -\frac{\Delta p}{2nl} r\mathrm{d}r = \frac{\Delta p}{4\eta l}(R^2 - r^2) \tag{8.7}$$

The velocity has its maximum at the axis where $r = 0$; $v = \Delta p R^2/4\eta l$, which is larger by a factor of 2 than the average velocity

$$\bar{v} = \frac{1}{\pi R^2} \int_{r=0}^{r=R} \frac{\Delta p}{4\eta l}(R^2 - r^2) 2\pi r\mathrm{d}r = \frac{R^2 \Delta p}{8\eta l} \tag{8.8}$$

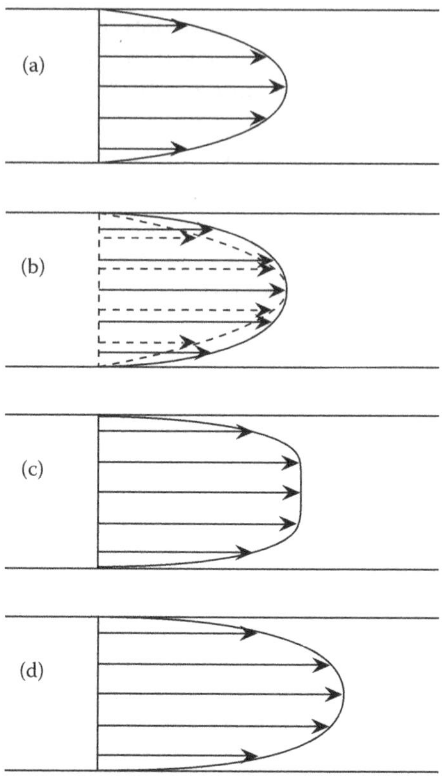

Figure 8.2 *Position of fluid particles a short time after being in the same cross-sectional plane. (a) Newtonian fluid; (b) power law fluid n < 1 (full line), n > 1 (dotted line); (c) Bingham model; (d) fluid slipping at the wall of the tube. (Redrawn from Whorlow, R.W.,* Rheological Techniques, *Ellis Horwood, Chichester, 1992.)*

The volume flow rate Q (m^3 s^{-1}) is obtained by integrating v over the cross section of the tube.

$$Q = \int_{r=0}^{r=R} v 2\pi r \, dr = \frac{\pi R^4 \Delta p}{8 \eta l} \tag{8.9}$$

Equation 8.9, the so-called *Poiseuille's law* (also known as the Hagen–Poiseuille law), allows the calculation of η from the volume flow rate.

An alternative starting point for deriving Equations 8.1 through 8.9 is the *Navier–Stokes equation* for incompressible Newtonian fluids (van Vliet and Lyklema 2005). The approach that has been followed above is mathematically somewhat simpler although less universal.

8.1.1.1 Non-Newtonian liquids

For non-Newtonian liquids, the relationship between σ and $\dot{\gamma}$ has to be known to derive Equations 8.5 through 8.9. To obtain these relationships for non-Newtonian fluids, it is convenient to express v and Q in terms of σ rather than r. Then

$$\frac{dv}{d\sigma} = \frac{dv}{dr}\frac{dr}{d\sigma} = -\dot{\gamma}\frac{2l}{\Delta p} = -\frac{2l}{\Delta p}f(\sigma_r) \tag{8.10}$$

Integration of 8.10 from radius r to the wall, assuming no slip at the wall, results in

$$v = \frac{2l}{\Delta p}\int_{\sigma_r}^{\sigma_R} f(\sigma_r)\,d\sigma \tag{8.11}$$

The volume flow rate, Q, follows from integration over the cross section of the tube, using Equation 8.2

$$\begin{aligned} Q &= \int_{r=0}^{r=R} 2\pi r v\,dr = \int_0^{\sigma_R} 2\pi\frac{2l}{\Delta p}\sigma_r\left\{\frac{2l}{\Delta p}\int_0^{\sigma_R} f(\sigma_r)\,d\sigma\right\}\frac{2l}{\Delta p}\,d\sigma \\ &= \frac{16\pi l^3}{\Delta p^3}\int_0^{\sigma_R}\sigma_r\int_0^{\sigma_R} f(\sigma_r)\,d\sigma\,d\sigma \end{aligned} \tag{8.12}$$

Since $\Delta p/l = 2\sigma_R/R$ (Equation 8.3) and $\int \sigma\,d\sigma = (1/2)\sigma^2$, we obtain

$$\frac{Q}{\pi R^3} = \frac{1}{\sigma_R^3}\int_0^{\sigma_R}\sigma_r^2 f(\sigma_r)\,d\sigma \tag{8.13}$$

Because the right-hand side of Equation 8.13 is only a function of σ_R and $f(\sigma_r)$, a graph of $Q/\pi R^3$ as a function of σ_R gives a unique curve that is characteristic for a given material and independent of the pressure gradient and tube radius.

Equation 8.13 can be integrated when $f(\sigma_r)$ is known and has a manageable form. The simplest case is that for a Newtonian fluid where $f(\sigma_r) = \dot{\gamma} = \sigma/\eta$, where η is constant, resulting in

$$\frac{Q}{\pi R^3} = \frac{1}{\sigma_R^3}\int_0^{\sigma_R}\frac{\sigma^3}{\eta}\,d\sigma = \frac{\sigma_R}{4\eta} = \frac{R\Delta p}{8l\eta} \tag{8.14}$$

and finally in Equation 8.9.

For a power law fluid, it holds that $f(\sigma_r) = \dot{\gamma} = (1/k)\sigma^{(1/n)}$ (see Equation 4.10), where k and n are constants. Substitution of this relation in Equation 8.13 and integration gives

$$\frac{Q}{\pi R^3} = \frac{1}{\sigma_R^3} \int_0^{\sigma_R} \frac{\sigma_r^{(2n+1)/n}}{k} \mathrm{d}\sigma = \frac{n\sigma_R^{1/n}}{k(3n+1)} \tag{8.15}$$

For a power law liquid, the flow profile will deviate from the parabolic form observed for Newtonian liquids; the extent and direction depend on n (**Figure 8.2b**). For a shear thinning fluid ($n < 1$) the velocity profile is more blunted, whereas for a shear thickening fluid ($n > 1$) the profile is more sharp. For a shear thinning fluid, $\dot{\gamma}$ increases at low σ less rapidly with σ than for a Newtonian fluid at the same Q resulting in the more blunted profile. Moreover, it appears that the flow rate increases more strongly with pressure than that for a Newtonian fluid (Equation 8.15). For both cases, the reason is the decreasing apparent viscosity with increasing shear stress.

For a Bingham liquid, one can derive from Equation 4.13 $f(\sigma_r) = \dot{\gamma} = (\sigma - \sigma_y)/\eta_B$ for $\sigma > \sigma_y$, and $\dot{\gamma} = 0$ for $\sigma < \sigma_y$. Therefore the, contribution of $f(\sigma_r)$ to the integral in Equation 8.13 will be zero for $\sigma < \sigma_y$ so,

$$\frac{Q}{\pi R^3} = \frac{1}{\sigma_R^3} \int_{\sigma_y}^{\sigma_R} \sigma_r^2 \frac{\sigma - \sigma_y}{\eta_B} \mathrm{d}\sigma = \frac{1}{\eta_B}\left[\frac{\sigma_r}{4} - \frac{\sigma_y}{3}\left\{1 - \frac{1}{4}\left(\frac{\sigma_y}{\sigma_r}\right)^3\right\}\right] \tag{8.16}$$

This equation was originally derived by Buckingham (1921). It does not lead to a simple linear plot of experimental data if $Q/\pi R^3$ is plotted as a function of $\Delta p/l$ or in another form. For an ideal Bingham liquid, $Q/\pi R^3$ will become proportional to $\sigma_R/(4\eta_B)$ when σ_R becomes $\gg \sigma_y$. The flow profile (**Figure 8.2c**) has a truncated parabolic form, and the more so when σ_R is closer to σ_y. Remember that σ is zero in the center of the tube (Equation 8.2) and shear flow only occurs when σ exceeds σ_y. The material near the axis moves as a solid plug, whose radius decreases as the pressure gradient increases. For σ_R approaching σ_y, the flow profile will approach that for plug flow (the flow profile will be nearly the same as that of a solid plug, which slips along the wall). Well-known examples are mayonnaise, tomato ketchup, and toothpaste squeezed out of a tube.

In liquids for which the relationship between the shear stress and shear rate is not known, an expression for the shear rate at the wall of the tube can be derived starting from Equation 8.13. Rearranging and differentiating gives the well-known Rabinowitsch–Mooney equation (Whorlow 1992; Steffe 1996).

$$f(\sigma_R) = \frac{1}{\sigma_R^2}\frac{\mathrm{d}}{\mathrm{d}\sigma_R}\left(\frac{\sigma_R^3 Q}{\pi R^3}\right) = \frac{Q}{\pi R^3}\left(3 + \frac{\mathrm{d}\ln(Q/\pi R^3)}{\mathrm{d}\ln\sigma_R}\right) \tag{8.17}$$

For liquids that deviate only slightly from Newtonian behavior, this equation can be simplified by introducing an effective shear rate at the wall $\dot{\gamma}_{R,\mathrm{N}}$, which can be obtained by combining Equations 8.5 and 8.9

$$\dot{\gamma}_{R,\mathrm{N}} = \frac{4Q}{\pi R^3} \tag{8.18}$$

Strictly speaking, Equation 8.18 only applies for Newtonian fluids. Then, by defining $b = \mathrm{d}\ln\dot{\gamma}_{R,\mathrm{N}}/\mathrm{d}\ln\sigma_R$, Equation 8.17 can be written as

$$\dot{\gamma}_R = \dot{\gamma}_{R,\mathrm{N}}\left(\frac{3+b}{4}\right) \tag{8.19}$$

The parameter b may depend on the range of shear rates considered. For a power law liquid, $b = 1/n$.

8.1.2 Instruments

A profound discussion on different types of capillary viscometers and the manipulation of these instruments has been given by Whorlow (1992) and Steffe (1996). Below, we will only discuss the most popular ones in the food industry and laboratories, and their major sources of inaccuracies. The discussion will be limited to viscometers in which flow is driven by gravity and are used for determining the viscosity of liquids. Measuring apparatuses in which flow is induced by gas or hydrostatic pressures or due to a piston are much less used in the food industry and laboratories. However, several empirical apparatuses are in use that involve flow through a short tube (e.g., Ford funnel). In those cases, flow is often dominated, or at least strongly affected, by entrance and exit effects. In industrial practice, this is also the case for extrusion through a die. On the other hand, transport by flow through a long pipe involves well-developed tube flow. By measuring the pressure fall over a certain pipe length by using adequate pressure transducers and the volume flow rate, a good impression of the (apparent) viscosity can be obtained, provided that the shear stress–shear rate relation is known.

Well-known examples of tube (or capillary) viscometers in which flow is due to gravity are the *Ostwald* and the *Ubbelohde viscometers* (**Figure 8.3**), which essentially consist of two bulbs connected by a capillary. In these viscometers, an accurately known volume of liquid flows, driven by gravity, through the capillary, and the required time is measured. The pressure drop Δp over the capillary is equal to $\rho g l$, where ρ is the density of the liquid. The required time is a direct measure of the kinematic viscosity $\upsilon = \eta/\rho$ (flow velocity $\bar{v}$ is proportional to $1/t$). Hence (see Equation 8.8),

$$\upsilon = Ct, \tag{8.20}$$

where C is a constant depending on the geometry of the capillary, and especially on its radius R, because of Poiseuille's law (Equation 8.9), $C \propto R^4$. The

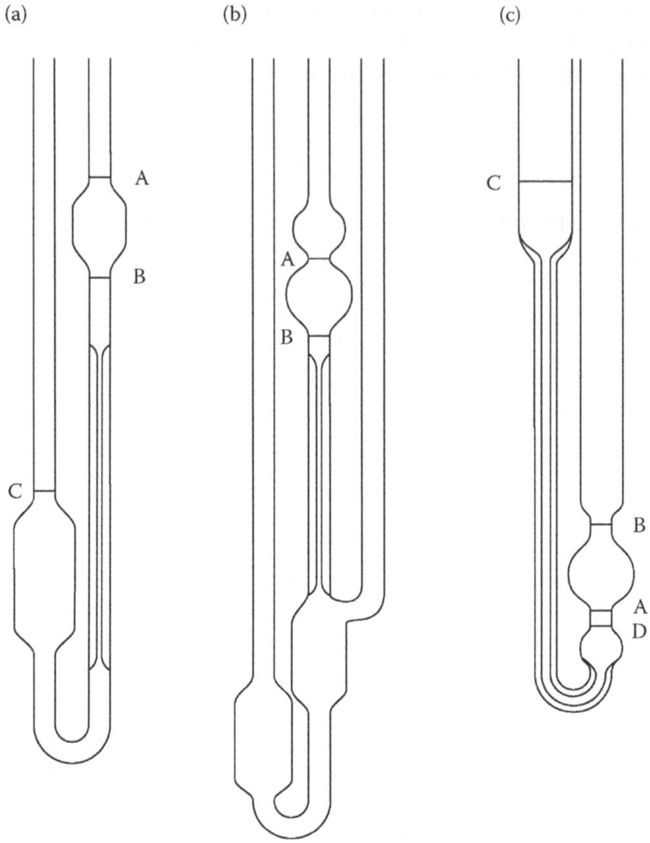

Figure 8.3 *(a) Ostwald, (b) Ubbelohde viscometers, and (c) reverse flow viscometer (not to scale). A, B are timing markings and C, D are filling markings. (Reprinted from Whorlow, R.W.,* Rheological Techniques, *Ellis Horwood, Chichester, 1992.)*

two bulbs should have the same diameter to minimize errors owing to surface tension. In the Ubbelohde viscometer, the pressure at the exit of the capillary is always atmospheric in contrast to the Ostwald viscometer, where it increases during the measurement. Reverse flow viscometers are available for opaque liquids that often does not allow good reading of the timing markings because of the presence of residual liquid (e.g., milk) on the tube wall (**Figure 8.3c**).

Capillary viscometers are popular for several reasons. First, the viscosity of Newtonian liquids can be determined very accurately. Therefore, they are very well suited for calibrating standard liquids to be used to calibrate various rheometers. Second, thermostating can be done accurately and in a relatively simple manner. Third, they are relatively cheap and easy to handle.

The main sources of errors in tube viscometers are discussed in the following subsections (for a more extensive discussion, see Whorlow 1992).

8.1.2.1 Entrance effect

The velocity profiles in the bulb and the capillary are different; a certain length of the capillary is required to obtain the equilibrium velocity profile. Moreover, at the entrance of the capillary the flow is strongly convergent; it contains a large elongational component. Both the constriction of the flow and the velocity rearrangement increase the effective length of the tube. For Newtonian liquids this increase is somewhat less than the tube radius R. Its effect is incorporated into the calibration of viscometers for Newtonian liquids, but for non-Newtonian liquids with a Trouton ratio larger than 3 (e.g., polymer solutions), this may lead to a larger energy dissipation at the entrance than is accounted for in the calibration. The increase in effective length may be more than $10R$.

8.1.2.2 Kinetic energy correction

This concerns an exit effect. Depending on the construction of the viscometer and the properties of the liquid, part of the driving pressure is converted into kinetic energy of the liquid as it leaves the end of the capillary. The kinetic energy generated per second is of the order of $\alpha Q\rho\bar{v}^2$, where α is a factor of order unity, which depends on the velocity profile and on the exit conditions. To correct for this, Δp in the Poiseuille equation (Equation 8.9) should be corrected by a term proportional to $\alpha\rho\bar{v}^2$, which modifies it into

$$Q = \frac{\pi R^4}{8\eta l}(\Delta p - \alpha\rho\bar{v}^2) \tag{8.21}$$

and Equation 8.20 into

$$v = C_1 t - \frac{C_2}{t} \tag{8.22}$$

where C_1 and C_2 are calibration constants for a particular viscometer. The kinetic energy correction is, in practice, only important for low viscosity fluids at high flow rates. For long flow times the correction is small. At high flow rates, turbulence for low viscosity fluids and temperature increase due to energy dissipation for high viscosity liquids are usually more serious problems than kinetic energy effects.

8.1.2.3 Turbulence

The flow velocity must be small enough to avoid turbulence, which starts for tube flow at a Reynolds number ($2R\rho\bar{v}/\eta$; **Table 3.2**) above about 2000. In addition, turbulence may develop at quite low flow rates at the entrance of the tube or, in general, in the presence of abrupt changes in the tube radius, but then laminar flow will develop at a short distance from the inlet.

8.1.2.4 Particle migration

Particles in a dispersion flowing through a tube exhibit a tendency to move to the center of the tube, creating a particle-depleted layer with reduced viscosity along the wall. This may partly be attributed to an excluded volume effect, because the particle centers cannot become closer to the wall than their radius, R_p. Another effect is the so-called "tubular pinch" effect (Segré and Silberberg 1963). According to this effect, particles move radially to concentrate eventually at 70% of the inner cylinder radius. However, even an extreme redistribution of particles only produces a change in flow rate equivalent to the presence of a particle-free layer of thickness 0.7 R_p.

8.1.2.5 Wall slip

This is the result of the formation of a lubricated layer along the wall. The most common is the apparent wall slip owing to, for example, particle migration, or exudation of a low viscosity liquid, and alignment of polymer molecules. Effects due to wall slip may lead to large effects on the finally calculated (apparent) viscosity.

8.1.2.6 Viscous heating

This will only play a role for highly viscous liquids at high shear rates as may occur in pressure-driven tube viscometers.

8.2 Rotational rheometers

Rheological measurements are often performed in rotational rheometers in which the test material is deformed between two coaxial cylinders, cones, plates, or cone and a plate. This group of instruments has several fundamental advantages over capillary viscometers. First, for an appropriate geometry, the shear strain and shear rate are almost uniform over the test material. Second, the sample can be sheared for as long as desired, allowing the study of time-dependent behavior. Third, flow curves as a function of shear rate or shear stress can be determined in a more convenient way. On the other hand, these types of instruments are less accurate for determining the viscosity of low, or moderate, viscosity liquids, and are more liable to viscous heating at higher shear rates for moderately and highly viscous materials. We shall use the term, "rotational rheometers" instead of rotational viscometers, because these instruments can be, and are, used for more than just the determination of viscosity. They are well suited for studying such properties as gel formation by oscillatory measurements, and other gel properties, in addition to creep tests.

The most frequently used geometries are those in which the sample is sheared between two coaxial cylinders, or a cone and a plate, or two parallel plates (**Figure 8.4**). Other variants are discussed by, for example, Whorlow (1992),

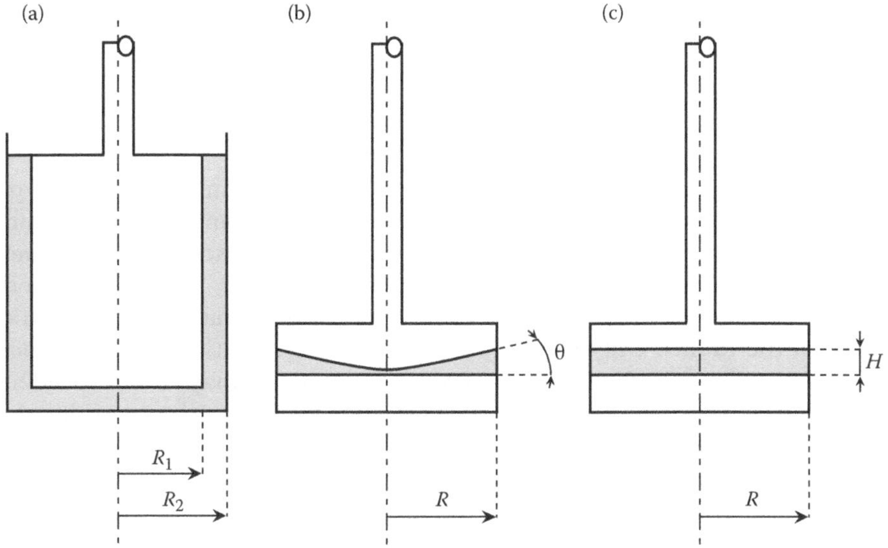

Figure 8.4 *(a) Concentric cylinder; (b) cone and plate; (c) plate–plate geometry. R_1 radius inner cylinder, R_2 radius outer cylinder, θ angle between cone and plate, H distance between the plates.*

and Ferguson and Kemblowski (1991). Most instruments are based on the relative rotation about a common central axis of the two parts forming the measuring geometry.

8.2.1 Concentric cylinder geometry

In these types of measuring devices, the sample is confined between two cylinder surfaces, of which one can rotate. In the *Couette system*, the outer cylinder rotates, whereas in a *Searle system* the inner one rotates. The equations relating rotation to torsion are the same for the two systems.

In the following derivation, it is assumed that end and inertial effects may be neglected and that there is no slip at the cylinder surfaces. The relation between the net applied torque, being the product of the force times the lever arm at which it is measured, and the local shear stress can be evaluated by considering the balance of moments acting on any cylindrical surface with radius r in the material. The torque T (N m) must be balanced by the moment owing to the shear force developed within the material

$$T = 2\pi r l \sigma r, \tag{8.23a}$$

where l is the length (height) of the cylinder. Equation 8.23a applies for all values of r for which $R_1 < r < R_2$, where R_1 is the radius of the inner cylinder and R_2 that of the outer one (**Figure 8.5a**). Rearrangement of Equation 8.23a gives for the shear stress σ at any point r in the test material

$$\sigma = \frac{T}{2\pi r^2 l} \tag{8.23b}$$

This equation implies that in contrast to T, σ is not constant over the gap between the two concentric cylinders, but decreases from the inner cylinder to the outer one. Just as for tube flow, the shear stress distribution in Couette geometry is determined only by the equation of motion, and is hence independent of the test material properties. It is proportional to the reciprocal square of the radius. Only if there is a very small gap between the cylinders, compared with the cylinder radii, say a few percent, may σ be considered constant over the gap. Then, the two cylinder walls may be considered as parallel, and in the absence of wall slip the shear strain in the gap can be approximated by (**Figures 3.2 and 3.3**)

$$\gamma = \frac{R_a}{R_2 - R_1}\alpha \tag{8.24}$$

and for laminar flow, the shear rate $\dot{\gamma}$, by

$$\dot{\gamma} = \frac{R_a}{R_2 - R_1}\Omega, \tag{8.25}$$

where α is the angular displacement of the rotating cylinder (rad), Ω is the angular velocity (rad s^{-1}), and R_a is an average of R_1 and R_2, the radius of the inner and outer cylinders, respectively. As a first approximation, $R_a = (R_1 + R_2)/2$. A very small gap between the cylinders requires that they are very precisely positioned because the gap width has to be constant. In addition,

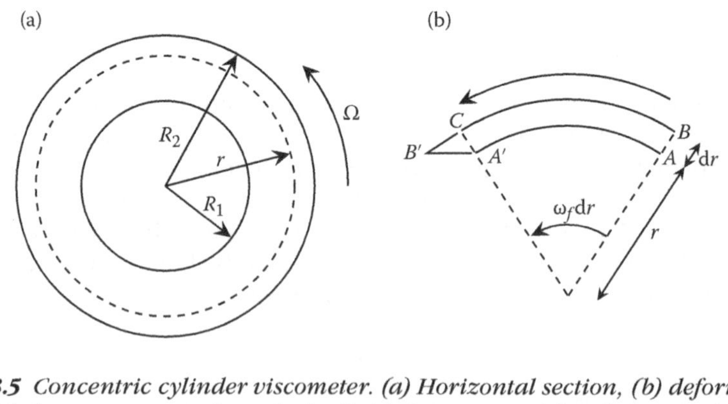

***Figure 8.5** Concentric cylinder viscometer. (a) Horizontal section, (b) deformation of a fluid element. ω_f, angular velocity fluid element considered; Ω, angular velocity rotating cylinder; r, radius; R_1, radius inner cylinder; R_2, radius outer cylinder.*

when studying dispersions, the gap width has to be large compared with the dimensions of the suspended particles (see below). For many suspensions (e.g., drinks containing fruit cells), this would require very large cylinders.

In reality, the flow in the gap between two concentric cylinders is curve linear. For obtaining an expression for the strain (rate) in Couette geometry, we consider two material points A and B, a distance dr apart (**Figure 8.5b**). During shearing over a time dt, the radial line AB moves to $A'B'$, whereas if the material had rotated as a rigid body around the central axis, it would have remained radial at $A'C$. Because

$$BC = (r + dr)\omega_f \, dt \tag{8.26}$$

and

$$BB' = (r + dr)\,(\omega_f + d\omega_f)dt \tag{8.27}$$

the shear strain becomes

$$\gamma = B'C/A'C = (r + dr)\, d\omega_f \, dt/dr, \tag{8.28}$$

where ω_f is the angular velocity of the fluid element. It increases with the distance from the inner cylinder and thus with r. In the limit where $dr \to 0$, the strain rate $d\gamma/dt$ is given by

$$\dot{\gamma} = r\frac{d\omega_f}{dr} \tag{8.29}$$

The next expressions can be obtained for the shear rate as a function of r, for $R_1 < r < R_2$ by combining Equations 8.29 and 8.23b with an expression relating σ to $\dot{\gamma}$.

For a Newtonian fluid, $\dot{\gamma} = \sigma/\eta$, so we obtain

$$r\frac{d\omega_f}{dr} = \frac{\sigma}{\eta} = \frac{T}{2\pi\eta l r^2} \tag{8.30}$$

Because $\omega_f = 0$ for $r = R_1$ and $\omega_f = \Omega$ for $r = R_2$,

$$\int_0^{\Omega} d\omega_f = \frac{T}{2\pi\eta l}\int_{R_1}^{R_2} r^{-3}\, dr \tag{8.31}$$

Hence,

$$\Omega = \frac{T}{4\pi\eta l}\left(\frac{1}{R_1^2} - \frac{1}{R_2^2}\right) \tag{8.32}$$

So, for a Newtonian fluid, a plot of T versus Ω will result in a straight line through the origin. The viscosity is given by

$$\eta = \frac{T}{4\pi\Omega l}\left(\frac{1}{R_1^2} - \frac{1}{R_2^2}\right) \tag{8.33}$$

By combining Equations 8.23b and 8.33, one obtains for the shear rate as a function of r for $R_1 < r < R_2$

$$\dot{\gamma} = \frac{\sigma}{\eta} = \frac{2\Omega}{r^2\left(R_1^{-2} - R_2^{-2}\right)} = \frac{2R_1^2R_2^2\Omega}{r^2\left(R_2^2 - R_1^2\right)} \tag{8.34}$$

In a similar manner, an expression can be obtained for the shear strain as a function of r for a Hooke (linear elastic) solid for $R_1 < r < R_2$. This reads as

$$\gamma = \frac{\sigma}{G} = \frac{2\alpha}{r^2\left(R_1^{-2} - R_2^{-2}\right)} = \frac{2R_1^2R_2^2\alpha}{r^2\left(R_2^2 - R_1^2\right)} \tag{8.35}$$

Expressions for the shear rate and shear strain at the inner and outer cylinders are obtained by replacing r^2 by R_1^2 and R_2^2, respectively, in the appropriate equation. From Equations 8.34 and 8.35, it follows that γ as well as $\dot{\gamma}$ decreases over the gap between the inner and outer cylinders with r^2, as σ does. Both γ and $\dot{\gamma}$ are higher at the inner cylinder wall by a factor R_2^2/R_1^2. If $R_2 = 1.1 \cdot R_1$, this already implies a difference of 21%.

An expression similar to Equation 8.34 can be derived for non-Newtonian fluids starting with Equation 8.30 if the relationship between σ and $\dot{\gamma}$ is known. For a power law fluid, $\sigma = k\dot{\gamma}^n$ (Equation 4.10) or $\dot{\gamma} = (1/k)\sigma^{1/n}$. Inserting in Equation 8.30 gives

$$r\frac{d\omega_f}{dr} = \frac{1}{k}\left(\frac{T}{2\pi l r^2}\right)^{1/n} \tag{8.36}$$

Integration, as before, gives

$$\Omega = \frac{n}{2k}\left(\frac{T}{2\pi l}\right)^{1/n}\left(\frac{1}{R_1^{2/n}} - \frac{1}{R_2^{2/n}}\right) \tag{8.37}$$

Combining Equations 8.36 and 8.37 gives for the shear rate

$$\dot{\gamma} = \frac{2\Omega}{nr^{2/n}\left(R_1^{-2/n} - R_2^{-2/n}\right)} \tag{8.38}$$

Substitution of R_1 or R_2 for r gives the shear rate at the inner and outer walls, respectively. From Equation 8.38, it follows that for shear thinning fluids ($n < 1$), $\dot{\gamma}$ decreases more strongly with r than for Newtonian fluids (**Figure 8.6**), and for shear thickening fluids it decreases less strongly. For instance, for a fluid characterized by $n = 0.5$, $\dot{\gamma} \propto r^{-4}$. If $R_2 = 1.1 \cdot R_1$, this implies a difference of 46% between the shear rate at the inner and outer cylinders. This stronger decrease is because for a shear thinning fluid, η_{app} is lowest where σ is highest, which is near the inner cylinder wall.

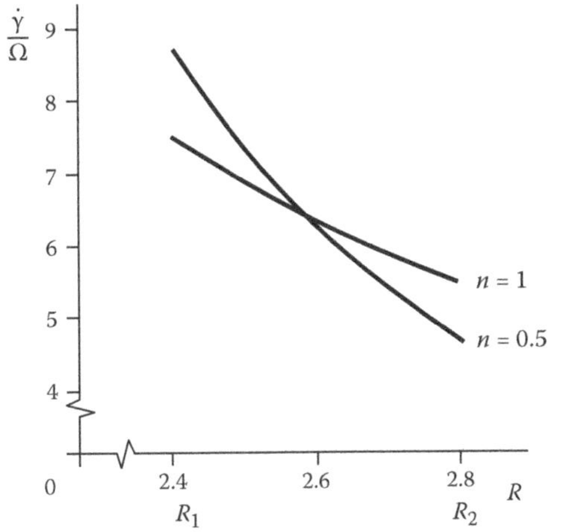

***Figure 8.6** Shear rate $\dot{\gamma}$, divided by angular velocity of rotating cylinder Ω, in the gap between two coaxial cylinders as a function of distance r from central axis for $R_2 = 2.8$ cm and $R_1 = 2.4$ cm, for a Newtonian fluid ($n = 1$) and for a shear thinning fluid characterized by a power law index $n = 0.5$. (Reprinted from van Vliet, T., in* Food Texture: Measurement and Perception, *ed. Rosenthal, A.J., Aspen Publishers, Gaithersburg, MD, 1999. With permission.)*

For a material behaving according to the Bingham model, $\sigma = \sigma_y + \eta_B\dot{\gamma}$ (Equation 4.13), implying $\dot{\gamma} = (\sigma - \sigma_y)/\eta_B$ for $\sigma > \sigma_y$, and $\dot{\gamma} = 0$ for $\sigma < \sigma_y$. Three distinct situations can be distinguished. For $\sigma > \sigma_y$ everywhere in the gap between the cylinders, flow will occur over the whole gap width. On the other hand, for $\sigma < \sigma_y$ everywhere in the gap, no flow occurs at all. The third case refers to situations where flow takes place only in that part of the gap where $\sigma > \sigma_y$. For the first case, substitution of Equations 8.23b and 8.29 into $\dot{\gamma} = (\sigma - \sigma_y)/\eta_B$ gives

$$r\frac{d\omega_f}{dr} = \frac{T}{2\pi l \eta_B r^2} - \frac{\sigma_y}{\eta_B} \tag{8.39}$$

Equation 8.39 can be integrated to give

$$\Omega = \frac{T}{4\pi l \eta_B}\left(\frac{1}{R_1^2} - \frac{1}{R_2^2}\right) - \frac{\sigma_y}{\eta_B}\ln\frac{R_2}{R_1} \tag{8.40}$$

Hence, a plot of Ω against T will be linear with a slope $\left(R_1^{-2} - R_2^{-2}\right)/4\pi l\eta_B$, provided that $T > 2\pi l R_2^2\sigma_y$. As soon as T falls below this value, the yield stress has not been exceeded at the outside of the gap, that is, the material near the outer cylinder remains solid and flow occurs at best in only a part of the gap between the two cylinders. In this situation, integration of Equation 8.39 only makes sense between the limits $\omega_f = 0$ at $r = R_1$ and $\omega_f = \Omega$ at $r = \sqrt{T/2\pi l\sigma_y}$, giving:

$$\Omega = \frac{T}{4\pi l \eta_B}\left(\frac{1}{R_1^2} - \frac{2\pi\sigma_y}{T}\right) - \frac{\sigma_y}{2\eta_B}\ln\left(\frac{T}{2\pi l\sigma_y R_1^2}\right) \tag{8.41}$$

Figure 8.7 shows the angular velocity as a function of the applied torque for a material that behaves according to the Bingham model. As this model

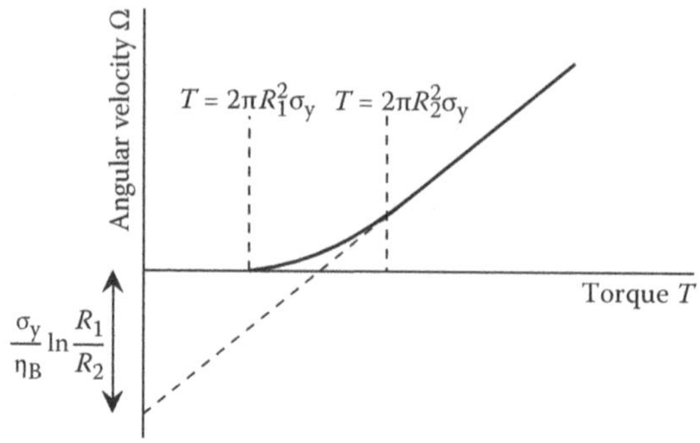

Figure 8.7 *Plot of angular velocity Ω as a function of applied torque T for a Bingham liquid. (After Whorlow, R.W.,* Rheological Techniques, *Ellis Horwood, Chichester, 1992.)*

describes idealized behavior, such a relationship is seldom, if ever, observed. The concept of a unique yield stress, independent of flow conditions and time, does not hold for most (or all) materials (Sections 4.1.1 and 4.2), which explains why the curved portion described by Equation 8.41 is not observed. However, the formation of a stationary layer of material near the outer layer has been observed for a material having a high enough yield stress under the prevalent conditions.

Several torque versus angular velocity relationships have been derived for other types of flow equations. Moreover, approximate solutions have been given for the case where the flow curve is not known. For a further discussion of these aspects, readers may refer to textbooks, for example, those written by Whorlow (1992) and Darby (1976).

8.2.1.1 Sources of errors

The main sources of errors in using concentric cylinder geometry are discussed in the following subsections.

8.2.1.1.1 End effects The resistance against deformation of the material below (and above) the inner cylinder is often neglected compared to the contribution of deformation between the two cylinders. Below the cylinder, the velocity gradient will be different from that between the concentric cylinders, the extent of which depends on the shape and dimensions of the gap below the cylinder. A correction for the end effect can be obtained by determining the torsion as a function of the immersed height of the inner cylinder. The end effect will be quantitatively different between Newtonian and non-Newtonian liquids, that is, in general larger for shear thinning liquids and smaller for shear thickening ones. For a typical example, see **Figure 8.8**. The reason is that in the gap below the inner cylinder, $\dot{\gamma}$ is usually lower than in the gap between the concentric cylinders and thus for shear thinning liquids the accompanying η_{app} is higher. This results in a larger contribution of σ in the gap below the inner cylinder to the measured σ than is observed for Newtonian liquids.

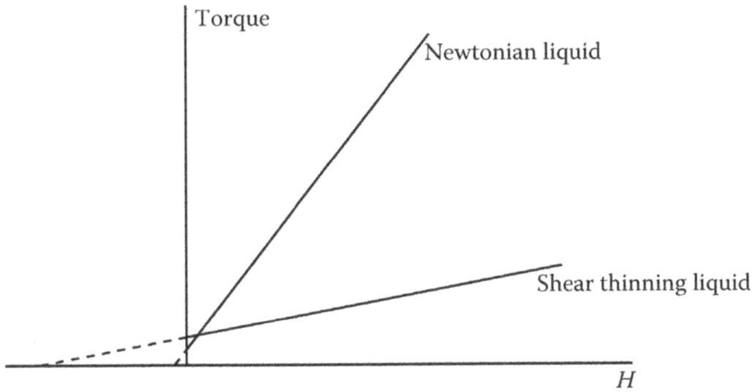

Figure 8.8 *Determination of the end effect by extrapolation of measured torque as a function of immersed height H for a concentric cylinder measuring geometry.*

The end effect can be reduced by using specially designed measuring bodies, for example, double concentric cylinders or an inner cylinder below which an air bubble will be entrapped by inserting it (**Figure 8.9**). Another solution is to use an inner cylinder with a cone at the bottom end, where the angle of the cone has been chosen in such a way that $\dot{\gamma}$ below the cone is the same as that in the gap between the concentric cylinders. In a double concentric cylinder geometry, the end surface is very small compared to the total surface. However, cleaning of the measuring geometry is more difficult, and it requires more precise lining of the inner and outer cylinders.

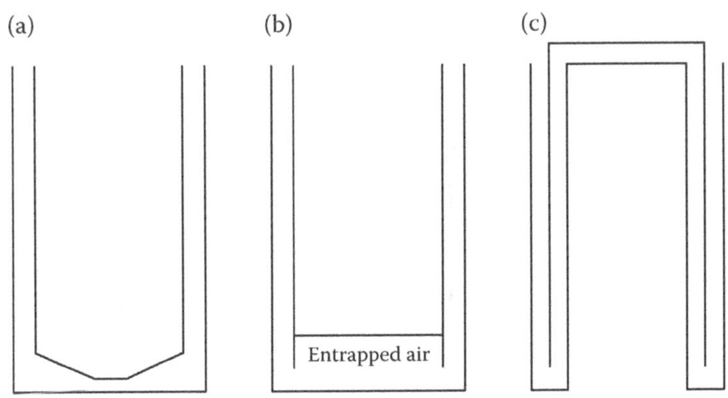

Figure 8.9 *Examples of concentric cylinder geometries with a reduced end effect. (a) Cylinder with a cone below, (b) entrapped air bubble under inner cylinder, and (c) double concentric cylinder. (Reprinted from van Vliet, T., in* Food Texture: Measurement and Perception, *ed. Rosenthal, A.J., Aspen Publishers, Gaithersburg, MD, 1999. With permission.)*

8.2.1.1.2 Departures of streamlines from circular geometry In the section above, we have assumed that the flow is laminar and occurs in a circular path around the axis of rotation of the instrument. Departure from such a flow regime may arise from inertia resulting from centrifugal forces. If the inner cylinder rotates (Searle system), centrifugal forces cause the (relatively) fast-moving liquid near the inner cylinder to move outward. The velocity v of the liquid near the rotating inner cylinder is $\Omega \cdot R_1$ and 0 near the outer cylinder. Because the liquid cannot move outward en masse, localized secondary flow patterns will develop, the so-called *Taylor vortices* (Taylor 1923) (**Figure 8.10**). The development of these vortices is opposed by viscous forces. Taylor vortices will not develop if the outer cylinder rotates (Couette system), because centrifugal action then stabilizes the flow: v is zero near the inner cylinder and at its maximum near the outer one.

For Newtonian fluids, Taylor vortices will develop in a Searle system when

$$Re\left(\frac{R_2 - R_1}{R_2}\right)^{1/2} > 41.3, \tag{8.42}$$

where Re is the Reynolds number ($\equiv (R_2 - R_1/R_2) Rv\rho/\eta$, where $v = \Omega \cdot R_1$ is the liquid velocity at the wall of the inner cylinder (Section 3.3). Using $\rho/\eta = 1/\upsilon$, where υ is the kinematic viscosity, gives:

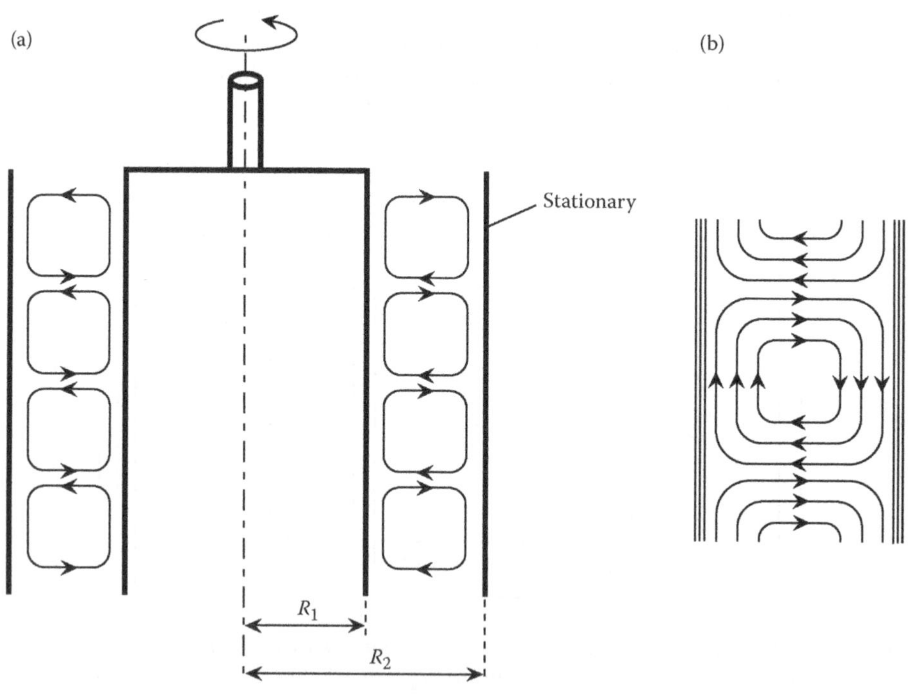

Figure 8.10 *Schematic representation of Taylor vortices. R_1, radius inner cylinder; R_2, radius outer cylinder. Right panel, more detailed picture of one of the vortices.*

$$\frac{v(R_2 - R_1)^{3/2}}{\upsilon R_1^{1/2}} > 41.3 \quad (8.43)$$

When the left side of Equation 8.43 is just larger than Equation 4.13, the shape of the vortices is rectangular. The flow is still ordered. Turbulence will occur at still higher Reynolds numbers, depending strongly on R_1/R_2 (**Table 3.2**).

Taylor vortices cause extra energy dissipation, resulting in an increase of the torque and with that of the measured apparent value of η. Moreover, it gives the occurrence of elongational and colliding flow in the regions where the vortices meet. Elongational flow may lead to disaggregation of aggregates of particles, whereas colliding flow may lead to the opposing effect (usually a much smaller effect). This formation and breakup of aggregates may strongly affect the apparent viscosity of the dispersion, resulting in a measured η_{app} largely determined by the measuring conditions.

8.2.1.1.3 Wall slip due to a slip layer The occurrence of wall slip due to a slip layer has to be checked by using various measuring geometries and, if possible, by visual inspection. It will often occur when measuring the viscosity of the material with a yield stress that exhibits some expulsion (syneresis) of liquid. A solution may be the use of roughened cylinders or covering the cylinders with emery papers with an appropriate roughness. For shear rate thinning liquids with a rather high apparent viscosity, the use of an inner spindle equipped with vanes may provide a good solution (Section 8.4.1.2).

For dispersion of particles, one has to be aware that the center of a particle cannot come closer to the wall than its radius (excluded volume effect). This leads to a layer with a thickness of about the diameter of the particles that has a lower particle concentration and thus a lower η. The concentration of particles in the middle part between the two walls will become slightly higher as a result of this demixing effect. However, the effect of this on η measured is generally much smaller than that due to the layer with the lower η. The latter effect will be larger for a lower ratio of gap width over particle diameter d. The effect may be serious as long as $(R_2 - R_1)/d < 20$ (**Figure 8.11**).

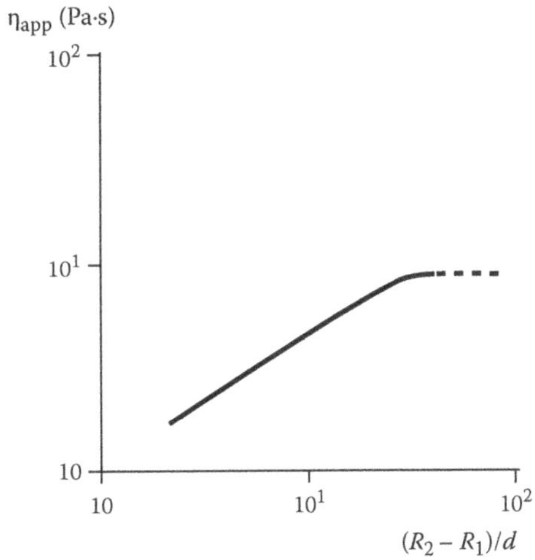

Figure 8.11 *Measured apparent viscosity of a dispersion of polystyrene particles in glycerol–water mixtures as a function of gap width $R_2 - R_1$ over particle diameter d ratio. (Redrawn from Cheng, D.C.-H.,* Powder Technol., *37, 255–273, 1984.)*

8.2.1.1.4 Viscous heating This will only play a role for materials with a high η measured at a high $\dot{\gamma}$.

8.2.1.1.5 Effects due to instrument inertia Concentric cylinders cannot be used easily for dynamic measurements at high frequencies because the inertia of the oscillating body inhibits that. The moment of inertia of a cylinder oscillating around its axis is proportional to mR^2. Generally, for simple shear, Equation 7.20 for $\sigma(t)$ has to be replaced by

$$\sigma(t) = \frac{m}{A}\frac{d^2x}{dt^2} + G'\gamma + \frac{G''}{\omega}\frac{d\gamma}{dt} \tag{8.44}$$

where m is the mass of the moving measuring body and A the adjoining area of the test piece, x the displacement of the mass. For an oscillating cylinder x is equal to αR. The inertia effect (first term on the right hand side) increases with ω^2, setting an upper limit to ω at which Equation 7.20 may be used.

8.2.2 Cone and plate geometry

This measuring device consists of a circular plate and a cone, with a radius R, having its axis perpendicular to the plate and its vertex in the plane of the surface of the plate (**Figure 8.12**). The cone or the plate may rotate, resulting in a shearing deformation of the material in between. Liquids are normally retained in the gap between the cone and plate by surface tension forces. Generally, the point of the cone is flattened to avoid direct contact with the plate. If the angle θ between the cone and plate is small (<5°), σ, γ, and $\dot{\gamma}$ are about uniform over the material in between. This is the main advantage of this measuring geometry compared to concentric cylinders or plate–plate geometries. It makes the cone and plate geometry well suited for the study of

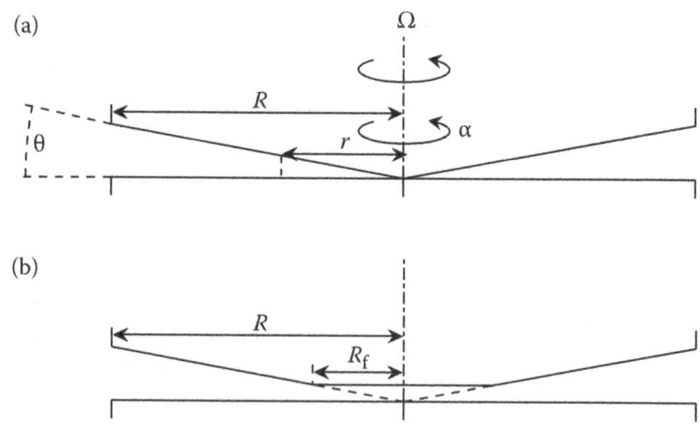

Figure 8.12 *(a) Vertical section through an ideal cone and plate viscometer; (b) section when cone is truncated. R, radius cone; R_f, radius flattened tip; θ, angle between cone and plate; α, angular displacement; Ω, angular velocity.*

strongly shear thinning (or thickening) liquids and of materials with a yield stress.

With respect to expressions for γ and $\dot{\gamma}$, we restrict ourselves to values of θ that are so small that the difference between sin θ and θ may be neglected. Moreover, we assume that the material is moving in concentric circles around the axis of rotation of the cone. This is normally the case for θ smaller than 5°. Then, the strain γ at radial distance r will be

$$\gamma = \frac{r\alpha}{r\sin\theta} \approx \frac{\alpha}{\theta} \tag{8.45}$$

and the strain rate

$$\dot{\gamma} = \frac{r\Omega}{r\sin\theta} \approx \frac{\Omega}{\theta}, \tag{8.46}$$

where α is the angular displacement (rad) of the rotating part and Ω is the angular velocity (rad s^{-1}). Equations 8.45 and 8.46 show that both γ and $\dot{\gamma}$ are independent of the radial distance from the center of the cone. Because both are uniform over the cone (plate) area, the shear stress exerted on the cone (plate) (σ, in spherical coordinates) will also be constant everywhere. The total torque on the cone is obtained by summing up the contributions δT of narrow rings between radii r and $r + \delta r$, being $\sigma r 2\pi r \delta r$. This gives for the total torque:

$$T = \int_{r=0}^{r=R} 2\pi r^2 \sigma \mathrm{d}r = \frac{2}{3}\pi R^3 \sigma \tag{8.47}$$

When the flattened tip has a radius R_f, Equation 8.47 becomes

$$T = \int_{r=R_f}^{r=R} 2\pi r^2 \sigma \mathrm{d}r = \frac{2}{3}\pi\left(R^3 - R_f^3\right)\sigma \tag{8.48}$$

As long as R_f is less than $0.2R$, the torque is reduced by less than 1% compared to the unflattened cone device.

The cone and plate geometry also permits the determination of the first normal stress difference (Whorlow 1992; Walters 1975). Other advantages of the cone and plate geometry over concentric cylinders are that they are well suited for measuring small samples and that end effects do not play a role. On the other hand, thermostatting is more difficult. Moreover, drying out of the samples at the edges will have a big effect on the measured torque. For solids, covering the edge with grease is often a good solution to prevent (diminished) drying out. To ensure that γ and $\dot{\gamma}$ are uniform, gap setting has to be done very carefully. In addition, one should be aware of the possibility of secondary flow due to inertia effects at high $\dot{\gamma}$ and viscous heating for liquids with a high η.

Just as for concentric cylinder systems, one has to be aware of the occurrence of demixing effects due to the excluded volume effect. Because of the small distances between the cone and plate, the effect of a layer near the wall with a lower particle concentration giving apparent slip, will be relatively more important. Moreover, a large particle may become trapped between the cone and plate. In such cases, a plate–plate geometry may be a useful alternative.

8.2.3 Plate–plate geometry

This measuring device consists of two parallel plates of radius R at a distance H from each other (**Figure 8.4c**) or an upper plate of radius R and a larger lower plate. For thixotropic materials, in particular, it can be advantageous to use a plate–plate geometry because the test materials are subjected to smaller stresses while they are introduced between parallel flat plates than between a cone and a plate or between concentric cylinders. Moreover, one may set the distance between the plates freely, within certain limits, whereas for a cone and plate geometry this has to be done very precisely. On the other hand, for many solid materials, care has to be taken to ensure that the sample has two parallel surfaces. A height difference of 0.1 mm between two sides of a solid test piece 10 mm high implies a compressive strain of ≥0.01 of the higher part before the upper plate is in contact with the lower plate. Such a height difference cannot be seen visually and can hardly be determined instrumentally. A strain larger than 0.01 is outside the linear region of, for example, margarines. Below, first the use of a parallel plate geometry for studying flow behavior of liquids and semiliquids will be discussed, and afterward its use for determining large deformation and fracture behavior of (elastic) solids will be clarified.

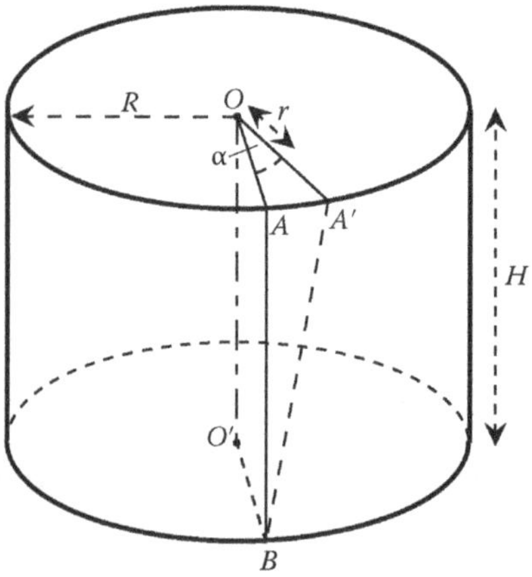

***Figure 8.13** Deformation of a cylindrical piece of material between two parallel plates in torsion. H, height test piece; R, radius of test piece; α, angular displacement. (Redrawn after Whorlow, R.W.,* Rheological Techniques, *Ellis Horwood, Chichester, 1992.)*

The main disadvantage of the plate–plate geometry compared with a cone and plate geometry is that γ, $\dot{\gamma}$, and σ are not constant in the test piece during deformation, but vary with the distance from the center, r. During a measurement, one of the plates is rotated around the axis through its center. On rotating the upper surface of a cylindrical piece of material over an angle α, a point on the line OA at radius r is sheared relative to the line OB to a point on the line OA' (**Figure 8.13**). The shear strain γ at radius r will be

$$\gamma = \frac{\alpha r}{H}, \tag{8.49}$$

where H is the distance between the plates (height test piece). The strain rate is given by

$$\dot{\gamma} = \frac{\Omega r}{H}, \tag{8.50}$$

where Ω is the angular velocity of the rotating plate. Equations 8.49 and 8.50 illustrate that both the shear strain and the strain rate increase with r. They are at maximum at the outside of the test piece, being $\alpha R/H$ and $\Omega R/H$, respectively.

The total torque on each of the plates is

$$T = \int_{r=0}^{r=R} 2\pi r r \sigma(r)\,\mathrm{d}r \tag{8.51}$$

However, because γ and $\dot{\gamma}$ depend on the radius, σ is now a variable depending on r. For a Hooke solid characterized by a shear modulus G and a Newtonian fluid, $\sigma(r)$ changes linearly with γ and $\dot{\gamma}$, respectively. Integration of Equation 8.51 gives then the next expressions

$$T = \int_{r=o}^{r=R} 2\pi r^2 G \frac{r\alpha}{H}\,\mathrm{d}r = \frac{\pi R^4}{2H} G\alpha \tag{8.52}$$

and similarly,

$$T = \frac{\pi R^4}{2H} \eta\Omega \tag{8.53}$$

for Hooke- and Newtonian-type materials, respectively. By determining the torsion at which a semiliquid yields, it is possible to calculate its yield stress from Equation 8.52 providing that σ increases linearly with γ below σ_y ($\sigma_y = G\gamma_y$) by

$$\sigma_y = \frac{2T_y}{\pi R^3} \tag{8.54}$$

where T_y is the torsion at which yielding occurs.

For non-Newtonian liquids, of which the viscosity depends only on the strain rate and not on shearing time, an expression for the torque can be obtained by taking the shear rate as the variable in Equation 8.51. Using Equation 8.50 gives for the torque

$$T = \int_0^{\dot{\gamma}_R} 2\pi \frac{H^3}{\Omega^3} \dot{\gamma}^3 \eta(\dot{\gamma})\,\mathrm{d}\dot{\gamma} \tag{8.55}$$

or

$$\frac{\dot{\gamma}_R^3 T}{2\pi R^3} = \int_0^{\dot{\gamma}_R} \dot{\gamma}^3 \eta(\dot{\gamma})\,\mathrm{d}\dot{\gamma} \tag{8.56}$$

This gives for the viscosity at $\dot{\gamma}_R$

$$\eta(\dot{\gamma}_R) = \frac{1}{2\pi R^3}\left(\frac{\mathrm{d}T}{\mathrm{d}\dot{\gamma}} + \frac{3T}{\dot{\gamma}}\right) = \frac{3T}{2\pi R^3 \dot{\gamma}_R}\left(1 + \frac{1}{3}\frac{\mathrm{d}\ln T}{\mathrm{d}\ln \dot{\gamma}_R}\right) \tag{8.57}$$

However, for non-Newtonian fluids, it is usually better to use a concentric cylinder or a cone and plate geometry, because shear rate variations throughout the sample and edge effects are larger for the plate–plate geometry. On the other hand, the latter geometry has advantages for the determination of normal stress differences (Whorlow 1992) and, as mentioned above, for the study of strain-sensitive materials, for example, in oscillatory measurements. The plate–plate geometry also has several other disadvantages: (1) thermostatting is difficult, except if the distance between the plates is small, and (2) drying out of the test pieces at the edges will have a big effect on the measured torque.

8.2.4 Torsion tests

During torsion tests, specimens are deformed by twisting them along their axial axis. Test pieces are usually cylindrical in cross section and fixed to the measuring body by gripping or by gluing, for example, with cyanoacrylate ester glue to two flat parallel plates. In addition, rectangular strips or square rods can be used, but then stress and strain calculation becomes more complicated. For the case where the test piece is cylindrical and glued to parallel plates, the torsion test is very similar to the test using the parallel plate geometry described above. The main distinction is that the height over diameter of the specimen is usually (much) larger. On twisting a cylindrical rod, its deformation is simple shear. The strain will vary from zero on the axial axis to a maximum at the surface. The maximum strain γ_{max} can easily be calculated from the relative angular rotation of both ends of the specimen by

$$\gamma_{max} = \frac{\alpha R}{H} \tag{8.58}$$

The measured torque can be calculated by Equation 8.52 as long as stress and strain are proportional to each other (linear region). The corresponding shear stress σ_{max} is

$$\sigma_{max} = \frac{2T}{\pi R^3} \quad (8.59)$$

Its calculation is similar to that of the yield stress by Equation 8.54.

Torsion experiments can be used for studying large deformation and fracture behavior of materials that are (about) linear up to fracture. To ensure that fracture happens at the middle of the specimen, the cylinders are often carved into a capstan shape by milling a smooth annular groove at its midlength (**Figure 8.14**). Fracture will occur at the basis of the groove. The shape of the groove will affect both σ_{max} and γ_{max}. The possible presence of disk-like end pieces with radius R will have no effect on σ_{max}, because the torque will be the same over the whole length of the test piece. On the other hand, the deformation of the end pieces will contribute to the total strain of the specimen. Equations 8.58 and 8.59 should be modified by incorporating shape-dependent factors K_γ and K_σ for the calculation of γ_{max} and σ_{max}, respectively,

$$\gamma_{max} = \frac{\alpha R K_y}{2H} \quad (8.60)$$

$$\sigma_{max} = \frac{2K_\sigma T}{\pi r_{min}^3}, \quad (8.61)$$

where r_{min} is the radius of the thinnest part of the test piece, K_σ is a function of r_{min} and the radius of the cutout (groove), and K_γ is a function of the same parameters and, in addition, of the length and radius of the end pieces.

In torsion, the magnitude of the shear, tensile, and compressive stresses are equal, but with different directions (Hamann et al. 2006). There is an angle of 45° between the shear stress and the tensile and compressive stresses. The result is that the fracture mode can easily be seen visually. When the fracture plane angle is perpendicular to the longitudinal axis of the specimen, fracture is in shear, and when the angle is 45° fracture is in tension.

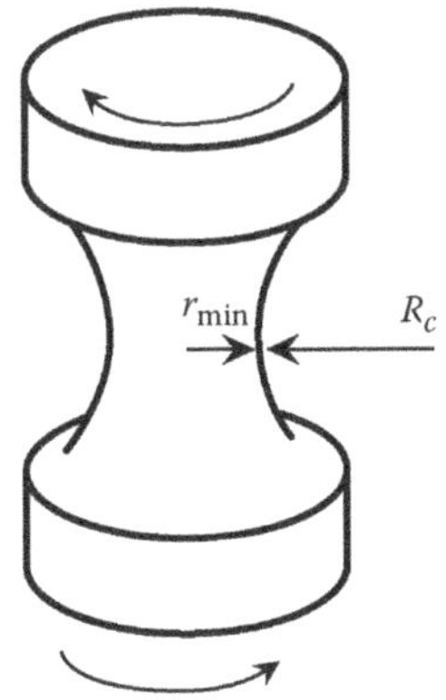

Figure 8.14 *Test piece used in torsion tests. r_{min}, radius of the thinnest part of test piece; R_c, radius groove.*

Advantages of torsion test are (1) deformation is in shear, implying no (significant) change in specimen volume; (2) specimen shape is maintained during the test, minimizing geometric considerations; (3) no friction between the specimen and the test fixture; (4) fracture in the tensile and shear mode can easily be distinguished; and (5) internal bulk stresses remains low, so a minimal amount of fluids will be forced from the specimen during testing. Its drawbacks are as follows: (1) specimen shaping and preparation are more complex than for various other tests and (2) strong attachment of the specimen to the machine is required. Moreover, calculation of fundamental fracture parameters can only be done easily for linear elastic behavior.

8.3 Tension compression apparatus

Tension compression or the so-called *general material testing instruments* are often used to determine large deformation and fracture behavior of materials. Essentially, these apparatuses consist of a fixed base and a moving bar, to each of which a part of the testing attachment (measuring geometry) can be fixed (**Figure 8.15**). A load cell is placed in the moving bar or in the base, by which the forces, required to deform the test piece, can be registered. For most instruments, the measuring range of the force can be varied by replacing the load cell with one suited for a lower or higher force range.

The apparatuses are available in two designs: one with the moving bar supported at one side by a column and one in which the bar is supported at both sides. The displacement of the moving bar can be registered in various ways. In many apparatuses, the displacement is determined by measuring the rotation of the screw(s) that drive the moving arm. For these apparatuses

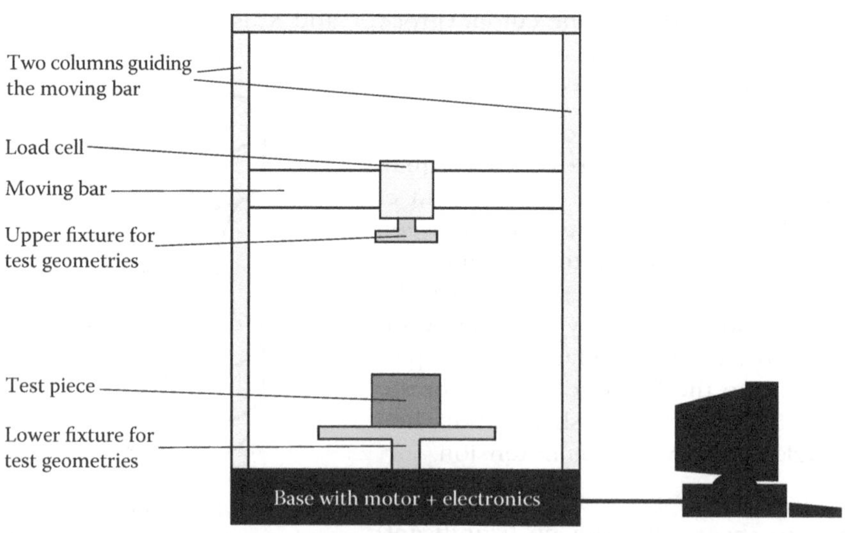

Figure 8.15 *Schematic illustration of a tension compression apparatus with a double screw driving system. For further explanation, see the text.*

equipped with a single column, the moving arm and its fixture to the column may deform to such an extent that on measuring stiff materials, the registered deformation of the test piece becomes unreliable (Rohm et al. 2000). The two-column frame allows the application of substantially higher forces. For most apparatuses, the moving bar can be operated at varying speeds allowing the determination of the strain rate-dependent properties of the tested materials.

Results are generally obtained in the form of a force versus displacement or time curve. Different types of tests can be performed by varying the measuring geometry. Depending on the geometry used and the shape of the test piece, results can be recalculated to stress versus strain curves, allowing the determination of fundamental parameters in SI units that characterize the material unequivocally. Tension compression instruments can be used both for studying large deformation and fracture behavior of predominantly solid materials and for determining the yield stress and flow behavior of semiliquid foods. Fundamental tests suited for solids include uniaxial compression tests, uniaxial extension (tensile) tests, (three-point) bending tests, wedge tests, and wire cutting tests, whereas for liquid-like foods biaxial extension tests are a useful option. Besides these tests, various (semi-)empirical tests can be executed depending on the measuring geometry used, for example, (cone) penetration tests and back extrusion tests. A popular test in the food industry for solids is the so-called Texture Profile Analysis (TPA) test, which essentially involves two successive uniaxial compression tests. A complication of empirical tests is that the deformation is not just one of the three simple types discussed in Section 3.1, but a combination of these simple types, which makes the description of the deformation in terms of a strain (tensor) very difficult or even impossible.

There is no single test by which the complete large-deformation and fracture behavior of a material can be characterized in terms of fundamental or empirical characteristics. In general, the methods that are easier to perform provide less information. In most popular tests on predominantly solid materials, a parameter is determined related to the fracture force or stress. Moreover, in many tests, information is obtained on the deformation at fracture (fracture strain) and on the modulus (in the fracture mechanics literature, this is often called *stiffness*) of the material as a function of deformation. In other tests, such as the so-called wedge tests and cutting tests (with a knife or wire), the critical strain energy release rate (in J m^{-2}) can be determined.

Below, a series of tests will be discussed.

8.3.1 Uniaxial compression tests

Uniaxial compression between flat parallel plates is the simplest and most popular technique for determining large deformation and fracture properties of foods. Usually, a constant compression rate is applied and the required force is measured as a function of the displacement. An alternative is to compress the material with a constant force and measure the deformation in time [determination of a creep curve (Section 7.2) in uniaxial compression]. The latter test will not be discussed separately, because the calculation of stress and strain is not fundamentally different from the constant compression rate test.

During uniaxial compression of a test piece, the material will not only be compressed, but also extended in the direction perpendicular to the compressive strain to an extent depending on the Poisson ratio. This results in an increase of the area on which the force acts. For the calculation of the stress, the actual surface area A_t should be used, which for a cylindrical test piece with a Poisson ratio of 0.5 is equal to $A_0(l_0/l_t)$, where the subscript 0 denotes the original cross section and length l of the test piece and t the actual one (Equation 3.21).

Also, the calculation of the strain from the displacement is less straightforward than is often thought (Peleg 1984). Except for brittle materials that fracture at small strain (≤0.1), the use of the Hencky or natural strain ($\varepsilon_H = \ln(l_t/l_0)$) (Equation 3.22) is strongly recommended above using the Cauchy strain ($\varepsilon_C = (dl/l_0)$) (Section 3.4).

The main advantage of uniaxial compression tests is the simple execution of the test, including the relatively simple preparation of the test pieces. In most modern instruments, the strain rate can be varied in a simple way by varying the speed of the moving bar. Be aware that a constant compression speed implies that the strain rate $\dot{\varepsilon}_H$ increases during the experiment according to $\dot{\varepsilon}_H = (v/l_t)$, where v is the compression speed (Equation 3.24). The dependence of $\dot{\varepsilon}_H$ on l_t implies that v has to be adjusted if one wants to compare measurements on test pieces with different heights at the same $\dot{\varepsilon}_H$.

Young's modulus E may be calculated from the stress–strain curve if the test piece has flat parallel ends. In general, this may be the case for materials with a large enough linear region. However, be aware that for a linear region of 2% and a test piece height of 1 cm, the unevenness of the surface must be less than 0.2 mm, a condition that is hardly possible to obtain.

Fracture stress σ_{fr}, fracture strain ε_{fr}, and the total energy required for fracture W, can in principle be obtained from the stress–strain curve. For σ_{fr} and ε_{fr}, usually the values are taken for the stress and strain, respectively, at the maximum of the curve and for W the area below the curve up to this point (Section 5.1). However, often there is no clear fracture point. When the test piece is compressed, its circumference will increase, giving rise to tensile stresses. When fracture occurs due to these tensile stresses, its start can mostly be noticed easily. Often, the start of visual crack formation (fracture) is observed before the maximum in the stress–strain curve. When fracture occurs due to the shear stresses in the test piece (Section 5.2), visible fracture at the outside of the test piece can often only be seen after the maximum. Despite this, the maximum in the stress–strain curve gives, for most purposes, a workable measure of σ_{fr}, ε_{fr}, and W. Other parameters such as the critical strain energy release rate, the critical stress intensity factor, and notch sensitivity cannot be determined in uniaxial compression.

A complication for the calculation of the stress–strain curve is the occurrence of friction between the test piece and the plates. It will lead to bulging of the middle part of the test piece and thus to a varying cross section of the test piece as a function of its height (**Figure 8.16**). The bulging leads to a larger tensile strain along the circumference of the test piece, which can be estimated for the case where the contact area between the test piece and the plate is constant (Gent and Lindley 1959). In general, the effect of friction will be larger for a lower height/diameter ratio. On the other hand, a too high ratio (>1.5) may lead to buckling of the test piece. Friction can often be diminished by applying a lubricant between the test piece and the plates or by using plates giving a low friction coefficient [Teflon (coated) plates for many food materials].

For very brittle materials or, in general, for materials that fracture at low relative deformation, ε_{fr} cannot be determined accurately. This is especially the case if the surface of the food is rough (e.g., many biscuits and cookies). In such cases, the products may already fracture, whereas the force only acts on the unevenness of the surface, making the determination of ε_{fr} and σ_{fr} impossible. For these products, three-point bending tests are more appropriate (**Figure 8.17**).

Summarizing, uniaxial compression is a suitable method to determine routinely large deformation and fracture behavior of foods with the exception of

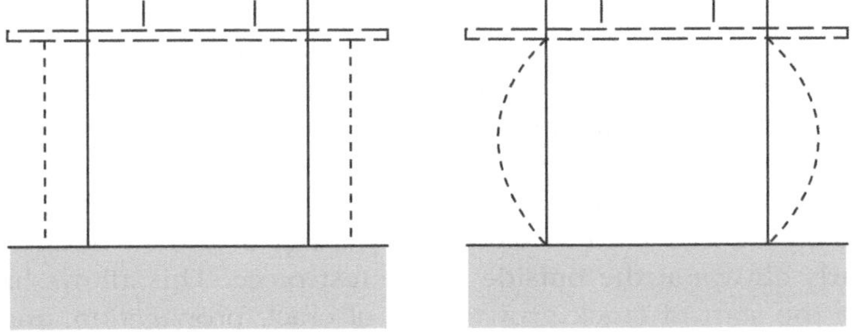

Figure 8.16 *Change in shape of a specimen during uniaxial compression. (a) No friction between test piece and plates. (b) Strong friction between test piece and plates. (Redrawn after van Vliet, T., Luyten, H., in* New Physico-Chemical Techniques for the Characterization of Complex Food Systems, *ed. Dickinson, E., Blackie Academic & Professional, Glasgow, 1995.)*

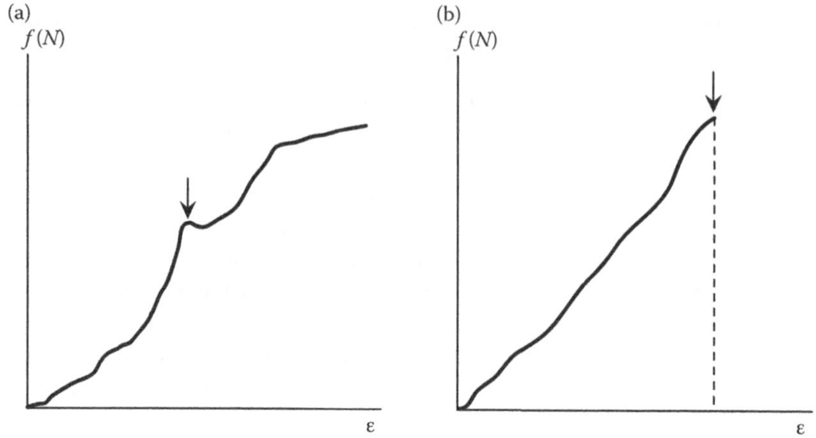

Figure 8.17 *Force f versus deformation curves for chocolate bars in (a) uniaxial compression and (b) three-point bending. Points where fracture was visible are indicated by an arrow. (Reproduced with permission from van Vliet, T., Luyten, H., in* New Physico-Chemical Techniques for the Characterization of Complex Food Systems, *ed. Dickinson, E., Blackie Academic & Professional, Glasgow, 1995.)*

products that fracture at very small deformations (most brittle products) as long as the test is performed in a reproducible way, taking into account the disadvantages of this test method.

8.3.2 Uniaxial extension tests

Uniaxial extension tests (often called tension tests) largely parallel uniaxial compression tests. Both within and especially outside food science, uniaxial tension tests are widely used to determine material properties. For these tests, usually a strip of material is clamped at both ends or fixed by gluing and pulled apart at a fixed deformation rate (or strain rate). The force is measured as a function of the displacement and next recalculated in stress versus strain data in the same way as has to be done for uniaxial compression tests. From the data obtained, E, σ_{fr}, ε_{fr}, and W can be calculated. An advantage of tensile testing above uniaxial compression tests is that the start of crack growth can usually easily be observed visually or by using a camera because fracture starts nearly always at the outside of the test piece. This allows handlers to determine the start of crack growth and of crack propagation, and for non-brittle products characterized by a low crack growth speed, the speed of crack growth. In addition, by applying artificial notches of varying length to the test piece, notch sensitivity and inherent defect length can be determined (Section 5.2.1.2). Another advantage of this type of tests over compression tests is that the problem of friction is avoided.

The specific critical strain energy release rate G_C or the specific fracture energy R_s can also be determined in experiments on notched test specimens from the energy available in the volume of material around a throughout notch per unit fracture area by using Equation 5.7 ($G_C = 2\pi l_c W_e$), where l_c is the critical notch length for crack propagation and W' is the average elastic (stored) energy per unit volume in the whole test specimen.

The main disadvantage of tension tests is that they are more difficult to execute than uniaxial compression tests. In general, test pieces have to be rather large, so that large homogeneous products are required from which they can be cut. This is especially difficult for many foods. A second problem is that for notch-sensitive materials, the fracture properties of the test pieces are sensitive for small scratches on the surfaces formed as result of preparation and handling. A third problem is fixing of the test pieces to the apparatus without damaging them. In addition, fracture may occur at the grips instead of in the middle of the test piece. Moreover, highly deformable test pieces may partly slip out of the grips. To avoid the latter, test pieces may be glued to the grips. For less deformable products such as Gouda and Cheddar cheese and many gels, specially shaped test pieces can be used that are thin in the middle, for example, dog bone-shaped samples (**Figure 8.18**). In addition, to prevent slipping, the inner part of the grips can be made, for instance, from

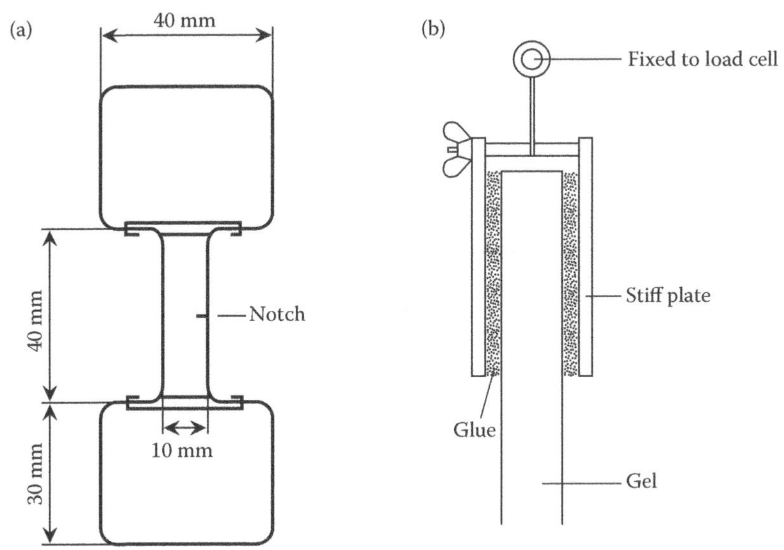

Figure 8.18 *(a) Shape and size of test pieces used for tensile testing of Gouda cheese and (b) a possible way of fixing them to the testing instrument. (Reprinted with permission from van Vliet, T., Luyten, H., in* New Physico-Chemical Techniques for the Characterization of Complex Food Systems, *ed. Dickinson, E., Blackie Academic & Professional, Glasgow, 1995.)*

high-friction, soft, and pliable plastic foam material or the test pieces can be glued to the grips.

8.3.3 Bending tests

Three-point bending tests are normally done (**Figure 8.19**), but for very long (nonfood) specimens four-point bending tests have also been described (Dobraszczyk and Vincent 1999). In the latter case, the ratio of the test piece length to the diameter must be larger than is feasible for most food products. As with uniaxial tension tests, it allows the determination of more fracture characteristics than uniaxial compression tests do. Its significant advantage over tension tests is that it avoids gripping and clamping problems. Moreover, if test pieces are available in the form of a reasonable rigid rod of circular or rectangular cross section, deflections for a given force are generally much larger in bending than in extension or compression, making the determination of the deflection much easier (**Figure 8.17**). It makes this technique especially suited for studying brittle materials such as chocolate bars, biscuits, and many other dry crispy materials. Bending tests are easy to perform, and it is not necessary to fix the specimen to the measuring body.

The major disadvantage of bending tests is that they do not produce a uniform strain in the test piece. They involve a combination of compression, extension, and some shear. This combination of different strains generally makes the analysis of the local strain impractical for nonlinear materials. The inner curved surface is deformed in compression and the outer surface in tension with a neutral axis in between where deformation is zero. The shear component will be larger for specimens with a lower length/rod thickness ratio. For three-point bending experiments, its contribution can be neglected for length/thickness ratios greater than 10. Tensile stress is at maximum at the outer surface and compressive stress at the inner surface. Fracture mostly starts at the outside of the tensed part of the test piece, and therefore it can be observed. The high tensile stress at the outer surface makes fracture behavior sensitive to scratches at this surface that may have been formed during specimen preparation.

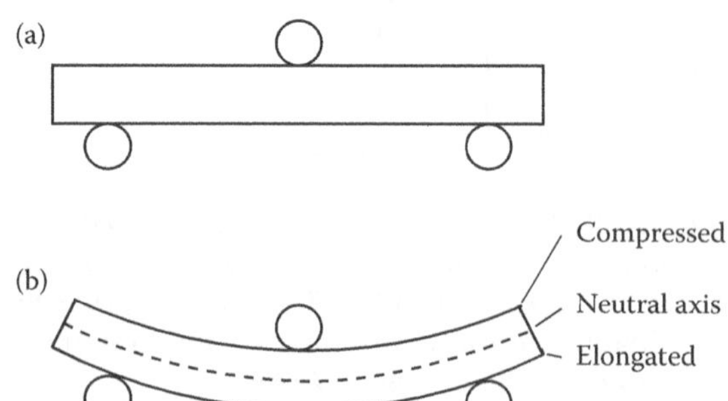

Figure 8.19 *Deformation of a test piece during a three-point bending test: (a) before bending; (b) after bending.*

The stress and the strain in a test piece can be calculated from the measured force versus deflection curve, assuming that (i) the material is isotropic and has the same modulus in tension and compression, (ii) strains are in the linear or Hookean region, (iii) the beam is straight and has a uniform cross section, (iv) the load is applied perpendicular to the neutral axis, and (v) the ratio of the distance between the supporting bars over specimen thickness is large—ideally ≥10, although it has been reported for Gouda cheese that there was no longer any effect of this ratio for values larger than 3.5 (van Vliet and Luyten 1995).

Then, the deflection y (vertical displacement) of the center of the beam due to a force f is given by (Whorlow 1992; Young 1989)

$$y = \frac{fL^3}{48EI}, \tag{8.62}$$

where L is the distance (span) between the supports (**Figure 8.20**) and I is the second moment of area of the beam cross section. For a solid circular beam with diameter d, $I = (1/64)\ \pi d^4$, and for a solid rectangular beam with a thickness (height) H and width w, $I = H^3w/12$, giving for Young's modulus

$$E = \frac{4fL^3}{3\pi d^4 y} \tag{8.63}$$

$$E = \frac{fL^3}{4wH^3 y} \tag{8.64}$$

for a solid circular beam and a solid rectangular beam, respectively. Note that the ratio of the modulus for the circular rod over that for the rectangular rod

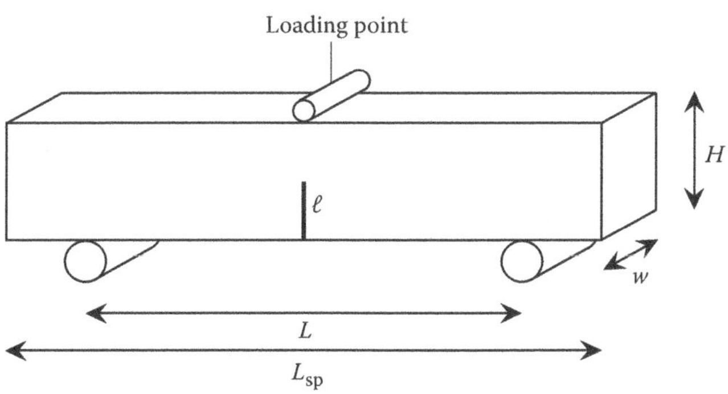

Figure 8.20 *Test specimen complying with requirements for a single-edge notch bending (SENB) test (for explanation symbols, see text).*

is equal to the ratio of the second moment of area of the beam cross section of both rods. The maximum stress, σ_{max}, for a circular rod is given by

$$\sigma_{max} = \frac{8fL}{\pi d^3} \tag{8.65}$$

and for a rectangular rod

$$\sigma_{max} = \frac{3fL}{2wH^2} \tag{8.66}$$

The combination of Equation 8.62 with the expression for σ_{max} gives for the maximum strain, $\varepsilon_{max} = \sigma_{max}/E$, for a circular rod

$$\varepsilon_{max} = 6dy/L^2 \tag{8.67}$$

and for a rectangular rod

$$\varepsilon_{max} = 6Hy/L^2 \tag{8.68}$$

Requirement (i) for the derivation of Equations 8.62 through 8.68 requires that the test pieces must be homogeneous, a requirement difficult to fulfill for layer products and products that are inhomogeneous over relatively large lengths (≥ a few millimeters, depending on test piece height).

There should be no substantial bending of the test piece due to its own weight; so the modulus of the tested material should be rather high. The maximum strain resulting from the weight of a circular rod is given by

$$\varepsilon_{max} = \frac{5g\rho L}{4Ed} \tag{8.69}$$

where g is the acceleration due to gravity and ρ is the material density. For stiff food products ($E > 10^6$ N m^{-2}), bending due to the weight of the test piece can normally be neglected. Furthermore, Young's modulus should be high enough to prevent substantial compression of the beam by the plunger and both supports. It prevents the use of knives as plunger and support. On the other hand, the radius of the supports and the plunger should be small compared to L.

For materials where fracture takes some time to occur, fracture stress and strain are usually taken to be equal to σ_{max} and ε_{max} at the point that start of fracture can be observed (Luyten et al. 1992). In addition to these parameters and E and W (area below the stress–strain curve up to fracture), notch sensitivity, inherent defect length, and critical strain energy release rate of the material can be determined by applying artificial notches of varying lengths at the side of the specimen that will form the outer tensed part. For stiff products such as chocolate bars, it is possible to determine all the major fundamental fracture parameters with this type of test.

A well-established bending test for determining the critical stress intensity factor in tensile deformation, K_{1C}, is the so-called single-edge notch bending (SENB) test, which is based on the use of standardized rectangular measuring geometries. An artificial notch is made in the center of the bottom side of the specimen (the side that will be deformed in tension) over the whole width (breadth) of the test piece, for example, by using a razor blade. The test specimen (**Figure 8.20**) has to comply with the next requirements: (1) length L_{sp} of the test specimen ≥4.5 times its height H; (2) width $w \sim 0.5H$; (3) distance L between the supports $4H$; (4) notch depth l (0.45–0.55)H. For linear elastic materials, the critical stress intensity factor for mode 1 (in tension) fracture K_{1C} can then be calculated from the data by using

$$K_{1C} = \frac{fL}{wH^{3/2}} f\left(\frac{l}{H}\right) \tag{8.70}$$

where $f(l/H)$ is a geometrical factor that accounts for the size effects of the specimen and f is the force at the onset of crack propagation (Ewalds and Wanhill 1985; Anderson 1991). For $(l/H) = 0.45$ or 0.5, $f(l/H)$ is equal to 2.29 and 2.66, respectively. In addition, the critical strain energy release rate can also be obtained for linear elastic behavior of the test material. For these materials, it can be calculated by

$$G_C = \frac{W}{wh\phi}, \tag{8.71}$$

where W is the energy (in J) represented by the area under the force displacement curve between zero and the crack initiation load and ϕ is a calibration factor that depends on l/H and L/H ($\phi = 0.26$ and 0.235 for $l/h = 0.45$ and 0.5, respectively ($L/H = 4$)) (Williams and Cawood 1990).

Summarizing, bending tests are very useful for studying rather brittle materials, and they are easy to carry out, but preparation of test specimens has to be done with great care.

8.3.4 Comparison of compression, tension, and bending tests for determining fracture behavior

A summary of the three most common and versatile types of tests that can be executed using a tension compression (or general material testing) instrument suited for determining large deformation and fracture behavior of materials is given in **Table 8.1**. It is clear that all methods have their strong and weak points. The easiest test to perform provides the lowest amount of information, but for many purposes such as quality control, it is in general enough. Regarding the determination of fracture parameters by using a bending test, these tests are very useful for stiff materials, but they are less suited for rather soft and deformable materials; therefore, determination of most fracture

Table 8.1 Advantages (+) and Disadvantages (–) of Compression, Tension, and Bending Tests for Determining Deformation and Fracture Properties of Food Materials

Aspect	Compression	Tension	Bending
Sampling	+	–	±
Fixing to apparatus	+	–	+
Weight of sample	+	±	–
Execution of method	+	–	+
Friction	–	+	+
Homogeneous deformation	±	±	–
Young modulus	+[a]	+	+
Determination of Fracture Parameters			
Fracture stress	±	+	±
Fracture strain	±	+	±
Total fracture energy	±	+	±
Critical strain energy release rate	–	+	±
Critical stress intensity factor	–	±[b]	±[b]
Notch sensitivity	–	+	±
Inherent defect length	–	+	±
Initiation–propagation	–	+	±

Source: Luyten, H. et al., *J. Texture Stud.*, 23, 245–266, 1992. With permission.
Note: For further explanation, see Sections 8.3.1 through 8.3.4.
[a] For large enough linear region.
[b] Requires specially shaped specimen (Ewalds and Wanhill 1985; Anderson 1991).

parameters is marked with ±. For a more detailed discussion, see the sections above on the various methods.

8.3.5 Controlled fracture tests

Various types of experiments have been developed to obtain data for the specific fracture energy at an imposed constant rate of fracture. Examples are wedge tests, cutting tests (with a knife or a wire), the "trouser tear" test, loading–unloading test (or the double cantilever beam test), and instrumental scissors (Dobraszczyk and Vincent 1999).

A controlled fracture test that has been applied successfully for determining the fracture energy of various food materials such as potatoes, apples, cheeses, and carrots is the so-called wedge test (van Vliet and Luyten 1995). By driving a wedge through a piece of material, the sides of it are bent outward (**Figure 8.21**). Initially, the material will deform below the wedge tip, and after some time a (small) crack will be formed below the wedge tip. The energy stored in the material by the ongoing pressing of the wedge into the material can be transported to the tip of the growing crack. The result is that the crack propagates ahead of the tip of the wedge. By adjusting the size and shape of the test piece and the driving speed of the wedge, the rate of crack growth can be controlled, resulting in a crack speed equal to the driving speed of the wedge. If so, the force f_w required for wedge penetration is constant, and the fracture energy (J m^{-2}) can be calculated directly from f_w/w, where w is the width of the test specimen. For obtaining the critical strain energy release rate, various corrections have to be applied, among others, for the curling of the test material next to the wedge and for friction between the wedge and the test specimen. A typical force–time curve for a wedge experiment is shown in **Figure 8.22**.

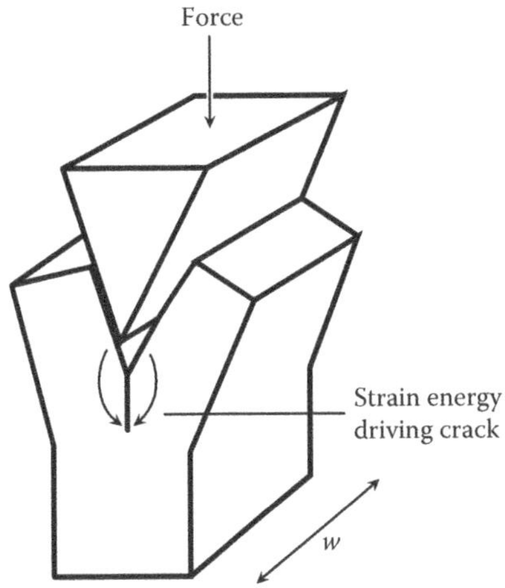

Figure 8.21 *Diagram showing principle of wedge test. w, width of test piece. (After Vincent, J.F.V.,* Biomechanics—Materials: a Practical Approach, *Oxford University Press, Oxford, 1992.)*

Wedge experiments offers several advantages for determining fracture energy compared to tension tests (van Vliet and Luyten 1995): (1) The test specimen can be very small. (2) The execution of the method is very simple; the test piece does not have to be clamped. (3) Fracture is in tension ahead of the wedge. This is probably very similar to biting food with the front teeth. Good correlations have been observed between results of a wedge test and fracture behavior of pieces of cheese during biting (Vincent et al. 1991). (4) Very small differences in fracture energy within a test piece, for instance, due to (layered) structural inhomogeneities, can be recognized.

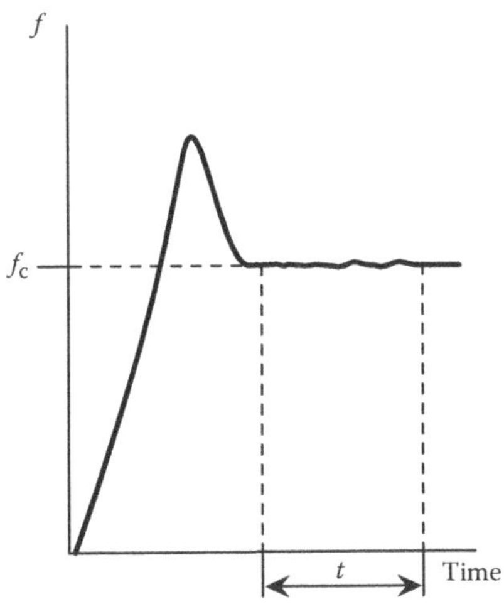

Figure 8.22 *Example of a force f versus time curve as observed in a wedge test or in a cutting test experiment. f_c is constant value of f over time interval t during controlled rate of crack growth. (Reprinted from van Vliet, T., Luyten, H., in* New Physico-Chemical Techniques for the Characterization of Complex Food Systems, *ed. Dickinson, E., Blackie Academic & Professional, Glasgow, 1995. With permission.)*

On the other hand, the wedge test has also several disadvantages. (1) The main one is that only fracture energy can be determined, so additional tests have to be performed for determining stress–strain curves and other fracture properties. (2) Side cracks may be formed because of inhomogeneities in the structure of the material. This is especially the case for brittle materials. The result is that no constant force is obtained, and the size of the fractured area cannot be determined. So, for these cases, no fracture energy can be calculated. (3) There may be friction between the wedge and the test material, leading to a too high measured force. Lubricating the wedge may help.

Another controlled fracture test used for food materials is cutting with a knife or wire attached to, for example, a tension compression instrument (**Figure 8.23**). This type of test is related to the wedge test. A sharp knife (e.g., for fruit) or a thin wire (for soft materials such as cheese) is forced through a material, resulting in the formation of a crack ahead of the knife or wire.

The measured force F_c is the sum of various components. (1) The force needed to fracture an area wvt, where w is the width of the test piece, v is the cutting speed, and t is the cutting time. (2) The force needed to deliver the energy that is in the off cut, the so-called curling energy. Its contribution can be determined by cutting slices of different thicknesses and extrapolating to zero slice thickness. (3) The friction force between the knife and the test material. (4) The force needed to deliver the energy that dissipates during deformation of the test material below and around the knife tip or wire. This gives for cutting off a slice of zero thickness

$$f_c vt = G_C vtw + (W_e + W_d)Crwvt, \tag{8.72}$$

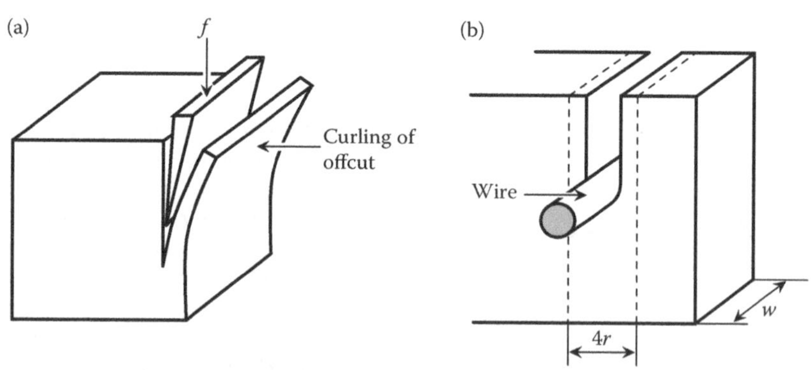

Figure 8.23 *Cutting experiments. (a) Cutting with a knife. (b) Cutting with a wire. (Reprinted from van Vliet, T., Luyten, H., in* New Physico-Chemical Techniques for the Characterization of Complex Food Systems, *ed. Dickinson, E., Blackie Academic & Professional, Glasgow, 1995. With permission.)*

where W_e and W_d are the elastic and dissipated energy due to the cutting process, C is a constant accounting for the volume of material deforming around the wire or knife, and r is the wire radius or a factor accounting for the sharpness of the knife. By performing experiments using wires of different thicknesses, f_c $(r = 0)$ can be calculated by extrapolating the measured f_c to zero wire thickness. The critical strain energy release rate is then given by

$$G_C = f_c\ (r = 0)/w \tag{8.73}$$

The wire should be fixed tautly to prevent strain energy from being stored in bending the wire that cannot be released into the cut.

For knife cutting, there is no extrapolation procedure that allows handlers to calculate f_c $(r = 0)$. This modification is more appropriate for stiff elastic materials such as carrots and apples where the friction between the material and the knife as well as the width of the area below the knife, where the material is deformed, is low. Then the measured force is representative of the area just ahead of the knife tip, yielding that information about the variation in structure can be obtained from the force–deformation curve.

In general, compared to other tests, the advantages and disadvantages of wire and knife cutting tests are similar to those mentioned above for wedge tests.

8.3.6 Biaxial extension tests

These tests are usually not performed for the purpose of measuring the fracture properties of the investigated material, but to study its biaxial flow behavior. Regarding test setup, there are many similarities between these types of tests and the uniaxial compression tests (discussed above). When a uniaxial compression test is performed in such a way that there is no friction between the sample and the plates, the test piece is not only uniaxially compressed but also biaxially extended (**Figure 3.7**) (Chatraei et al. 1981). Determination of biaxial flow properties can be relevant, because this type of flow occurs in various practical situations, such as around growing gas bubbles during fermentation and baking of bread and cakes, cheese ripening, and in the mouth when pressing the tongue against the palate.

A clear advantage of biaxial extension tests over measuring flow properties in a concentric cylinder instrument is that test specimens can be prepared much more easily without extensive structural damage. The latter is inherent to bringing a liquid-like material in the narrow gap between two coaxial cylinders. Slip between the test material and the plates is essential for a good execution of the tests, and not a nuisance as it is during measurements in rotational instruments such as the plate–plate viscometer. This can be advantageous for characterizing products that exhibit slip during shear measurements, such as mayonnaise. To avoid friction, often plungers are used with the same diameter as the initial diameter of the cylindrical test piece (constant area arrangement) and made of material that favors slip (e.g., Teflon for many food materials) or by covering the plates with a lubricating oil of low viscosity. An alternative is to use plates that have a larger diameter than the test material so that the test is executed on a constant volume of the material. Because it is difficult to form a regular (cylindrical) and stable test sample of semiliquid materials, the constant area and changing volume arrangement is usually preferred.

The biaxial stress σ_{be} that the test material experiences, is equal to the uniaxial compressive stress σ_e and given by $f_t/\pi R^2$, where f_t is the force recorded at time t and R is the radius of the plate. For a material with a Poisson ratio of 0.5, the biaxial Hencky strain $\varepsilon_{b,H}$ is half the uniaxial compressive strain ε_H (Equation 3.25), which reads for test specimen with initial height H_0 as

$$\varepsilon_{b,H} = -\frac{1}{2}\varepsilon_H = -\frac{1}{2}\ln\frac{H_t}{H_0} \tag{8.74}$$

The biaxial extensional strain rate $\dot{\varepsilon}_{b,H}$ is given by

$$\dot{\varepsilon}_{b,H}\frac{d\varepsilon_{b,H}}{dt} = \frac{1}{2}\frac{dH}{H_t dt} = \frac{v}{2H_t} \tag{8.75}$$

and equal to half the uniaxial strain rate (Equation 3.24) (be aware that Equations 3.24 and 3.25 are given for test specimens of length l). Note that for a constant deformation speed v, the strain rate increases during the tests. From the results, a biaxial extensional (apparent) viscosity $\eta_{b,app} \equiv \left(\sigma/\dot{\varepsilon}_{b,H}\right)$ can be calculated, which is 6η for Newtonian liquids.

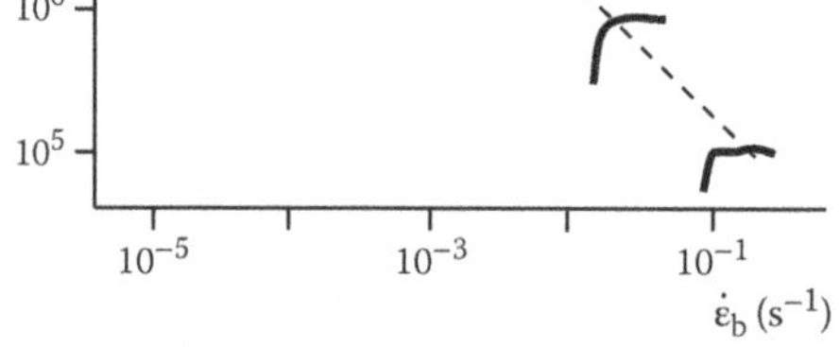

***Figure 8.24** Apparent biaxial extensional viscosity $\eta_{b,app}$ as a function of biaxial strain rate of Gouda cheese for four displacement speeds of the moving plate. Dashed line connects points deformed to the same biaxial strain $\varepsilon_{b,H}$.*

Products such as bread dough, peanut butter, and mayonnaise will exhibit stress overshoot. As discussed in Section 7.3, the deformation up to the maximum in the stress is mainly governed by elastic effects, implying that a real viscosity can only be determined for strains larger than the strain corresponding with the maximum. One should be aware of this when interpreting results from biaxial extensional tests presented in the form of those in **Figure 8.24**.

A useful modification of the biaxial extension test is described by Damreu and Peleg (1997). It is based on compressing a liquid(-like) product in a shallow container by a wide (circular) plate (**Figure 8.25**). The test material does not need to have a perfect circular shape. In principle, the specimen can be formed in the test container, allowing handlers to study strain-sensitive materials. Because of the imperfect geometry, measured forces are affected by entry effects and buoyancy. The entry effect is attributable to the time delay before squeezing flow has fully developed. Buoyancy results from the increase in height of the liquid level in the gap between the upper plate and the container with respect to the position of the upper plate level. The size of these factors is considerably reduced for a larger upper plate radius, provided that the gap between the upper plate and container wall remains sufficiently large. A danger of a large radius of the upper plate is the occurrence of some tilt, resulting in uneven specimen height. Equations have been deduced both for the case of lubricated flow between the liquid and the plates and for nonlubricated flow. These read as

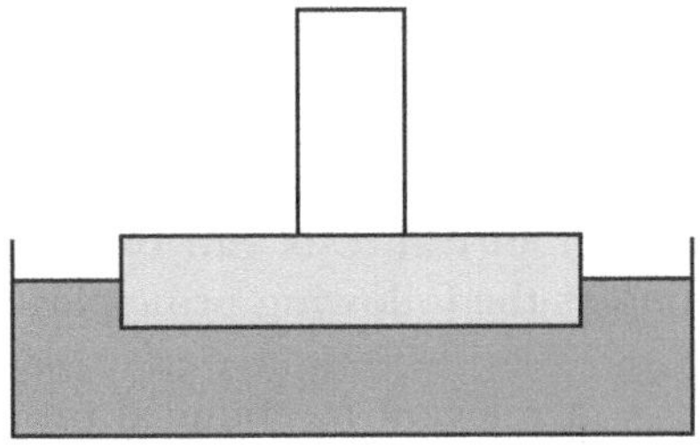

Figure 8.25 *Imperfect squeezing flow geometry. (After Damrau, E., Peleg, M.,* J. Texture Stud., *28, 187–204, 1997.)*

$$\sigma_e = 3\frac{\eta v}{H_t} + \rho g(H_0 - H_t)\frac{R^2 R_c^2}{R_c^2 - R^2} \tag{8.76}$$

for lubricated flow and as

$$\sigma_e = \frac{3}{2}\frac{\eta v R^2}{H_t^3} + \rho g(H_0 - H_t)\frac{R^2 R_c^2}{R_c^2 - R^2} \tag{8.77}$$

for nonlubricated flow (Damreu and Peleg 1997). In these equations, v is the displacement speed of the upper plate, ρ is the liquid density, R is the disk radius, R_c is the container radius, and H is the specimen height below the upper plate. The term at the far right is the extra stress needed to overcome the extra hydrostatic pressure due to the liquid in the annular gap between the upper plate and the container.

8.4 Empirical tests

Empirical tests for determining mechanical properties of food are very popular in food and other industries. They are often very well suited for quality control tests whereby their main advantages are that they are relatively cheap and can be performed easily, even by people who are only trained to a limited extent. Results are mostly expressed in one parameter, whose value can be compared with the standard quantity. The main disadvantage is that the stresses and strains involved are not well defined, which makes the calculation of fundamental rheological parameters impossible. Moreover, it strongly limits the comparison of the results obtained with data obtained by other tests. At best, results can be correlated to those from other tests. In spite of this, empirical tests can be of great value, especially for quality control, as long as the following principles are followed:

- A relevant mechanical property is determined or one that is unequivocally related to it. Measurements based on an accidental correlation may turn out to be dangerous if the product, the raw material properties, or the manufacturing conditions are changed.
- The experiment is done over a time scale (deformation rate and time) that is relevant for the property being investigated.
- The measurements are reproducible in terms of test sample preparation and test execution.

Many empirical test methods are variations on fundamental tests; for example, the determination of the torque required to let a cylinder, a vane, or a helical path rotate at a set speed in a "liquid" in a large container. In addition to the type of apparatus used, the measured torque will depend on the (apparent) viscosity of the liquid and on the characteristics (e.g., dimensions) of the apparatus. In general, for the interpretation of the results, it may be useful to analyze roughly which fundamental parameter(s) (or combination thereof) determine(s) the reading obtained.

For several test methods, it is a matter of taste whether one rates them as a fundamental test or as an empirical test. An example is the determination of the viscosity of thick liquids by using a vane rheometer. In this book, this method is rated as an empirical test because the exact flow around the vanes tips can often only be described globally. However, as discussed below, the use of a vane viscometer allows the determination of some fundamental rheological parameters.

Below, first an overview will be given of tests predominantly suited for determining the mechanical properties of liquid and semisolid materials and second of those suited for solid materials. No attempt is made to give an extensive overview of the whole variety of tests that exist. The discussion will be limited to a series of typical examples of empirical test methods.

8.4.1 Empirical tests primarily suited for liquids and semisolids

8.4.1.1 Flow funnels

A very simple instrument that is often used in industry to get an impression of the flow behavior of a liquid is the flow funnel (**Figure 8.26**). Depending on the country (region) where it is used, it is variously known as Ford cup, Marsh funnel, consistency cup, and Posthumus funnel. The instrument consists of a circular container with a conical end in which there is a small orifice leading to a short exit tube (length/diameter ratio <1.5). The driving pressure for flow is gravity. One simply determines the time required for a certain amount of liquid to flow out of the funnel. Sometimes, pins are placed at the inside of the container to allow for a better determination of the time when the required amount of liquid has flowed out. Often, the conical part can be replaced by another one with an orifice with a smaller or larger diameter to allow accurate measurement of liquids with lower or higher viscosities.

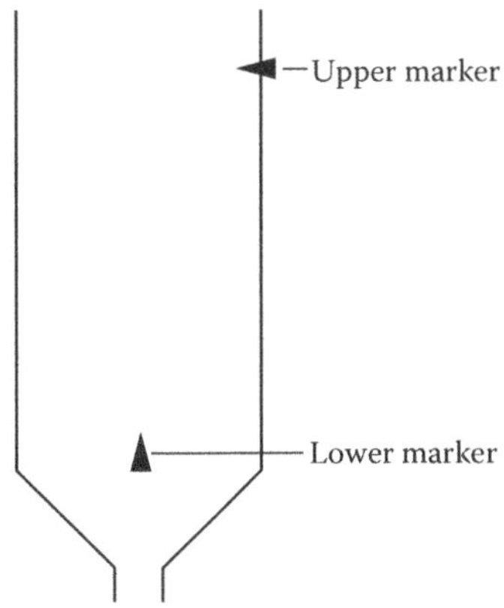

Figure 8.26 *Illustration of a flow funnel.*

Because the small exit tube is often short or even absent, flow is to a large extent determined by the convergent flow through the conical part, causing a strong contribution of the elongational viscosity to the resistance against flow. For long exit tubes with a small diameter, the shear flow through it may contribute substantially to or even dominate flow resistance. The shear component will be zero in the middle of the funnel and maximum at the walls just as in tube viscometers. Flow rates will decrease as the funnel empties. This makes the flow pattern very complicated. For Newtonian liquids and for liquids whose flow behavior can be described by a power law, the (apparent) viscosity may be estimated from the flow times (Barnes 1980).

Liquids showing marked thixotropic behavior over long shearing times will give efflux times that are (much) longer than expected based on steady-state shear viscometry because flow times in a flow funnel are mostly relatively short. In addition, because of the elongation flow component present, liquids with a large Trouton ratio (η_E/η) will have a long efflux time compared to Newtonian liquids with the same viscosity in shear flow. A combination of these effects may explain, for instance, the relatively large efflux times observed for yogurt and Dutch custard (**Figure 8.27**).

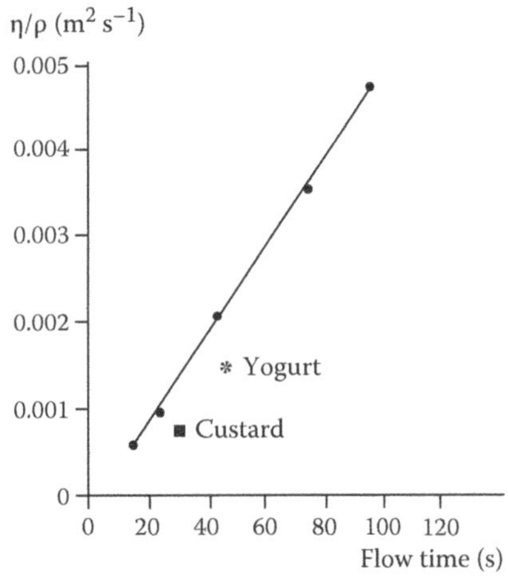

Figure 8.27 *Relation between kinematic viscosity (η/ρ) and efflux time of a flow funnel (cylinder diameter 60 mm, height 98 mm, cone height 50 mm, orifice diameter 8mm). Data for Newtonian oils. Data for stirred yogurt and custard indicated.*

8.4.1.2 Vane rotational measuring geometry

Vane rotational viscometers are a kind of rotational instruments with a spindle equipped with thin vanes (often four or six) as inner measuring body (**Figure 8.28**). Such spindles can in principle easily be attached to available standard rotational instruments. It provides, especially in combination with a roughened outer cylinder, an effective way to measure the rheological properties of highly shear rate thinning non-Newtonian liquids that otherwise would show large slip effects at a smooth inner cylinder. Moreover, it allows insertion of the inner spindle with minimal disruption of the sample. Measurements do not require a narrow gap between the vane and the outer cylinder (in contrast to the concentric cylinder geometry). It also makes the measurements less sensitive for artifacts because of the presence of large particles. The vane measuring body is therefore suited for testing foods that are sensitive for structure breakdown and/or slippage such as mayonnaise, tomato paste, and apple sauce. In addition, the instrument allows the determination of the yield stress of these types of products.

The simple formula usually used for the calculation of the yield stress is based on the assumption that the yield stress acts uniformly on the sides and both ends of an equivalent cylinder formed when a vane completely immersed in the material is slowly rotated at a constant rotation speed. The total torque on the sides of an equivalent cylinder with radius R and length l in a large vessel is $2\pi R^2 l\sigma$ and on both ends together $4\pi R^3\sigma/3$. Then the yield stress is given by

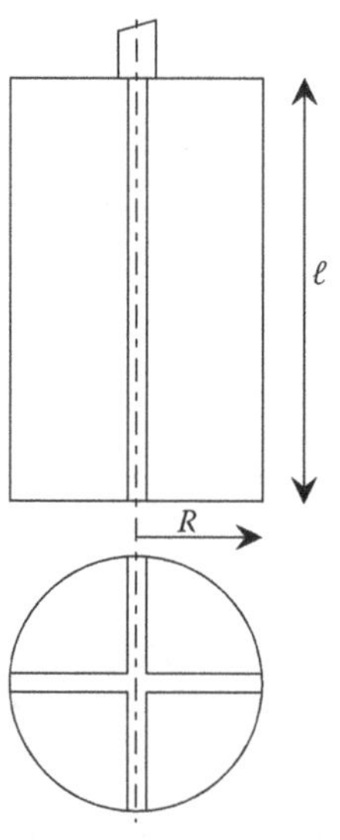

__Figure 8.28__ Schematic diagram of a four-bladed vane. R, radius of vane blades; l, length of vane blades.

$$\sigma_y = \frac{T_y}{2\pi R^3}\left(\frac{l}{R}+\frac{2}{3}\right)^{-1} \tag{8.78}$$

where T_y is the measured torque at yielding, and is often assumed to be equal to the maximum in a torque versus deformation/time curve. Various studies have shown the suitability of Equation 8.78 for calculating the yield stress (Barnes and Nguyen 2001).

The vane geometry is also suited for determining the flow properties of shear thinning liquids (power law index of the power law model <0.5). Then the vane shears the liquid outside a circular cylinder prescribed by the vane blade tips, whereas there is little or no shearing within this cylinder. For a Newtonian liquid, the streamlines are not circular, but dipped into the space between the vanes. The vane geometry is also not suited for studying low viscosity liquids.

Inertial effects will become important, leading to distortion of the streamlines and a too high (calculated) viscosity.

8.4.1.3 Spreading consistometers

With these instruments, the spreading of a material under its own weight is measured after a certain time. Examples are the Bostwick and the Adam's consistometers. With the Bostwick consistometer, the distance is measured over which a certain amount of material, initially contained behind a sliding door, flows through a rectangular horizontal trough in a given time after the sliding door is opened. For instance, the standard method for characterizing the consistency of tomato paste is based on a measurement with the Bostwick consistometer. The Adam's consistometer is regularly used as a standard tool for testing of, for example, apple puree. It consists of a flat horizontal plate with a series of concentric equally spaced rings. The sample is placed in a short tube positioned on the central ring. After lifting the tube, the sample will radially flow out over the rings. For both types of instruments, the distance traveled by the material within a certain time will primarily be determined by a combination of fundamental characteristics such as yield stress and apparent shear viscosity at decreasing strain rates (stresses). In addition, surface tension may play a part if the serum expelled by the material moves in front of the concentrated pastes.

Rough information about the yield stress of a thick paste can often be obtained visually by estimating (or measuring) the height to which a pile of the material levels off. The yield stress over the given time can then be calculated simply by using

$$\sigma_y = \rho g H, \tag{8.79}$$

where ρ is the density of the material, g is the acceleration due to gravity, and H the height of the pile after leveling. One should be aware that for heights lower than ~1–3 mm, surface tension effects may make the calculation invalid. This height will be higher for materials with a larger surface tension that do not wet the solid plate.

8.4.1.4 Penetrometer tests

With these tests, the penetration depth is measured of a semisolid material by a usually conically shaped measuring body that falls from a predetermined height onto the product. The penetrometer is a relatively simple and cheap instrument used for applications such as the determination of the consistency of fats and the spreadability of butter.

Penetration depth depends on the properties of the cone such as weight, cone angle, and surface roughness, on falling height, and on a combination of test material properties. In the classical test, the tip of the cone has to be placed exactly on the surface of the test material. The falling cone will first deform the material and at large deformations the material may yield (e.g., in the case of butter and margarine) or may fracture (e.g., in the case of elastic gels). Materials that yield will subsequently flow upward through the gap between the cone and the unyielded material. This flow contains both a shear and an elongation flow component (**Figure 8.29**); their relative contribution will depend on the material and test conditions. The material outside the yielding/fracture zone will be compressed. For an elastic material, the modulus and fracture strain of this material will determine the penetration depth to a large extent. Thus, next to instrument characteristics, penetration depth is determined by a combination of the following fundamental rheological characteristics: elastic modulus in shear and compression, yield or fracture stress, and shear and elongational viscosity.

Instead of using a cone, penetration can also be measured using a rod or a disk. However, this does not affect the basic advantages and drawbacks of the test to a large extent. Therefore, they will not be discussed separately.

Summarizing, a penetrometer test may give an unequivocal characterization of the mechanical properties of a product as long as the relative contribution of these properties to the penetration depth does not vary and the tests are executed in a standardized way in terms of the apparatus, cone properties, and measuring conditions. Its main disadvantage is that results cannot be or can only be roughly translated in fundamental rheological properties and therefore quantitatively be compared with results of other tests.

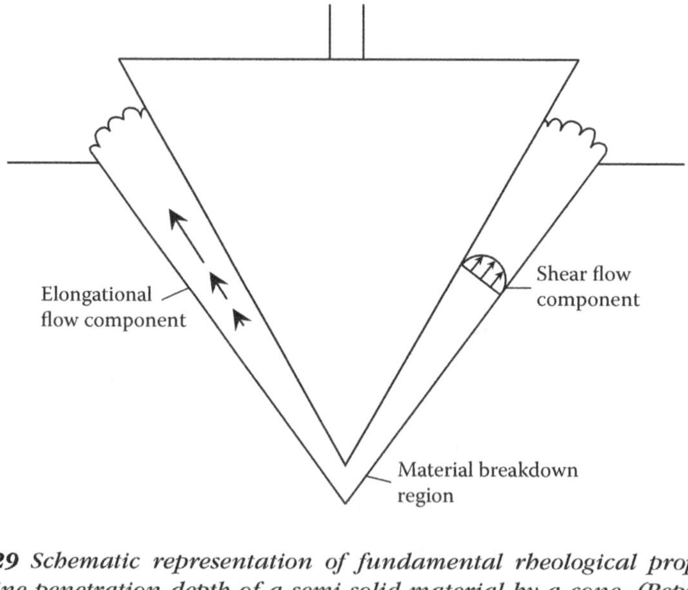

***Figure 8.29** Schematic representation of fundamental rheological properties that determine penetration depth of a semi solid material by a cone. (Reprinted from van Vliet, T., in* Food Texture: Measurement and Perception, *ed. Rosenthal, A.J., Aspen Publishers, Gaithersburg, MD, 1999. With permission.)*

8.4.2 Empirical tests primarily suited for solids

By far, most empirical tests suited for predominantly solid products are based on compression of the test specimen and determination of the required force. For these types of tests, in principle the instruments can be used as described in Section 8.3. In addition, a large range of relatively cheap apparatuses can be bought in which mostly the compression speed is fixed or the measuring body is pressed into the test material by hand. A large variety of compression geometries is used; examples include flat cylindrical plates of various diameters, cylindrical rods with a flat or rounded tip, needles, perforated plates, and artificial teeth. The choice strongly depends on the goal of the test: does it involve puncturing of the test specimen, deformation of the specimen up to fracture or yielding, or compressing and extruding the specimen through holes in the container or the plunger?

8.4.2.1 Puncture tests

An extensive overview of puncture tests and the theory behind this test has been given by Bourne (2002). Here, we will only discuss several principles. Puncture tests are often used for testing the maturity of fruits, although other kinds of food can also be tested in this manner. During a puncture test, the force required to push a probe into the product is measured. Penetration of the specimen results in irreversibly damaging its structure as a result of crushing and/or yielding. Usually, the penetration depth is held constant. For cheaper instruments, results are reported as the maximum force measured. For more sophisticated instruments, a force–distance or force–time curve is provided. The latter offers more possibilities to determine the fracture or yield force. The force f at the yield point has been found to depend both on the area A below and the perimeter P of the probe. Below the probe the test material is compressed, whereas it will be deformed in shear around the edges of the punch (**Figure 8.30**). Above relationship can be written as (Bourne 2002)

$$f = K_A A + K_P P + C, \qquad (8.80)$$

where K_A and K_P are related to the (uniaxial) compression and shear modulus, respectively, of the test material. C is a constant (N) depending on the apparatus and the test material.

In view of this, the stress below a (flat) probe will be highest at its perimeter and lowest in the center under the probe. The exact stress distribution will depend on the shape of the tip of the probe, being flat or rounded. For a needle, shear deformation will generally dominate.

In view of the deformation involved, the force at the yielding or fracture point will primarily depend on the fracture force and modulus of the material in compression and shear. After this point, in addition, friction forces between the probe and the test material will play a role. The size of the test specimen has to be much larger than the probe diameter and penetration

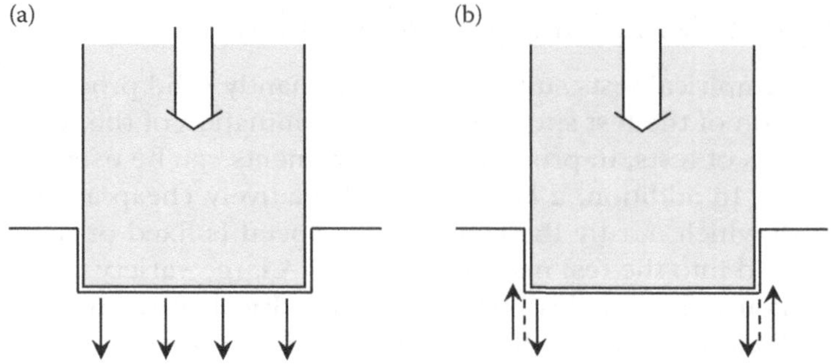

***Figure 8.30** Schematic representation of compression and shearing deformation in a material due to penetration by a flat rod: (a) compression ∝ cross section rod and (b) shear deformation ∝ perimeter rod. (Redrawn after Bourne, M.C.,* Food Texture and Viscosity: Concept and Measurement, *2nd ed., Academic Press, New York, 2002.)*

depth (at least more than three times) to prevent the effects of its size on the force measured.

The advantages of puncture tests are as follows:

- Test apparatus can be relatively simple and cheap. Many of them can easily be transported to a test location.
- Tests are easy and fast to execute.
- The test can be used for many types of food as long as they show a clear yielding or fracture point, and the probe diameter and penetration depth is small compared to food dimensions.
- Tests are suited for foods that are heterogeneous over distances larger than the probe diameter. Puncture test can then be performed on the different components separately.

8.4.2.2 Compression extrusion tests

During a compression extrusion test, the test material is compressed at a certain speed and extruded through an outlet in the form of slots or holes in the test cell. Usually, the maximum force required to extrude the food is measured or, in more advanced apparatuses, the force versus compression distance. The test is not suited for testing materials that cannot be extruded easily. In its most simple form (the so-called back extrusion test), the measuring geometry consists of a (circular) test cell with an open top, containing the test material. Next, this is compressed by a (circular) plunger with a smaller diameter than the test cell. On lowering the plunger, the material will first be compressed and finally be extruded through the (annular) slot between the plunger and the container. The required force will primarily depend on the yielding force and the apparent shear and elongational viscosity of the product for the given measuring geometry dimensions. The same product characteristics will determine the measured force by extrusion through a (circular) hole or a slit in front of the moving plunger. For products that do not flow easily, friction forces between the material and measuring geometry may also play a role. An extensive overview of compression extrusion tests is given by Bourne (2002).

8.4.2.3 Texture profile analysis (TPA)

This test, developed by Szczesniak and coworkers (Friedman et al. 1963) comprised, in essence, double compression of "bite-size" samples between two flat plates. Double compression is applied to better simulate the mastication of food. Moreover, it allows obtaining information on the mechanical properties of broken down products.

In the original test scheme, the specimen was compressed to 25% of its initial height at a sinusoidally varying speed that roughly agrees with the speed of the teeth during mastication. The product was compressed between a plunger that moves with an arc motion and a horizontal plate. The disadvantage of this arc motion is that during the initial stage, the area of contact between the plunger and the test specimen increases with ongoing compression. Moreover, the direction of the forces on the test specimen changed with the rotation of the lever. Consequently, the results cannot be recalculated to stresses and strains, but can only be compared mutually when the sizes of the test pieces are standardized. At present, TPA is usually performed using a tension compression apparatus with constant compression–decompression speed. In addition, the definitions and calculations of the TPA parameters have been adapted to some extent (Bourne 2002). Results are still mostly plotted as the required force for deforming the test piece versus time. **Figure 8.31** shows (schematically) a so-called TPA curve and illustrates the definition of some of the parameters that can be calculated from such a curve.

The original TPA test was clearly an imitative type of test. By using a tension compression instrument, parameters can be expressed in SI units and recalculated in stresses and strains. However, one has still to be careful when comparing TPA parameters with the parameters defined in the field of fracture mechanics and rheology. TPA hardness is defined as the force required to attain a given deformation, whereas in the official rheological nomenclature it denotes the fracture force. TPA fracturability (originally called brittleness) is defined as the force at which the material fractures. In fracture mechanics, this is denoted by the term "fracture force," whereas brittle behavior is defined as the behavior of a solid that fractures without appreciable permanent deformation.

Another possible complication is that during TPA measurements, compression of the test pieces is up to a given Cauchy strain (often somewhere between

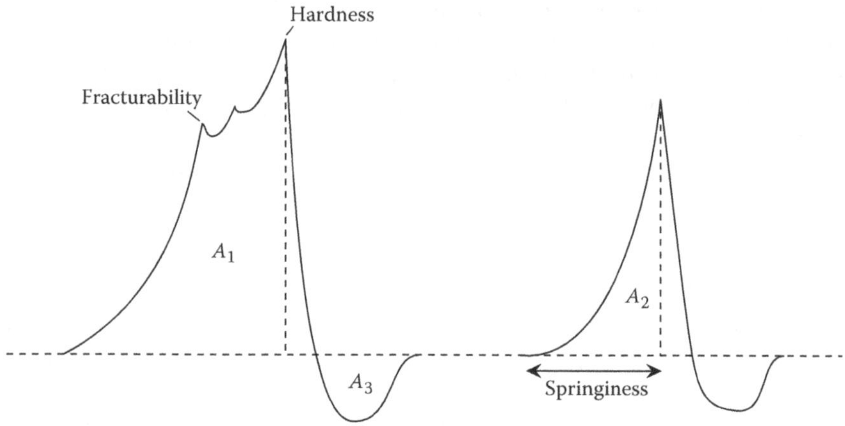

Figure 8.31 *A generalized TPA curve obtained using a tension compression apparatus. Indicated are the so-called first and second bite and some points on the curve corresponding to the mentioned TPA parameter. Cohesiveness is defined as A_2/A_1 and adhesiveness is given by A_3 and fracturability by the force at which the first significant break occurs during the first compression. Springiness is the extent to which food recovers its height during the time that elapsed between end of first bite and start of second bite.*

0.75 and 0.9). Some products will already fracture at much lower strains, whereas others fracture only at strains close to or even larger than the chosen strain. It is likely that while eating, people adapt the extent of deformation of the food during the first bite/chews depending on its fracture properties. Therefore, when comparing products that fracture at a widely varying fracture strain, TPA parameters obtained by deforming them to a constant deformation may have only a limited value for relating these data to sensory grading of the products.

When performing a TPA test, researchers should realize that during eating consumers close the gap between their teeth at a sinusoidal varying speed that, on average, is somewhere between 10 and 40 mm s^{-1}. This speed is much higher than the compression speed usually applied during an instrumental TPA test. In view of the clear dependency of the large deformation and fracture properties of most foods on deformation speed, this clearly limits the translation of the instrumentally determined parameters to sensorially determined characteristics of products.

Finally, the terms used by consumers or panel members are often not well defined. This implies that comparison of TPA parameters with sensory parameters should always be done very carefully, even when they are measured at the right deformation speed, relevant deformations, etc. For instance, it is still not well established if consumers mean the same when they speak about the hardness of a cracker, boiled egg, and cheese. In spite of this, some progress has been made in this field. Examples are discussed in **Chapter 18**.

In spite of all the criticisms, TPA tests are still popular in the food industry and related research and development laboratories. Its main advantage is that it includes double compression of the test piece, thereby imitating to some extent the processing of the food in the mouth. The latter factor is mostly not (well) considered during the development of instrumental tests intended to estimate or even predict the perception of a selected sensory attribute by humans. The main disadvantage of TPA tests is that results can mostly not be related to food characteristics and sensory properties in a causal way. One main reason is that the test is often performed in such a way that results cannot be translated to unequivocal product properties. Another reason is that many researchers do not take the effort to calculate the relevant characteristics from the data obtained, but just calculate the originally proposed texture parameters after executing the test under arbitrary test conditions.

References

Anderson, T.L. 1991. *Fracture Mechanics: Fundamentals and Applications*. Boca Raton: CRC Press.

Barnes, H.A. 1980. Detergents. In *Rheometry: Industrial Applications*, ed. K. Walters, 31–111. New York: Research Studies Press.

Barnes, H.A., and Q.D. Nguyen. 2001. Rotating vane rheometry. *J. Non-Newtonian Fluid Mech.* 98: 1–14.

Bourne, M.C. 2002. *Food Texture and Viscosity: Concept and Measurement*. 2nd ed. New York: Academic Press.

Buckingham, E. 1921. On plastic flow through capillary tubes. *Am. Soc. Test Mater. Proc.* 21: 1154–1161.

Chatraei, S.H., C.W. Makosko, and H.H. Winter. 1981. Lubricated squeezing flow: A new biaxial extension rheometer. *J. Rheol.* 25: 433–443.

Cheng, D.C.-H. 1984. Further observations on the rheological behaviour of dense suspensions. *Powder Technol.* 37: 255–273.

Damreu, E., and M. Peleg. 1997. Imperfect squeezing flow viscometry of Newtonian liquids—theoretical and practical considerations. *J. Texture Stud.* 28: 187–204.

Darby, R. 1976. *Viscoelastic Fluids: An Introduction to Their Properties and Behavior*. New York: Marcel Dekker.

Dobraszczyk, B.D., and J.F.V. Vincent. 1999. Measurements of mechanical properties of food materials in relation to texture: The materials approach. In *Food Texture; Measurement and Perception*, ed. A.J. Rosenthal, Chap 5, 99–151. Gaithersburg, MD: Aspen Publishers.

Ewalds, H.L., and R.J.H. Wanhill. 1985. *Fracture Mechanics*. London: Edward Arnold.

Ferguson, J., and Z. Kemblowski. 1991. *Applied Fluid Rheology*, Chap. 4. New York: Elsevier.

Friedman, H.H., J.E. Whitney, and A.S. Szczesniak. 1963. The texturometer—a new instrument for objective texture measurements. *J. Food Sci.* 28: 390–396.

Gent, A.M., and P.B. Lindley. 1959. The compression of bonded rubber blocks. *Proc. Inst. Mech. Eng.* 173: 111–122.

Hamann, D.D., J. Zhang, C.R. Daubert, E.A. Foegeding, and K.C. Diehl. 2006. Analysis of compression, tension and torsion for testing food gel fracture properties. *J. Texture Stud.* 37: 620–639.

Luyten, H., T. van Vliet, and P. Walstra. 1992. Comparison of various methods to evaluate fracture phenomena in food materials. *J. Texture Stud.* 23: 245–266.

Peleg, M. 1984. A note on the various strain measures at large compressive deformation. *J. Texture Studies* 15: 317–326.

Rohm, H., Y. Noel, and G. Schleining. 2000. Stiffness of compression testing machines. *J. Texture Stud.* 13: 153–162.
Segré, S., and A. Silberberg. 1963. Non-Newtonian behavior of dilute suspensions of macroscopic spheres in capillary viscometer. *J. Colloid Sci.* 18: 312–317.
Steffe, J.A. 1996. *Rheological Methods in Food Processing Engineering*. 2nd ed. East Lansing, MI: Freeman Press.
Taylor, G.I. 1923. Stability of a viscous liquid contained between two rotating cylinders. *Philos. Trans. R. Soc.* A223: 289–343.
van Vliet, T. 1999. Rheological classification of foods and instrumental techniques for their study. In *Food Texture: Measurement and Perception*, ed. A.J. Rosenthal, 65–98. Gaithersburg, MD: Aspen Publishers.
van Vliet, T., and H. Luyten. 1995. Fracture mechanics of solid foods. In *New Physico-Chemical Techniques for the Characterization of Complex Food Systems*, ed. E. Dickinson, Chap. 7, 157–176. Glasgow, UK: Blackie Academic & Professional.
van Vliet, T., and J. Lyklema. 2005. Rheology. In *Fundamentals of Interface and Colloid Science*, Vol. IV. *Particulate Colloids*, Ch 6. Amsterdam: Academic Press.
Vincent, J.F.V. 1992 *Biomechanics—Materials: A Practical Approach*. Oxford: Oxford University Press.
Vincent, J.F.V., G. Jeronimidis, A.A. Kahn, and H. Luyten. 1991. The wedge fracture test, a new method for measurement of food texture. *J. Texture Stud.* 15: 317–326.
Walters, K. 1975. *Rheometry*. Ch 5. London: Chapman & Hall.
Williams, J.G., and M.J. Cawood. 1990. European group on fracture: K_C and G_C methods for polymers. *Polym. Testing* 9: 15–26.
Whorlow, R.W. 1992. *Rheological Techniques*. Chichester: Ellis Horwood.
Young, W.C. 1989. *Roark's Formulas for Stress and Strain*. New York: McGraw-Hill.

IV

Relation between Structure and Mechanical Properties

9

General Aspects

In the following chapters, the discussion will focus on the relationship between the structure and mechanical properties of materials relevant for understanding the behavior of food products. The emphasis will be on relatively simple materials. In this context, *structure* is defined as the spatial arrangement of structural elements, that is, the physical building blocks of the material. A structural element may be heterogeneous in itself, containing structural elements by itself. Mechanical properties do not depend only on the distribution in space of the structural elements, but also on the interaction forces between them and, in various cases, on the mechanical properties of the structural elements. Therefore, to establish the relationship between structure and mechanical properties, on one hand theoretical considerations based on models are required of the structure of the material and the relevant interaction between structural elements, and on the other hand, often a combination is required of rheological and fracture data with results from other measurements.

The starting point for the forthcoming discussion will be a Newtonian liquid containing a low volume fraction of dispersed spherical particles that are hard, smooth, and impermeable for the liquid. Unless explicitly mentioned otherwise, we will assume shear flow. The description of the relation between the structure of the material, and for liquids the rheological properties, will become more complicated when

- The shape and nature of the dispersed particles is no longer smooth, spherical, hard, or impermeable. Emulsion droplets can often be classified as smooth, spherical, hard, or impermeable. In contrast, for example, fat crystals are usually anisometric, aggregates of protein molecules or emulsion droplets are not smooth and often also not spherical, whereas gas cells are deformable. Dissolved polysaccharides are deformable and permeable and often not spherical. In the latter case, the hydrodynamic volume fraction (i.e., the volume fraction of smooth, spherical, hard, and impermeable particles giving the same effect on the viscosity as a very dilute dispersion of the macromolecules) may be lower than what would be calculated from the sum of volumes taken by the macromolecules.

- The volume fraction of dispersed particles is so high that the flow pattern around a given particle disturbs the flow pattern around the neighboring particles. This will already be the case at lower volume fractions when the particles are not spherical and smooth.
- Colloidal interaction forces between the particles cannot be neglected any more compared with hydrodynamic interactions. When attractive, they may lead to aggregates and even to gel formation.
- Friction between the (anisometric) particles should not be neglected with respect to hydrodynamic interactions.
- The continuous phase is no longer a simple Newtonian liquid. For many foods, it is not directly clear what the continuous phase is. For a dispersion of emulsion droplets in an aqueous macromolecular solution, water is the continuous phase for the macromolecules, whereas the macromolecular solution is the continuous phase for the emulsion droplets. In other cases, the continuous phase can be a (weak) macromolecular or particle gel with large particles dispersed (entrapped) that yield at a certain stress.

A short overview of the main factors determining rheological properties of a liquid dispersion is given in **Table 9.1**.

A particle may also be an aggregate of smaller particles. In this case, the volume fraction, shape, and size of the aggregate will depend on its structure and the colloidal interaction between the particles and, in addition, may be affected by the size and direction of the hydrodynamic forces exerted on it by the continuous phase relative to the colloidal interactions. The viscosity of dispersions of particles will be discussed in **Chapter 10**.

The viscosity of macromolecular solutions can partly be considered along the same lines as that of solid dispersions. However, since there are also several essential differences, it deserves separate consideration. First, many macromolecules are easily deformed owing to the flow and, second, permeation of the continuous solvent through the macromolecule can occur and, finally, macromolecules may mutually interpenetrate. These aspects are discussed in **Chapter 11**.

A huge amount of food products are not liquids but solids. For these materials, the mechanical properties are primarily determined by the structure of the material and the colloidal interaction forces between the structural elements. Hydrodynamic forces are less important, and for elastic materials they can be neglected completely. Many solid foods are not homogeneous, but contain particles, consist of several layers with different properties, or are bicontinuous systems (e.g., hard sponge structures such as toast). The solid.

Table 9.1 Main Factors Determining the Rheological Properties of a Liquid Dispersion of Particles

Type of Interactions		Factors of Importance	
Between	**Nature**	**Of the particles**	**Of the continuous phase**
Particles and continuous phase	Hydrodynamic	Volume fraction Size distribution Shape Surface smoothness Flow of solvent through particles Deformability	Rheological properties such as; Apparent viscosity Elasticity Yield stress
		Flow situations such as: laminar flow: flow rate elongational component turbulent flow: power density	
Particles	Hydrodynamic	Hydrodynamic interaction between particle and continuous phase, insofar as the particles affect the flow pattern around each other	
	Colloidal	Size Hamaker constant Zeta potential Charge and charge distribution Adsorbed amount of low and high molecular-weight substances	Via effect on particle–particle interaction Ionic strength Type of ions pH Dielectric permittivity Solvent quality In the presence of polymers: depletion flocculation

Source: Reprinted from van Vliet, T., Lyklema, J., in *Fundamentals of Interface and Colloid Science*, Vol. IV, *Particulate Colloids*, ed. Lyklema, J., Academic Press, Amsterdam, 2005. With permission.

matrix may be a macromolecular or particle gel, gel-like closed packed systems, cellular materials, glassy materials, etc. For filled systems (e.g., gels containing emulsion droplets or gas bubbles), the interactions between the dispersed particles and the continuous gel network will be important and, similar to liquid systems containing dispersed particles, how the deformation around a particle will be affected by the deformation around other ones. Solid and solid-like materials will be discussed in **Chapters 12** through **17**.

Reference

van Vliet, T., and Lyklema, J. 2005. Rheology. In *Fundamentals of Interface and Colloid Science*, Vol IV, *Particulate Colloids*, ed. J. Lyklema, Chap. 6. Amsterdam: Academic Press.

10

Viscosity of Dispersions of Particles

10.1 Dilute dispersions

Dispersion of hard particles in a liquid leads to an increase in viscosity. Then, part of the liquid is replaced by solid, nonflowing material. The sheared fluid must make a detour around the particles, resulting in higher shear rates in the liquid close to the surface of the particle (**Figure 10.1**). At low shear rates, the flow stays laminar. Even in that case, this leads to an extra energy dissipation which, in turn, is observed physically as an increase of viscosity.

As discussed in Section 3.3, the energy dissipation W_d per second due to flow in a liquid is equal to the applied stress σ times the change in strain per second. For shear flow, this gives

$$W_d = \sigma\dot{\gamma} = \eta\dot{\gamma}^2, \tag{10.1}$$

which is identical to Equation 3.19b. Equation 10.1 illustrates that η is a measure of energy dissipation, like $G'' = \eta/\omega$ in oscillatory tests (Section 7.4). Because shear rate $\dot{\gamma}$ is measured macroscopically for dispersions as a whole, every local increase in shear rate (e.g., around a dispersed particle) leading to extra energy dissipation will result in an increase in the calculated viscosity of the dispersion.

The calculation of the increase in viscosity due to the incorporation of a particle in a liquid is not simple and outside the scope of this book. Both the shear and the elongation component of the flow are affected, the extent of which depends on several factors including particle shape.

Einstein solved the basic problem for hard spherical particles in 1906 (Einstein 1906, 1911). In the first publication, he made an arithmetical error that was

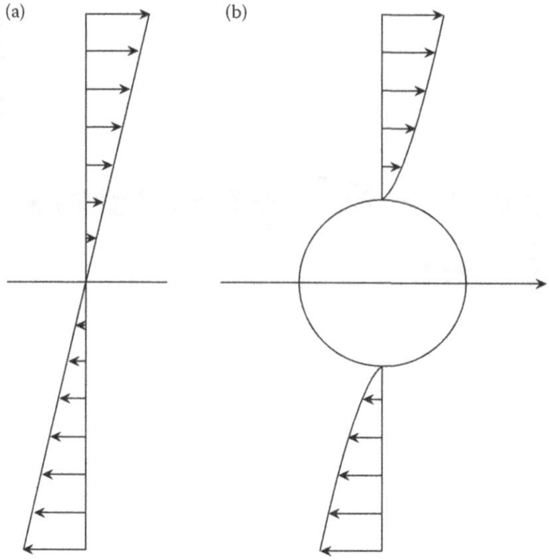

Figure 10.1 *Flow around a smooth, hard, spherical particle in a shear flow field at low shear rates (a) undisturbed flow field (b) disturbed flow field (schematic picture).*

improved in the 1911 paper. Assumptions made during the derivation were (1) particles are hard; (2) the fluid is incompressible; (3) there is no slip at the particle liquid interface; (4) inertial terms can be neglected (Reynolds number <1); (5) particles are large compared with the molecules forming the dispersing liquid, that is, they move in a continuum liquid; (6) thermal diffusion and gravitation effects can be neglected; and (7) particle volume fraction is so low that adjoining particles do not interact hydrodynamically. The final result was

$$\eta = \eta_c(1 + 2.5\varphi), \tag{10.2}$$

where η and η_c are the viscosity of the dispersion and of the continuous phase, respectively, and φ is the volume fraction of the colloid. Equation 10.2, the so-called *Einstein equation*, has been extensively tested experimentally and proven to be correct within a few percent for $\varphi \leq 0.01$.

In foods, particles are often not spherical, but fiber or disk-like, and they are anisometric. Such particles give a stronger increase in viscosity with φ than predicted on the basis of Equation 10.2. The reason for this is that anisometric particles will rotate in shear flow due to the rotational component of this flow. The disturbance of the flow pattern because of the presence of the particles will depend on their orientation, with respect to the flow direction. It will be higher when their long axis is perpendicular to the flow direction than when it is aligned in the direction of the flow. This effect promotes the alignment of the particles in the direction of flow, that is, it is stronger when the shear rate is higher. It induces rod-like particles to move in a tumbling motion. They rotate faster when their long axis is oriented perpendicular to the flow field than when it is aligned. Alignment is opposed by the rotation diffusion of the particles due to thermal motion. The extent to which alignment happens is determined by the *rotational Peclet number*:

$$Pe_r = \frac{\dot{\gamma}}{D_r}, \tag{10.3}$$

where D_r is the rotational diffusion coefficient (s^{-1}). D_r is proportional to $k_B T/\eta$ times a factor that depends on the shape of the particles and the direction of rotation. k_B is the Boltzmann constant ($J\ K^{-1}$) and T is the absolute temperature (K). For anisometric particles, D_r over the long axis will always be lower than that for spheres, implying that alignment of particles will occur.

In practice, it is common to adapt the Einstein equation empirically for non-spherical particles by replacing the factor 2.5 by υ giving

$$\eta = \eta_c(1 + \upsilon\varphi), \tag{10.4}$$

where υ depends on the shape of the particles and on the flow rate. For rod-like particles, in particular, the effect of an anisometric shape on the viscosity of the dispersion is considerable. Generally speaking, rods/prolate ellipsoids increase the viscosity more than disks/oblate ellipsoids do at equal concentrations (**Figure 10.2**).

Alignment of anisometric particles in the direction of the main flow component lowers the disturbance of the flow pattern and reduces the energy dissipation, and hence the viscosity. Because the extent of flow alignment increases with shear rate, this leads to shear rate thinning behavior. If changes in the extent of alignment with changing shear rate take longer than the reaction time of the apparatus, the dispersion will also behave in a thixotropic manner. For $Pe_r \to \infty$, particle alignment will be at a maximum, regardless of flow rate, and the viscosity will be low and independent of $\dot{\gamma}$ (**Figure 10.3**). On the other hand for $Pe_r \to 0$, there will be no flow-induced alignment of the particles and the viscosity will be at maximum and also independent of $\dot{\gamma}$. So, both at low and high Pe_r, flow behavior is apparently Newtonian with a shear rate thinning region in between, roughly starting at $Pe_r \approx 1$.

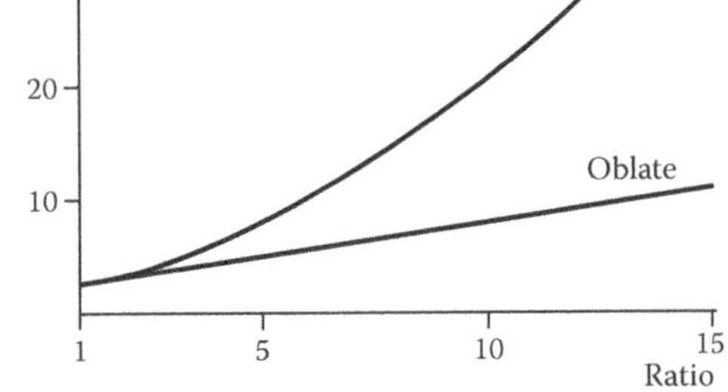

Figure 10.2 *Theoretical values of factor υ for oblate and prolate ellipsoids as a function of ratio of long over short axis for $Pe_r \to 0$. For spheres, υ is 2.5.*

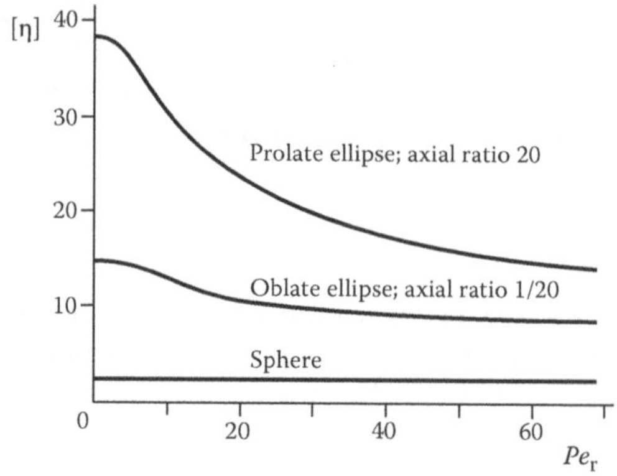

Figure 10.3 *Intrinsic viscosity [η] as a function of the rotational Peclet number* $Pe_r = \dot{\gamma}/D_r$. *(Reprinted from van Vliet, T., Lyklema, J., in* Fundamentals of Interface and Colloid Science, *vol. IV,* Particulate Colloids, *ed. Lyklema, J., Academic Press, Amsterdam, 2005.)*

In elongational flow, the effect of strain rate is different. For this flow type, an increase in elongational viscosity is observed with increasing strain rate starting at $Pe_r \approx 1$, in which the effect is larger for prolate than for oblate ellipsoids.

Finally, it should be emphasized that Equation 10.2 and its generalized form, Equation 10.4 for nonspherical particles, only holds for very dilute suspensions, that is, for $(\eta/\eta_c) < 1.03$.

For higher volume fractions of spherical particles, but still semidilute dispersions, a number of extensions to the Einstein equation have been proposed. These are usually based on a power series φ, such as

$$\eta = \eta_c(1 + 2.5\varphi + k_2\varphi^2 + \cdots) \tag{10.5}$$

The φ^2 term then accounts for binary hydrodynamic particle interaction, a possibly present φ^3 term for ternary ones, etc. In the literature, a series of equations with a form similar to Equation 10.5 have been proposed, varying in terms of the value of k_2. In the absence of, or for weak colloidal interaction forces, a main factor determining k_2 is the relative importance of the effect of hydrodynamic forces and of Brownian motion, as expressed in the *translational Péclet number*

$$Pe_t = \frac{R_p^2\dot{\gamma}}{D} = \frac{6\pi\eta_c R_p^3\dot{\gamma}}{k_B T}, \tag{10.6}$$

where D is the diffusion coefficient ($m^2\ s^{-1}$) of a single particle with radius R_p. Pe_t gives the ratio of the time it takes for a particle to diffuse over a distance equal to its radius divided by a time characterizing the flow field ($1/\dot{\gamma}$). For $Pe_t \ll 1$, the distribution of particles is not—or only slightly—altered by the flow. Then, for dilute hard-sphere homodisperse suspensions ($\varphi \leq 0.1$), the following equation holds (Batchelor, 1977)

$$\eta = \eta_c(1 + 2.5\varphi + 6.2\varphi^2) \tag{10.7}$$

On the other hand, for $Pe_t \gg 1$, only the effect of binary hydrodynamic particle interaction plays a role leading to

$$\eta = \eta_c(1 + 2.5\varphi + 5.2\varphi^2) \tag{10.8}$$

Comparison of Equations 10.7 and 10.8 shows the occurrence of shear thinning with increasing Pe_t, that is, increasing $\dot{\gamma}$. The factor 6.2 in Equation 10.7 stems partly from hydrodynamic interactions and partly from Brownian motion (thermal motion). Flow distorts the structure from its equilibrium form under rest, and Brownian motion acts as an entropic restoring force driving the structure back toward equilibrium. For high Pe_t (high shear rates), the equilibrium microstructure will be significantly affected by the flow, with the Brownian contribution becoming less important. This gives rise to shear thinning behavior, more strongly as the volume fraction is higher. This implies that k_2 and higher terms are smaller at high Pe_t than for low Pe_t. For uniaxial extensional flow, k_2 was found to be equal to 7.6.

For volume fractions higher than about 0.1–0.15, Equations 10.7 and 10.8 can no longer be applied. For these concentrated dispersions, a series of empirical and semiempirical equations are available that often contain one or more fitting parameters (Section 10.2). The effect of colloidal interaction forces between particles on the viscosity of dispersions will be discussed in Section 10.3. We will refrain from discussing the effect of the presence of an electric double layer or an adsorbed macromolecular layer on the viscosity of very dilute dispersions since these phenomena are not important for food systems.

10.1.1 Quantities characterizing viscosity increment due to particles added

For many practical purposes, it is important to know the extent to which the viscosity of a given solvent η_c increases if colloidal or other particles are added. As long as the rheological behavior is of the Newtonian type (laminar flow, spherical particles, no bonds between the particles, etc.), the viscosity is the sole quantity required to describe the system rheologically. However, knowing the viscosity of dispersions is not enough to quantify the contribution of added particles or molecules to it. **Table 10.1** gives an overview of the quantities used to characterize the viscosity of Newtonian liquids and the contribution of dispersed particles or molecules to it.

In **Table 10.1**, both the reduced viscosity and intrinsic viscosity are given on the basis of the volume fraction φ as well as the mass concentration of particles *c* present. These quantities are related to each other via the particle density. This relation reads for intrinsic viscosity as

$$[\eta]_\varphi = (c/\varphi)[\eta]_c = \rho_p[\eta]_c, \tag{10.9}$$

Table 10.1 Glossary of Definitions and Symbols for Viscosity of Newtonian Liquids

Symbol	Name	SI Unit	Notes
η	(Dynamic) viscosity	Pa s = N m^{-2} s	
$\eta_r = \eta/\eta_c$	Relative viscosity or viscosity ratio	–	
$\eta_{sp} = \dfrac{\eta - \eta_c}{\eta_c} = \eta_r - 1$	Specific viscosity or relative viscosity increment	–	
$\eta_{red} = \dfrac{\eta_{sp}}{c}$	Reduced viscosity, or viscosity number on concentration basis	m^3 kg^{-1}	
$\eta_{red} = \dfrac{\eta_{sp}}{\varphi}$	Reduced viscosity on volume fraction basis	–	
$[\eta]_c \equiv \dfrac{1}{\eta_c}\left(\dfrac{d\eta}{dc}\right) = \lim\limits_{c\to 0} \eta_{red}$	Intrinsic viscosity on concentration basis	m^3 kg^{-1}	a
$[\eta]_\varphi \equiv \dfrac{1}{\eta_c}\left(\dfrac{d\eta}{d\varphi}\right) = \lim\limits_{\varphi\to 0} \eta_{red}$	Intrinsic viscosity on volume fraction basis	–	a

Note: c, concentration dispersed particles (kg m^{-3}); φ, volume fraction dispersed particles.

[a] Usually, an intrinsic viscosity is given without the subscript c or φ since it is clear from the context on which basis it has been calculated. We will follow this custom except where it may give rise to confusion.

where ρ_p is the particle density (kg m^{-3}). The quantity $[\eta]_c$ (dimension m^3 kg^{-1}) is the rule in polymer science, whereas the quantity $[\eta]_\varphi$ (dimensionless) is normally used in particle science. The same applies for reduced viscosity.

To illustrate the meaning of the various characteristics, based on the Einstein equation (10.2) the relative viscosity is equal to $1 + 2.5\varphi$, $\eta_{sp} = 2.5\varphi$, $\eta_{red} = 2.5$, and $[\eta] = 2.5$ for $\varphi \to 0$. For this case, the distinction between reduced viscosity and intrinsic viscosity is irrelevant, because the Einstein equation only holds for φ close to zero. However, for extensions of the Einstein equation, the difference between both characteristics is very relevant.

The specific viscosity gives the increases in relative viscosity as a result of the dispersed particles. The reduced viscosity accounts for this increase, expressed per unit weight or volume fraction of the dispersed phase. In practice, this characteristic depends on the concentration/volume fraction dispersed particles at which it has been determined except for very low concentrations where Equation 10.2 holds. For a dispersion in which viscosity as a function of φ is described by Equation 10.5, $\eta_{red} = 2.5 + k_2\varphi$. This concentration dependence is eliminated by extrapolation to $\varphi \to 0$, giving a parameter that reflects a characteristic property of the dispersed particles. Therefore, the *intrinsic viscosity* $[\eta]$ is a very useful characteristic of dispersions. It is a measure of the "viscometric particle volume," which can be obtained if the particle number density, $\rho_N = N_p/V$, is known, where N_p is the number of particles. This is especially of importance for the characterization of the thickening action of macromolecules and of nonspherical particles and their aggregates.

10.2 Concentrated dispersions

Many empirical and semiempirical equations have been developed for describing viscosity as a function of φ beyond the range for which Equations 10.7 and 10.8 hold. At higher volume fractions, the viscosity increases much more steeply with increasing φ than described by these equations. The reason is the increase in simultaneous hydrodynamic interaction between various particles leading to a strong distortion of the flow field around the particles and with that, to extensive energy dissipation. This extensive energy dissipation results in a higher apparent viscosity.

For a concentrated dispersion of hard particles, viscosity will go to infinity when the particles form a close-packed three-dimensional network throughout the suspension that inhibits flow. It is therefore useful to consider the effect of particle concentration on the viscosity of a suspension in relation to the *maximum volume (packing) fraction*, φ_{max}. The parameter φ_{max} depends, among other things, on the arrangement of the particles on packing and on their size distribution and shape. For example, for homodisperse spheres it is 0.637 for random close-packing, but 0.52 and 0.74 for simple cubic and hexagonal close-packing, respectively. For suspensions of particles with a broad size distribution, φ_{max} can be substantially higher. The reason is that the smaller particles can fill the gaps between the bigger ones. Therefore, for concentrated systems, the ratio φ/φ_{max} is, next to φ, a relevant concentration characteristic to be considered in relation to dispersion viscosity.

Some equations in which this ratio occurs are those according to Eilers (1941)

$$\eta = \eta_c\left(1 + \frac{1.25\varphi}{1-(\varphi/\varphi_{max})}\right)^2, \tag{10.10}$$

Mooney (1951)

$$\eta = \eta_c \exp\left(\frac{2.5\varphi}{1-k\varphi}\right), \tag{10.11}$$

and Krieger and Dougherty (1959)

$$\eta = \eta_c\left(1 - \frac{\varphi}{\varphi_{max}}\right)^{-[\eta]_\varphi \varphi_{max}} \tag{10.12}$$

In Equation 10.11, η becomes infinite for $(1/k) = \varphi$, so $1/k$ can be identified with φ_{max}. The common characteristics of Equations 10.10 through 10.12 are that they reduce to the Einstein equation for $\varphi \to 0$ and that η goes to infinity for $\varphi \to \varphi_{max}$. Equations relating η to φ that do not have these characteristics should not be used for describing the concentration dependence of the viscosity of dispersions of hard particles.

The semiempirical Equation 10.12 can be derived from the Einstein equation (Equation 10.2) by writing the latter in a differential form

$$d\eta = \eta(5/2)d\varphi, \tag{10.13}$$

where $d\eta$ is the increment in viscosity on the addition of a small increment of the volume fraction dispersed particles $d\varphi$. When a particle is added to a relatively concentrated dispersion, it requires more space than $d\varphi$ because space will always remain in between spherical or ellipsoidal-like particles on packing. In view of this packing problem, $d\varphi$ should be replaced by $d\varphi/(1 - k\varphi)$. Integration then yields

$$\eta = \eta_c(1 - k\varphi)^{-5/(2k)} \tag{10.14}$$

The parameter k can be identified with $1/\varphi_{max}$. In addition, Equation 10.14 can be generalized by replacing the factor 5/2 by the intrinsic viscosity $[\eta]_\varphi$, which is dimensionless. This, in principle, provides a situation that allows particles of any shape to be accounted for (be aware of $\dot{\gamma}$ effects), and results in the so-called Krieger–Dougherty equation (Equation 10.12).

As mentioned above, the value of φ_{max} is strongly dependent on particle size distribution. It increases with increasing polydispersity. An increase in ϕ_{max} will lead to a reduction in viscosity. The effect of polydispersity is illustrated in **Figure 10.4**, in which the viscosity of dispersions is plotted as a function of the volume fraction of dispersed hard spherical particles for various maximum volume fractions. The effect is larger for larger volume fractions.

As indicated above, η goes to infinity for systems containing hard particles when φ approaches φ_{max}. This will no longer be the case if the particles

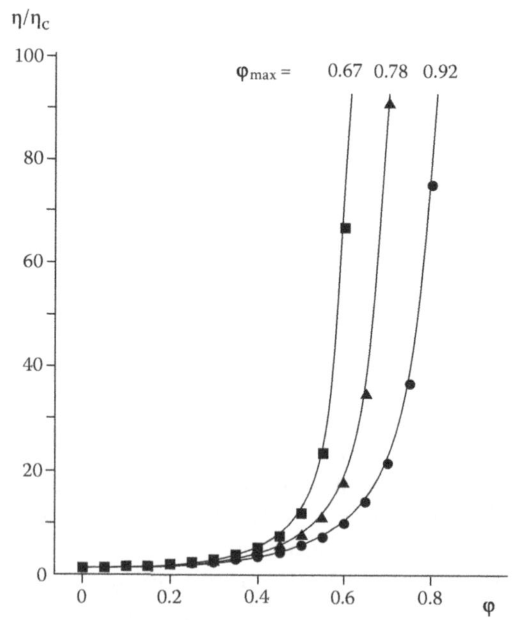

Figure 10.4 *Relative viscosity η/η_c of dispersions as a function of volume fraction dispersed hard spherical particles for various maximum volume fractions (indicated) (calculated data according to Equation 10.10).*

can be deformed by the stresses involved as, for example, is the case for tomato ketchup, apple sauce, and most mayonnaises. At low forces, these systems behave as solid-like materials with elastic properties. They exhibit a yield stress, and at higher stresses they start to flow. The yield stress is then a parameter characterizing the deformability (stiffness) of the particles. Similarly, for dispersions containing volume fractions of particles close to φ_{max}, the viscosity is largely determined by the stiffness of the particles, as has been illustrated, for example, for starch particles (**Figure 10.5**). In this figure, the concentration *c* times swelling volume *q* is equal to the volume fraction φ for $\varphi < 1$. For $cq > 1$, the starch granules will not be swollen maximally. The system will behave as a solid, but will yield above a certain stress. As can be seen already at volume fractions above ~0.5, the apparent viscosity depends both on φ and particle stiffness.

Just as for semidilute dispersions, shear thinning may occur with increasing shear rate. At low shear rates and in the absence of gravity, the distribution

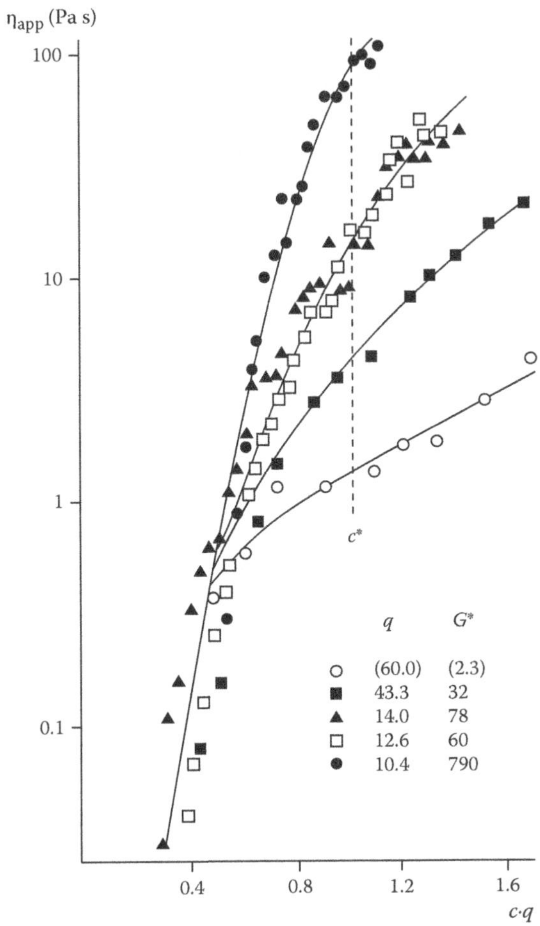

***Figure 10.5** Apparent viscosity of swollen native and modified starch particles as a function of concentration c (g/ml) times the swelling volume q under dilute conditions. q (ml/g) is the swelling volume at 95°C in excess of water (indicated), c* the concentration at which the fully swollen particles just fill up the available space and G* (N m^{-2}) the stiffness of the particles (indicated). ○ Native potato starch; ■ lightly cross-linked potato starch; □ native wheat starch; ▲ native maize starch; • highly cross-linked waxy maize starch. For further explanation, see text. (Redrawn from Steeneken, P.A.M.,* Carbohydr. Polym., *11, 23–42, 1989. With permission.)*

of nonaggregated particles in a dispersion of particles is random, owing to Brownian motion. This involves a large resistance to flow. With increasing shear rate (Pe_t), the equilibrium microstructure is affected by the flow, even more strongly than for semidilute suspensions. The flow induces an orientation of the particles with respect to each other, which allows them to move alongside each other more freely. At still higher shear rates, particles form layers separated by layers of the continuous phase. These layers act as a kind of slip layer between the solid-like particle layers. This arrangement of the particles in a layer-like structure proceeds faster and more extensively when φ is higher, and when the particles are smoother and more homodisperse (having the same size). This transition of the particles from a random distribution to an ordered layer structure gives rise to shear thinning behavior, which is more pronounced when the volume fraction of particles is higher. This effect has been experimentally observed clearly for homodispersed nonfood systems, but it will also play a role during, for example, shearing of concentrated emulsions. Because the formation of an ordered structure takes time, this phenomenon also leads to thixotropy. When shearing is stopped, the flow-induced structure will gradually disappear, owing to Brownian motion.

The formation of ordered structures might also lead to enhanced wall slip. Wall slip is a general phenomenon for suspensions because particles are depleted from the wall. For a suspension with $\varphi \leq 0.25$, wall slip can generally be neglected if the diameter/gap width ratio is more than 10 times the particle diameter (**Figure 8.11**). For higher φ, wall slip may become significant even for larger diameter/gap width ratios. In the extreme case, slip will result in plug flow.

At still higher shear rates than those involved in the shear rate thinning region (nearly), all concentrated dispersions of nonaggregating particles show shear rate thickening. A discussion of its cause, as so far known, is considered to be outside the scope of this book, because its practical relevance for food systems is limited.

A last aspect that is often neglected is the question, "What is the relevant volume fraction of particles in dispersions containing various kinds of particles of different sizes?" A clear example of such a product is cream, which consists of concentrated milk fat droplets (size 100–10^4 nm) in a dispersion of protein aggregates (so-called casein micelles, size range 10–300 nm), protein molecules (so-called serum proteins, average size ~7 nm), and sugar molecules (4%–5% lactose, size ~0.7 nm) in a roughly 1% solution of small molecules, mainly salts, with a viscosity 1.02 times that of water. As indicated above, one of the assumptions made during the derivation of the Einstein equation is that the dispersed particles should be large compared to those of the solvent (continuous phase). All extensions of the Einstein equation and those for concentrated dispersions (Equations 10.10 through 10.12) are implicitly based on this assumption. The lower limit is about a factor of 10 in volume, implying that the mentioned equations will hold for sugar molecules in a salt solution.

Below, we will illustrate the relevance of the assumption on the size of the dispersed particles compared to those of the solvent mole by discussing the relationship between viscosity and fat content of milk and cream. For these products, the assumption on the minimum size of the dispersed particles is not fulfilled for the fat droplets in a dispersion of the casein micelles, and for these in a dispersion of the protein molecules. For temperatures between 40°C and 80°C, cream was found to behave as a Newtonian liquid for volume fractions of fat $\varphi_f \leq 0.4$ at $\dot{\gamma} \geq 7\ s^{-1}$. Data measured for the viscosity ratio of creams over that of creams without fat (skimmed milk) as a function of φ_f could not be fitted by Equation 10.10 for any reasonable value of φ_{max} (**Figure 10.6**). However, by doing so, the contribution to the hydrodynamic interaction of the particles other than the fat globules is wrongly neglected. The volume faction φ_p of those particles can be estimated from the measured relative viscosity of skimmed milk ($\varphi_f = 0$) and was found to be ~0.16, which is in agreement with φ_p calculated on the basis on the voluminosity of the casein micelles, the other proteins, and the lactose and their concentrations. So, for φ to be filled in the relevant equation for calculating the relative viscosity, one should use $\varphi = \varphi_f + 0.16$, and for η_c the viscosity of the salt solution. Then, the measured results were found to be in excellent agreement with Equation 10.10, taking $\varphi_{max} = 0.88$ (**Figure 10.6**). A value of 0.88 is reasonable in view of the very broad size distribution of the various particles together. That curve *b* starts at η/η_c larger than 1 is attributable to the contribution of protein and lactose to η_r at $\varphi_f = 0$. An alternative to the approach followed here could consist of taking the lactose solution as a continuous phase in which a mixture of protein molecules, protein aggregates, and fat globules are dispersed. It leads to the same conclusion.

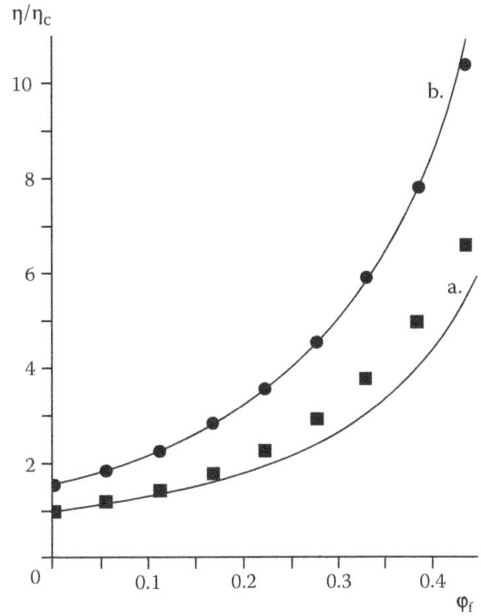

***Figure 10.6** Relative viscosity of milk and cream at 40°C as a function of its fat content. (■) Experimental results; (•) the same after correction for the presence of other dispersed particles in the system; full lines after Equation 10.10: curve a with $\phi_{max} = 0.67$ and $\phi = \phi_f$, curve b with $\varphi_{max} = 0.88$ and $\phi = \phi_f + 0.16$. (Redrawn after van Vliet, T., Walstra, P.,* J. Texture Stud.*, 11, 65–68, 1980.)*

10.2.1 Viscosity of concentrated dispersions of fruit cells

As an example of the difficulties that one may encounter in determining the relation between the structure of food dispersions and their apparent viscosity, a short overview will be given of the main factors determining the rheology of products such as tomato ketchup and apple sauce. Products such as these are prepared by a crushing and heating process followed by pulping/finishing, during which the mass is pressed through one or more sieves with pore sizes ranging from 0.5 to 1.5 mm. For tomato, the resulting juice is concentrated to a paste, whereas for apples the end product is apple sauce.

To a first approximation, both tomato paste and apple sauce may be considered as a concentrated suspension of tomato cells and apple cells, respectively, in a rather dilute macromolecular solution. Visually, by pouring these products on a plate it can be observed that they have a yield stress and flow above a certain stress. In the industrial practice of tomato ketchup preparation, the paste is often homogenized in a high-pressure homogenizer. Because it is difficult to determine the volume fraction of dispersed particles in these type of products, their concentration is often expressed as the % water insoluble (or unextractable) solid (WIS or WUS), which is considered to be a good measure of the amount of cell walls present and with that of the number of cells or clusters of cells. However, it is likely that this relation depends on the sample and on the type of fruit.

On determination of the apparent viscosity, η_{app}, of concentrated dispersions of fruit cells, a first pitfall is the occurrence of slip and a too low particle diameter gap width ratio (<10) (**Figure 10.7**). The differences in the σ versus $\dot{\gamma}$ curves in **Figure 10.7** are attributable to these two phenomena.

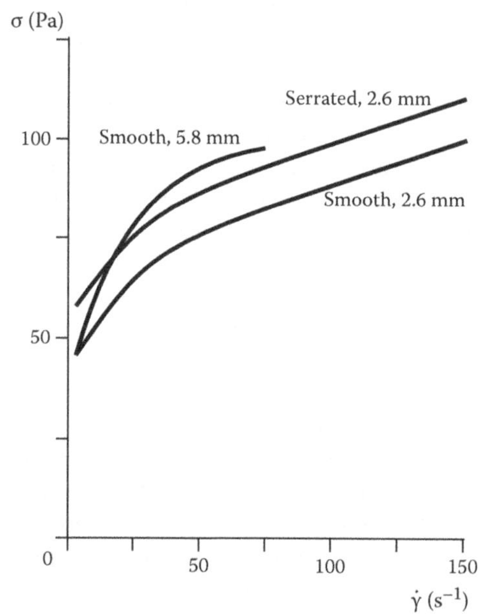

Figure 10.7 Shear stress σ as function of shear rate $\dot{\gamma}$ for one batch of apple sauce using various concentric cylinder geometries. Gap width and smoothness cylinder surface indicated. Volume surface averaged particle size diameter ~0.4 mm.

For higher concentrations, fruit cell dispersions exhibit solid-like behavior under low stresses. For WIS values up to about 1.5% (roughly the concentration in commercial apple sauce), yield stresses determined by using concentric cylinder systems and lubricated squeezing flow were about the same. However, at larger concentrations, yield stresses determined by the squeezing flow method were much higher (e.g., at 2% WIS by a factor 5). It is likely that this difference is primarily due to extensive slip in the concentric cylinder setup.

A plot of η_{app} versus concentration shows at low WIS% a relation, as expected for a particle dispersion (**Figure 10.8**). However, starting from a concentration just below that of the original sauce, the increase in η_{app} with WIS% became much less. In this region, it is primarily the deformability of the dispersed particles that determines the relation. This deformability will depend, among others, on the fruit particles involved and their history. For instance, concentrating a tomato suspension followed by diluting it back to its original concentration results in a suspension with a lower η_{app} (**Figure 10.9**). It is likely that the concentration process leads to damaging and microscopic fracture of the rigid-like, cellulosic microfibrillar network in the cell wall, resulting in cell walls that are more flexible, thereby causing a lower η_{app}. Its extent depends on the conditions during the concentration process. Moreover, it is less for homogenized tomato paste.

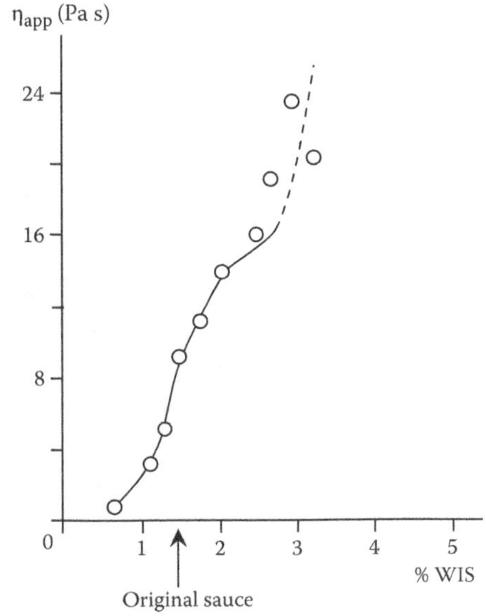

Figure 10.8 *Apparent viscosity η_{app} of apple sauce as a function of concentration water insoluble solids (% WIS). Shear rate $\dot{\gamma}$ 7.9 s^{-1}.*

The structure of a tomato paste can be affected substantially by homogenizing it (**Figure 10.10**). This treatment results in a paste with a higher η_{app} and yield stress. A similar effect is observed on homogenization of strawberry sauce, whereas for apple sauce (**Table 10.2**) and orange juice concentrate it results in a clear decrease in η_{app}. For all products, the surface-weighted mean diameter decreases upon homogenization. However, microscopy shows clear differences in the breakdown pattern. Particles in nonhomogenized apple sauce are

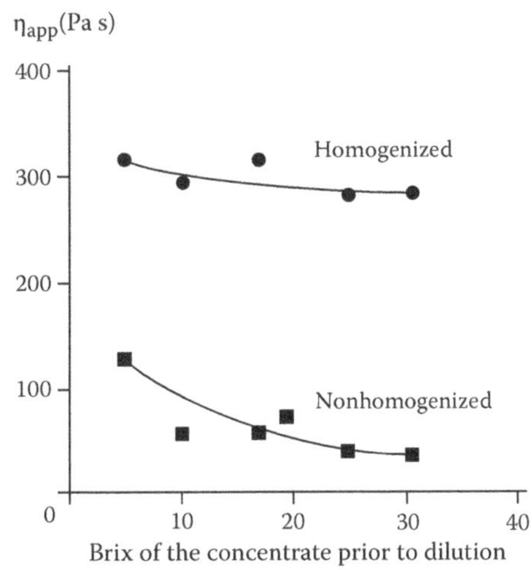

Figure 10.9 *Apparent viscosity η_{app} ($\dot{\gamma}$ 0.025 s^{-1}) as a function of Brix of tomato paste concentrates before dilution to a suspension with 0.65% WIS; nonhomogenized (■) and homogenized (•). (Reprinted from den Ouden, F.W.C., van Vliet, T.,* J. Texture Stud., *33, 91–104, 2002. With permission.)*

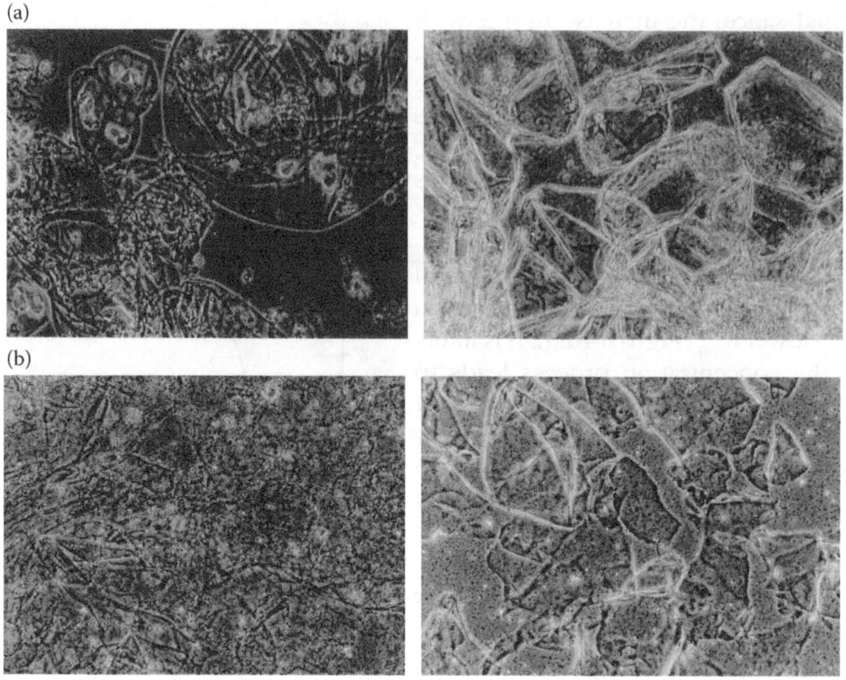

Figure 10.10 Phase contrast microscopic pictures of tomato suspensions (left panels) and of apple sauce (right panels). (a) Nonhomogenized; (b) homogenized (20 MPa). (Courtesy of F.W.C. den Ouden.)

often clusters of several cells with a size up to 700 μm. Upon homogenization at 20 MPa, the clusters fall apart into separated cells, and most cells become damaged and broken into regular pieces (**Figure 10b**, right panel). Polarized microscopy showed a significant brightening of the cell walls, revealing that (semi-)crystalline structures, probably mainly cellulose, are present in the relatively thick cell walls. For the orange juice concentrate, the number of long,

Table 10.2 Apparent Viscosity η_{app} ($\dot{\gamma}$ 10 s^{-1}) of Various Dispersions of Fruit Cells

	Apparent Viscosity (Pa s)	
Dispersion	**Nonhomogenized**	**Homogenized**
Tomato paste	2.8	5.5
Strawberry sauce	3.2	5.0
Apple sauce	11	5.5

Source: Based on data from Den Ouden, F.W.C., Physico-chemical stability of tomato products, Ph.D. thesis, Wageningen University, the Netherlands, 1995.

filamentous fibers decreased. In tomato and strawberry cells, no large (semi-) crystalline structures can be seen in the cell walls. After homogenization, the cells are completely ruptured and have fallen apart into long shaped filamentous cell fragments, partly with a fibrous brush-like appearance (**Figure 10.10b**, left panel). For these systems, products are formed with a very different structure resulting in a higher η_{app}.

10.3 Effects of colloidal interaction forces between particles

The discussion in Sections 10.1 and 10.2 was limited to dispersions of particles in which effects of colloidal and various short-range interaction forces could be neglected in comparison with those resulting from Brownian motion and mechanical forces due to imposed flow. For many liquid food systems, this is not a justified assumption. Various types of interaction forces between dispersed particles can be distinguished. A first rough distinction in interaction forces is between those acting over a short (1–2 nm) and over a long (>2 nm) distance. Short-distance interaction forces include covalent bonds, hydrogen bonds, hydrophobic bonds, and hard core or Born repulsion. These interactions play a role only between particles that are in close contact and determine, to a large extent, the force required to separate such particles. Above hard core, repulsion was implicitly taken into account by the assumption that the particles are mutually impermeable. Long-distance or so-called colloidal interaction forces include van der Waals forces, electrostatic interaction, and steric interaction forces. The latter two may lead to a repulsive as well as to an attractive interaction force between dispersed particles, whereas van der Waals forces always result in attractive interaction between the particles.

In Section 10.3.1, we will give a short overview of the main interaction forces acting between particles of colloidal (mesoscopic) size (0.1–100 μm) for those not familiar with these types of interaction forces. Their effects on the viscosity of dispersions of particles will be discussed in Sections 10.3.2 and 10.3.3.

10.3.1 Colloidal interaction forces

10.3.1.1 van der Waals attraction

For two spherical particles with a diameter d, the attractive van der Waals force, f_{vdW}, is given by

$$f_{vdW} = \frac{Ad}{24h_0^2} \tag{10.15}$$

where A is the Hamaker constant (J), a material property of the combination dispersed particle and solvent (for food: water or oil), and h_0 is the shortest distance between the two particles. The attractive force decreases with the distance squared and increases proportional to the particle diameter.

10.3.1.2 Electrostatic interaction

Electrostatic interaction is the result of charged groups on the surface of the particles. As a consequence of the presence of an excessive positive or negative charge at the particle liquid interface, it has an electrostatic surface potential ψ_0, the value of which depends on the number of charges and their valency. The interaction will be repulsive if the surface potential of both particles has the same sign and will be attractive when one particle is predominantly positively charged and the other is predominantly negatively charged. Electrostatic interaction is screened by the presence of the so-called counterions (i.e., ions with a charge sign opposite to that at the interface) in the surrounding aqueous solution. On the other hand, coions (i.e., ions with the same charge sign as at the interface) are excluded. In this way, a so-called *electric double layer* is formed. The charge at the surface is compensated for by the excess counterions and the depletion of coions. At a distance larger than a few nanometers, the surface potential ψ_0 decreases with the distance h from the surface as $\exp(-\kappa h)$, where κ is a screening parameter (m^{-1}). It is given by

$$\kappa = \sqrt{\frac{e^2 \sum_i m_i z_i^2}{\varepsilon_0 \varepsilon_r R_g T}}, \tag{10.16}$$

where e represents the charge of an electron, m_i is the molarity of the ions considered and z_i is their valency, ε_0 is the dielectric permittivity of vacuum, ε_r is the relative dielectric permittivity (dielectric constant) of the solvent, and R_g is the gas constant. The thickness of the double layer is usually characterized by $1/\kappa$.

When two particles approach each other, the double layers start to overlap, leading to electrostatic interaction. Its strength increases with decreasing distance between the particles and with their ψ_0, whereas it decreases with increasing concentration of the counterions and their charge (mono- or divalent counterions). For spherical particles of equal size and mutual distance $h \ll d$, their mutual electrostatic interaction force f is proportional to

$$f_{el} \propto \varepsilon_0 \varepsilon_r d \psi_0^2 e^{-\kappa h} \tag{10.17}$$

Equation 10.17 implies that electrostatic interaction between two particles of similar size increases proportionally to the particle diameter d and decreases with the distance h according to $e^{\kappa h}$.

10.3.1.3 Steric interaction

Steric interaction results from the interaction between adsorbed layers of macromolecules. The distances over which it acts depend on the thickness of these layers. Usually, the strength of the interaction increases strongly with decreasing distance. The interaction will be repulsive when the molecules repel each other and attractive when they like to mix or aggregate.

In foods, the adsorbed layer of macromolecules often consist of protein molecules, which contain charge groups. It causes that the interaction between the particles involves both steric and electrostatic interaction as well as the always-present van der Waals attraction. The balance between the electrostatic and steric interaction will depend on the protein and on conditions (solvent composition, pH, temperature, etc.) in an intricate way and, thereby, the interaction between the particles. In this section and the next, we will not discuss the effects of the various types of interaction forces separately, but will focus on the effect of repulsive or attractive interaction forces and their combinations for various situations.

10.3.1.4 Hydrodynamic force

In the absence of flow, the extent of aggregation of particles in a suspension will be determined completely by the ratio of the Brownian forces over the colloidal forces. *Hydrodynamic forces* will also start to play a role when a flow field is applied. A measure of the hydrodynamic shear force can be obtained by considering the frictional force of a spherical particle moving through a fluid with a velocity v, which for a dilute suspension is equal to $6\pi\eta R_p v$ (the so-called Stokes law, where R_p is the particle radius) (particle Reynolds number must be below ~1 (**Table 3.2**); otherwise, turbulence may develop in the wake of the particle, thereby increasing the frictional force). In a shear flow field, the particle velocity relative to the velocity of the surrounding fluid at particle radius distance is given by $v = \dot{\gamma} R_p$. This gives for a dilute suspension for hydrodynamic force, f_h

$$f_h = 6\pi\eta R_p^2 \dot{\gamma} \tag{10.18}$$

The hydrodynamic force increases linearly with shear rate and viscosity, and thus with shear stress, whereas colloidal interaction forces are roughly independent of these factors. The result is that the hydrodynamic force will be more important compared to colloidal forces at higher shear rates and for liquids with a higher viscosity.

10.3.2 Effect of colloidal forces on viscosity

The case where repulsion forces between the particles are dominant has the least consequences for the viscosity of dispersions. When they are subjected to shear flow, particles present in nearby trajectories (parallel streamlines) will be pushed toward each other because of their speed difference until hydrodynamic (thinning of the thin film between closely approaching particles) or colloidal repulsion prevents further approach. A transient, rotating doublet is formed (**Figure 10.11**). For a given shear rate, the minimum distance over which two particles will approach each other is determined by the balance between shear and, in this case, repulsive colloidal forces and the time that they are pressed together. An alternative way of saying this is that the balance between the energy (force × distance traveled in the

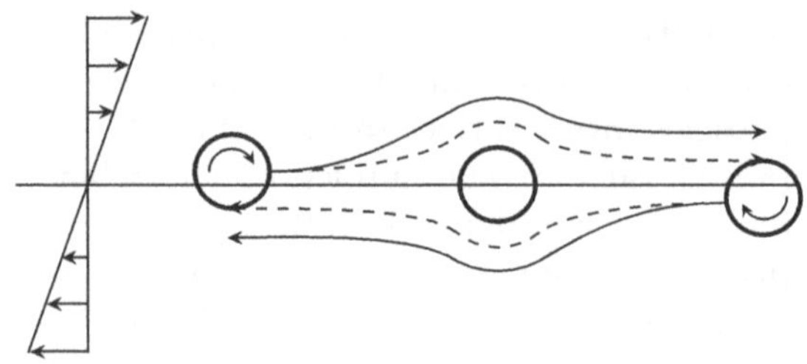

Figure 10.11 *Trajectories of two particles approaching each other in shear flow; dashed curve uncharged particles, full curves repulsive particles.*

direction of the second particle) provided by the hydrodynamic forces during the time that the particles are pressed against each other and the repulsive energy between the particles determines the minimum distance. Because of the presence of repulsive forces, the excluded volume of the particles during a collision is somewhat larger than that for uncharged particles. The particles stay at a greater distance from each other. As a result, during a collision, the trajectories of the particles that pass closely along each other will be altered (**Figure 10.11**). The paths of collision and recession become different from each other. This departure of the streamlines gives rise to extra energy dissipation observed macroscopically as a higher viscosity.

The effect of shear rate on the mutual approach between two particles is two-sided. Shear rate affects both the minimum distance of approach of the particles during an encounter and the time that they stay in each other's vicinity. Hydrodynamic forces increase with $\dot{\gamma}$, whereas colloidal forces do not. However, as discussed in Section.3.3, shear flow consists of a rotational part and an elongational part. So, if two particles in adjacent trajectories approach each other in a shear flow field due to the difference in velocity (**Figure 10.11**), they are not only pressed together, but also start to rotate around each other with a characteristic time roughly equal to $1/\dot{\gamma}$. As a result, the collision time is roughly proportional to $1/\dot{\gamma}$. These factors together give a shear rate-dependent contribution to the extent of approach of two particles when they meet and cause the dispersion to become (somewhat) shear thinning.

If the repulsion is of a steric nature, an additional effect on viscosity is caused by a mostly small increase in the effective volume fraction of the particles, φ_{eff}, due to the adsorbed macromolecular layer. Then, φ_{eff} becomes

$$\varphi_{eff} = \left(\frac{R_p + h}{R_p}\right)^3 \varphi, \tag{10.19}$$

where h is the thickness of the adsorbed macromolecular layer and R_p is the particle radius. At low φ, the effect of this factor on η is usually small except for small particles and/or very thick adsorbed layers. Still, in many cases, the effect is large enough that it can be determined and used for calculating the thickness of the adsorbed layer. At high φ, the effect on η of this small increase can be substantial in view of the steep increase of η with φ. For thick adsorbed macromolecular layers at φ close to φ_{max}, the dispersion may exhibit gel-like properties. Something similar to this will be the case for long-range electrostatic repulsion.

If the overall effect of the colloidal interaction is attractive, aggregates will be formed. In rest, the extent of aggregation will depend on the balance between Brownian motion and the net attractive forces and the possible presence of an activation energy for aggregation and/or disaggregation (**Figure 10.12**). Upon flow, hydrodynamic forces will affect the balance between Brownian motion and colloidal forces, and may push or pull particles over the activation barrier. It is outside the scope of this book to deal with all possible situations. Only some major trends will be discussed. The effect of flow on aggregate formation and breakup and its effect on the viscosity of dispersions will be discussed first without considering the structure of the aggregates formed explicitly. That effect will be considered in Section 10.3.3.

Regardless of whether doublet or aggregate formation is permanent or transient, its direct result is an increase in the effective volume fraction. Even in a doublet, part of the continuous phase is hydrodynamically trapped in the aggregate (**Figure 10.13**), leading to a substantial increase in φ. At the macroscopic scale this will lead to an increase in the viscosity, the extent of which depends on the particle volume fraction. This effect will, in general, be stronger for larger aggregates as these usually have a lower particle concentration

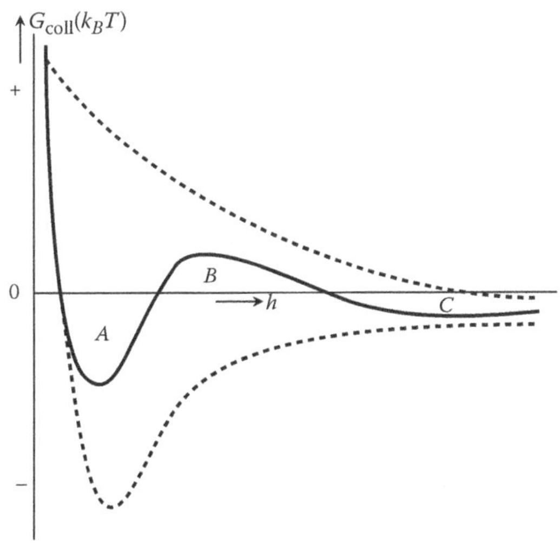

Figure 10.12 *Hypothetical example of colloidal interaction free energy G_{coll} between two particles as a function of interparticle distance h. In this example, interparticle interaction is attractive at long distances because of van der Waals attraction (region C). At intermediate distances, interaction is repulsive as a result of electrostatic repulsion (region B), and at short distances again it is attractive because of van der Waals forces (region A). At very short distances, interaction is repulsive because of steric repulsion and the so-called Born repulsion between the atoms. Upper dash curve illustrates dominant electrostatic repulsion, and lower dash curve denotes the case when only van der Waals attraction and short distance repulsion play a role. (Redrawn from Walstra, P.,* Physical Chemistry of Foods, *Marcel Dekker, New York, 2003.)*

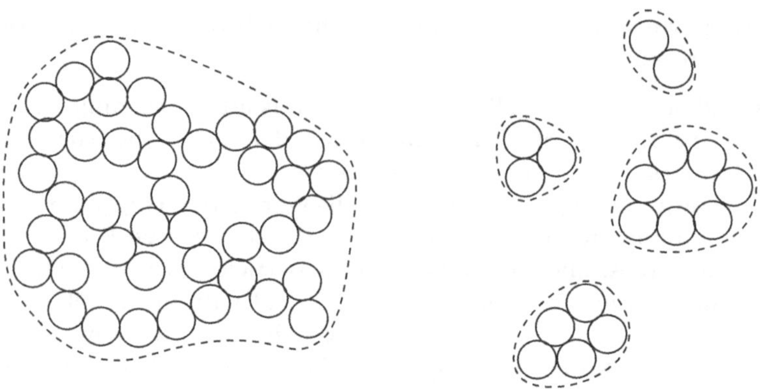

Figure 10.13 *Schematic indication of hydrodynamic volume (dashed lines) of various aggregates of spherical particles.*

per volume, implying a stronger increase in φ. A second effect mentioned in the literature is the breaking of bonds between the colloidal particles, which also expends energy and therefore leads to a higher viscosity. However, the amount of energy involved in breaking of bonds between the particles is usually much lower than the extra energy dissipation as a result of the flow process. For a suspension of particles with a diameter of 1 μm and $\varphi = 0.25$, the total energy in the bonds is 0.15 J m^{-3} if one assumes that there are, on average, three bonds per particle with an energy content of 50 kT per bond (75 kT per particle). For $\dot{\gamma} = 10\ s^{-1}$ and $\eta_{app} = 0.1$ Pa s, the energy dissipation by flow $\eta_{app}\dot{\gamma}^2 = 10\ J\ m^{-3}\ s^{-1}$, which is much higher than the energy dissipation caused by bond breaking. Moreover, during flow, only a fraction of the bonds will break, resulting in an even larger difference between the two contributions to the measured apparent viscosity. Therefore, the contribution of bond breaking to the viscosity increase can usually be neglected compared to effects due to the increase in φ on particle aggregation.

Because of the increase in hydrodynamic forces with increasing shear rate, particles aggregated by a relatively small attractive force will break up above a certain shear stress. The exact value of it will also depend on the aggregate structure, because, in general, more loose aggregates will break up at lower shear rates than compact ones (see below). Breakup of aggregates at higher shear rates will lead to a decrease in the effective volume fraction. The part of the continuous phase trapped in the aggregate becomes lower. The end result is a decrease of apparent viscosity with increasing $\dot{\gamma}$ (shear thinning behavior). This process will only stop when no further disaggregation of the aggregates with increasing $\dot{\gamma}$ takes place, resulting in apparently Newtonian behavior.

The prediction of aggregation and breakup is more complicated for suspensions in which the interparticle interaction is characterized by repulsion at intermediate distances and attraction at small distances between the particles (**Figure 10.12**). In addition to the size of the different forces, time also plays an important role, during which these forces act in a certain direction. As mentioned above, shear flow consists of a rotational part and an elongational part. So, two particles that approach each other are not only pressed together, but also start to rotate around each other with a characteristic time roughly equal to $1/\dot{\gamma}$. During rotation, the effect of the elongational flow component changes from pressing the particles together to pulling them apart (Section 3.3). If, during the time that the particles are pressed together, they do not approach each other so closely that they have passed the repulsion barrier, they will separate again.

In addition to the effect of the colloidal repulsion forces at intermediate distances, the rate of mutual approach of the particles will also be limited by the rate at which the continuous phase can flow out of the small gap between the particles. At small distances, the shear rate of the liquid from the gap between the particles will already be high for low approach rates, implying high energy dissipation. This phenomenon leads to a lower rate of aggregation than would be predicted based on the colloidal interaction forces alone. Increasing $\dot{\gamma}$ has two main effects on the rate of aggregation: (1) increasing the force with which particles are pressed together and (2) reducing the time they are pressed together. For most systems, the former effect, which is proportional to $\eta\dot{\gamma}$, is the most important. In the latter effect, aggregation as a result of a velocity gradient in the liquid (so-called orthokinetic aggregation) is proportional to $\dot{\gamma}^{0.8}$ rather than to $\dot{\gamma}$, as would be expected based on Equation 10.18.

Next, we will consider suspensions in which particle–particle interaction is characterized by a shallow, so-called secondary minimum at larger particle separations, a deep primary minimum at small separations, and a repulsion maximum in-between (**Figure 10.12**, full curve). In addition, we assume that, at rest, all particles have formed small aggregates because of the aggregation in the secondary minimum (interparticle distance C). For such systems, a plot of the apparent viscosity as a function of shear rate could exhibit a shape as shown in **Figure 10.14**.

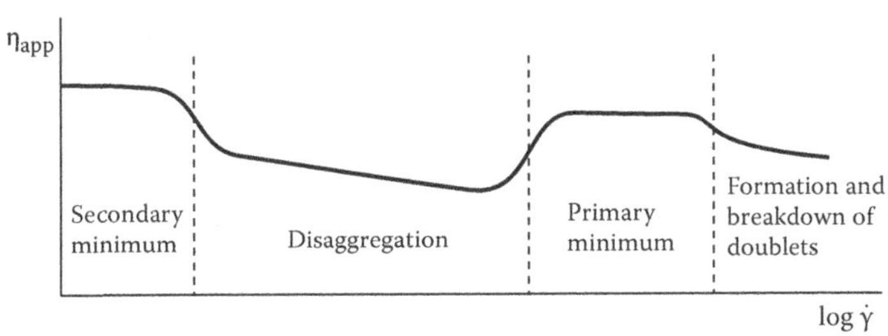

Figure 10.14 *Apparent viscosity, η_{app}, as a function of logarithm of shear rate, $\dot{\gamma}$, for a dispersion characterized by a shallow attractive force at larger distances (secondary minimum), a repulsive force at intermediate distances, and a large attractive force at short distances (primary minimum) (schematic picture).*

At low shear rates, apparent viscosity will be relatively high because of the increase in the effective volume fraction as a result of the aggregates formed. At higher shear rates, these aggregates will break up — the sooner the less deep shallow minimum at larger distances is. The apparent viscosity decreases to that of a dispersion containing nonaggregated repelling particles. With increasing shear rate, the apparent viscosity will continue to reduce somewhat (higher Pe_t) until hydrodynamic forces become large enough to cause aggregation in the primary minimum. This causes the apparent viscosity to increase again to an extent depending on the incorporation of the continuous phase in the aggregates formed. At very high shear rates these aggregates may be broken up again. Then there will be a continuous formation and breakup of aggregates. Mostly, shear rates required for disaggregation of particles aggregated in the primary minimum have to be much higher than those needed for causing aggregation, owing to the asymmetry in the force–distance curve. The free energy minimum at interparticle distance A is much deeper than the size of the maximum at distance B (**Figure 10.12**).

In practice, on shearing a dispersion at increasing shear rate, not all the transitions in $\eta_{app}(\dot{\gamma})$ shown in **Figure 10.14** will be observed. In the case of electrostatic repulsion and van der Waals attraction, the primary minimum is often so deep that no disaggregation is observed except at shear rates so high that flow will be turbulent. On the other hand, when part of the repulsion is of steric nature there may be no primary minimum. This may be the case for dispersions stabilized by (slightly or net uncharged) proteins that are aggregated at rest because of van der Waals and some electrostatic attraction. On shearing, these dispersions will exhibit a Newtonian flow behavior at very low and at high shear rates and shear thinning behavior in between (**Figure 10.15**).

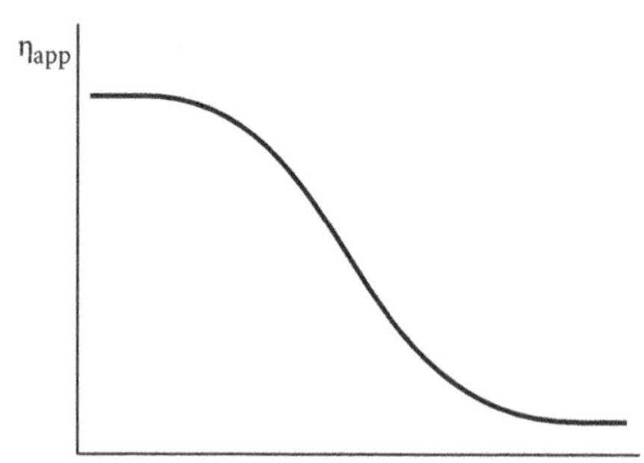

Figure 10.15 *Typical example of apparent viscosity η_{app} as a function of shear rate $\dot{\gamma}$ for a dispersion showing aggregate formation at low $\dot{\gamma}$ (schematic figure).*

For many systems at rest, not only aggregates will be formed, but even a *continuous particle network* may be formed, which provides the undisturbed system with gel-like properties. The chance that this will happen will be higher for systems containing larger volume fraction particles. These systems behave like solids at low shear stresses, yield at a certain stress, and exhibit shear thinning at still larger stresses. The yield stress is determined by the spatial structure of the network and the interaction between the particles (Section 13.3.1). A complicating factor is that the aggregate structure, and therewith the structure of the continuous particle network, depends on the aggregation process, and that may be affected by the flow regime during this process.

Disaggregation and aggregation processes usually do not proceed momentarily after a change in shear rate, but it takes some time before a new steady state is obtained with respect to these processes. The dispersion exhibits thixotropic behavior, and the measured η_{app} depends on the time that the dispersion is sheared at the set shear rate.

Modeling of the apparent viscosity as a function of shear rate, for liquid dispersions that do not show aggregation at high shear rates, requires at least three constants: the low shear rate viscosity plateau η_0, the high shear rate viscosity plateau η_∞, and a critical shear stress or shear rate. The last mentioned characteristic determines the region in which the transition from low shear rate to high shear rate behavior occurs. The resulting equations are often of the shape

$$\eta_{app} = \eta_\infty + \frac{\eta_0 - \eta_\infty}{1 + (\dot{\gamma}/\gamma_c)^m}, \tag{10.20}$$

where m is a constant and $\dot{\gamma}_c$ is a critical shear rate that proves to be a useful indicator of the onset $\dot{\gamma}$ for shear thinning. Parameter m is a measure of the dependence of the viscosity on shear rate in the shear thinning region. For a short overview of other equations, see, for example, Roberts et al. (2001). To include thixotropy, a memory function should also be included. When a continuous network is formed, the yielding and shear rate thinning part can be modeled by one of the equations discussed in Section 4.3.1.

10.3.3 Relation aggregate structure and shear thinning behavior

For aggregates with a specified structure, a theory has been developed for shear rate thinning behavior in which the aggregate structure is more explicitly accounted for. Computer simulation and experimental work have shown that aggregation of spherical particles at rest leads to aggregates with a *fractal structure*. The characteristics of fractal structures are that they are *self-similar* regardless of the length at which they are considered, and that the number of particles N_{agg} in the aggregate scales with the aggregate radius, R_{agg}, as

$$N_{agg} \propto \left(\frac{R_{agg}}{R_p}\right)^{D_f}, \tag{10.21}$$

where R_p is the particle radius and D_f is the so-called *fractal dimensionality*. It has a value smaller than 3, mostly between 1.8 and 2.4, depending on aggregation conditions such as the size of the repulsion barrier/activation energy for aggregation of the particles. In general, D_f will be larger if the activation energy is higher (slower aggregation). The fractal dimensionality is a measure of the change in packing density as a function of the size of the aggregate. For Equation 10.21 to hold, D_f must be independent of the aggregate size, that is, its structure must be self-similar. Otherwise stated, the packing must be scale invariant. The average radius of the primary particles R_p is taken as the lower cutoff length for the fractal geometry to apply. Real aggregates are, in general, not exactly self-similar, that is, they are not exactly scale invariant like deterministic fractals (**Figure 10.16a**), that is parts of the aggregates are not an exact copy of the aggregate itself (**Figure 10.16b**). However, it has been found by computer simulations and experimental work that, on average, Equation 10.21 describes the relation between the number of particles in an aggregate and its radius quite well.

The volume V of an aggregate will be proportional to R_{agg}^3 and thus the number of sites N_s in the aggregate that can be occupied by a particle will scale as

$$N_s \propto \left(\frac{R_{agg}}{R_p}\right)^3 \tag{10.22}$$

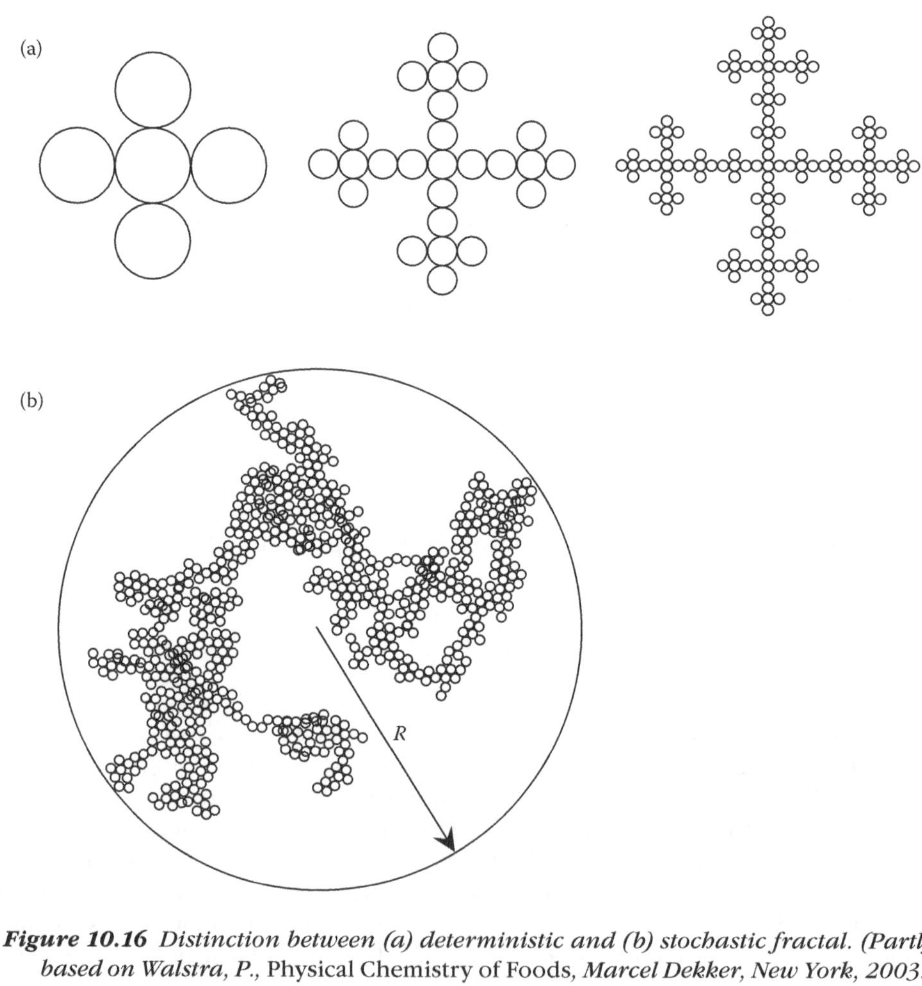

Figure 10.16 *Distinction between (a) deterministic and (b) stochastic fractal. (Partly based on Walstra, P.,* Physical Chemistry of Foods, *Marcel Dekker, New York, 2003.)*

The volume fraction of particles in a fractal aggregate, φ_{agg}, then scales as

$$\varphi_{agg} \equiv \frac{N_{agg}}{N_s} \propto \left(\frac{R_{agg}}{R_p}\right)^{D_f-3} \tag{10.23}$$

Because D_f is always smaller than 3 (in three-dimensional space), Equation 10.23 implies that φ_{agg} decreases with increasing values of R_{agg} or N_{agg}. The number of aggregates n_{agg} in the dispersion will be equal to

$$n_{agg} = \frac{N}{N_{agg}} = \frac{3\varphi}{4\pi R_p^3}\left(\frac{R_p}{R_{agg}}\right)^{D_f}, \tag{10.24}$$

where N is the number of primary particles in the system and φ denotes the volume fraction primary particles. Next, the hydrodynamic effective volume fraction of particles (aggregates) φ_{eff} in the dispersion can be derived given

$$\varphi_{eff} = n_{agg}\frac{4\pi}{3}R_{agg}^3 = \varphi\left(\frac{R_{agg}}{R_p}\right)^{3-D_f} \tag{10.25}$$

Equation 10.25 implies that φ_{eff} increases with ongoing aggregation and that its final value depends on R_{agg} and D_f. An overview of fractal aggregation is given by Meakin (1988).

For high interaction forces and sufficiently high φ, aggregation may go on and finally a gel may be formed (Section 13.3). In the case of weakly interacting particles, the system will exhibit a low shear rate Newtonian viscosity. In this section, we will consider the latter situation and, for strongly interacting particles, the situation where φ is not high enough for a gel to be formed or where the shear stress is larger than the yield stress. In these cases, breakdown of the aggregates as a result of hydrodynamic forces will lead to a lower aggregate radius and so to a lower φ_{eff} (Equation 10.25) and, as a result of that, to a lower apparent viscosity. The dispersion exhibits shear thinning behavior.

During shear, aggregates will not only break up, but their structure may also rearrange to a more compact one. This will result in a decrease in R_{agg} and an increase in D_f. In addition, small parts of the broken aggregates may aggregate again into new ones. Both experimental data and computer simulation data show that shear during aggregation leads to aggregates characterized by a higher D_f. The result of these processes is a decrease in φ_{eff} and, thereby, a lower η_{app}. These effects are usually irreversible. Because these processes proceed over time scales longer than the reaction time of standard viscometers, they lead to apparent thixotropic behavior. Apparent, because the measured viscosity will increase only to some extent when the shear rate is lowered, but it will not return to its original value before shearing (at the higher shear rate) has started.

A semiquantitative theory has been worked out for the shear rate thinning behavior of dispersions, containing aggregates with a fractal structure. During flow, a hydrodynamic stress σ_{hyd} is exerted on suspended aggregates of the size $\eta\dot{\gamma}$. The breakup stress σ_{fr} for fractal cluster scales as (Potanin et al., 1995)

$$\sigma_{fr} \propto \frac{2}{5\pi}\frac{f}{R_p^2}\varphi_{agg}^x, \quad (10.26)$$

where f is the bonding force between the primary particles and x is an exponent that depends on the structure of the fractal cluster. For brittle clusters, undisturbed after formation, x will be $4/(3 - D_f)$, whereby the numerical factor 4 depends on the precise way in which the fractal cluster deforms and fractures. The viscosity of the suspension can then be estimated by using, for example, Equation 10.12. For φ, the quantity φ_{eff} is taken. It can be calculated from the breakup criterion (Equation 10.26) in combination with Equation 10.25.

References

Batchelor, G.K. 1977. The effect of Brownian motion on the bulk stress in a suspension of spherical particles. *J. Fluid Mech.* 83: 97–117.

Den Ouden, F.W.C. 1995. Physico-chemical stability of tomato products. Ph.D. thesis, Wageningen University, the Netherlands, chap 4.

Den Ouden, F.W.C., and T. van Vliet. 2002. Effect of concentration on the rheology and serum separation of tomato suspensions. *J. Texture Stud.* 33: 91–104.

Einstein, A. 1906. Eine neue Bestimmung der Moleküle-Dimensionen. *Ann. Phys.* 19: 289–306.

Einstein, A. 1911. Berichtung zu meiner Arbeit: “Eine neue Bestimmung der Moleküle-Dimensionen”. *Ann. Phys.* 34: 591–592.

Eilers, H. 1941. Die Viskosität von Emulsionen hochviskoser Stoffe als Funktion der Konzentration. *Kolloid Z.* 97: 313–321.

Krieger, I.M., and T.J. Dougherty. 1959. A mechanisms for non-Newtonian flow in suspensions of rigid spheres. *Trans. Soc. Rheol.* 3: 137–152.

Meakin, P. 1988. Fractal aggregates. *Adv. Colloid Interface Sci.* 28: 249–331.

Mooney, M. 1951. The viscosity of a concentrated suspension of spherical particles. *J. Colloid Sci.* 6: 162–170.

Potanin, A.A., R. De Rooij, D. Van den Ende, and J. Mellema. 1995. Microrheological modeling of weakly aggregated dispersions. *J Chem. Phys.* 102: 5845–5853.

Roberts, G.P., H.A. Barnes, and P. Carew. 2001. Modelling the flow behaviour of very shear-thinning liquids. *Chem. Eng. Sci.* 56: 5617–5623.

Steeneken, P.A.M. 1989. Rheological properties of aqueous suspensions of swollen starch granules. *Carbohydr. Polym.* 11: 23–42.

van Vliet, T., and P. Walstra. 1980. Relationship between viscosity and fat content of milk and cream. *J. Texture Stud.* 11: 65–68.

van Vliet, T., and J. Lyklema. 2005. Rheology. In *Fundamentals of Interface and Colloid Science*, Vol. IV, *Particulate Colloids*, ed. J. Lyklema. Ch 6. Amsterdam: Academic Press.

Walstra, P. 2003. *Physical Chemistry of Foods*. New York: Marcel Dekker.

11

Viscosity of Macromolecular Solutions

Nearly all food products contain macromolecules (or polymers). The most important ones are proteins (e.g., gelatin, milk proteins, and soy proteins) and polysaccharides [e.g., starch (amylose and amylopectin), pectin, and xanthan gum]. Polysaccharides and—to a lesser extent—proteins are often used for increasing the viscosity of liquid or semiliquid food products whereby the choice depends also on factors other than their thickening efficiency, such as nutritional and price aspects. Their effect on the viscosity of a solution strongly depends on the concentration, size, and shape of the molecules. In its most simple form, a macromolecule is a linear chain of covalently bonded identical monomers (e.g., amylose). Much more frequent are chains of nonidentical monomers (e.g., proteins, alginate, and carrageenan). Others are heavily branched (e.g., xanthan and amylopectin).

In solution, macromolecules can assume a variety of shapes, depending on the nature of the molecule (e.g., stiffness and extent of branching and folding of the chain) and their interaction with the solvent (good or bad). Proteins, in particular, can be folded into a compact structure—the so-called globular proteins. Some shapes that macromolecules may adopt in solution are sketched in **Figure 11.1**. The differences in shape are reflected in the viscosity and other rheological properties. They are, to a large extent, responsible for the differences in viscosity increase per unit weight of the macromolecule added, that is, their thickening efficiency. The latter varies considerably between different macromolecules (**Figure 11.2**). In all cases, they are more efficient in raising the viscosity of a solution compared with nonaggregated spherical particles such as emulsion droplets. Globular proteins come closest to the line for emulsion droplets. They can be considered ellipsoidal shaped particles. Their effect on viscosity has been discussed briefly in Section 10.1 for dilute dispersions.

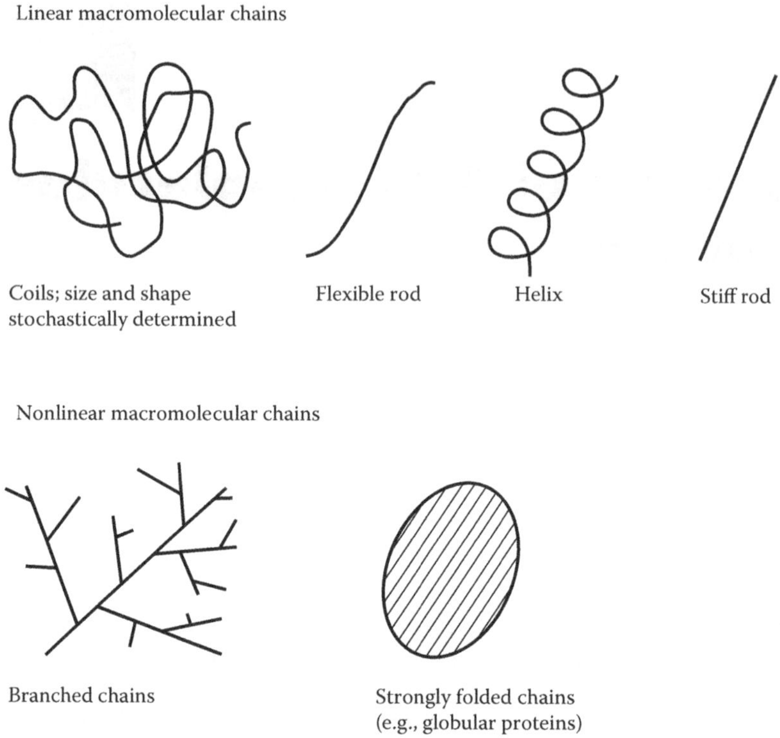

Figure 11.1 *Sketches of some structures that dissolved macromolecules can assume.*

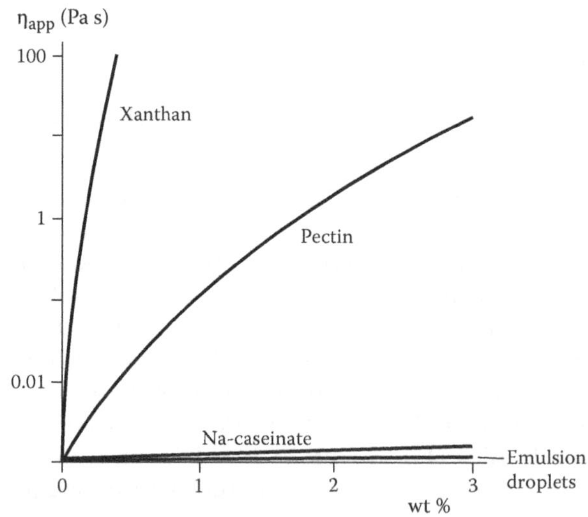

Figure 11.2 *Apparent viscosity η_{app} of dispersions of xanthan, pectin, sodium caseinate, and emulsion droplets as a function of their concentration in w/w%. For a dispersion containing 3 wt.% emulsion droplets, increase in viscosity is only about 8%–9%.*

In Section 11.1, solutions of macromolecules will be discussed for the very dilute and dilute range and in Section 11.2 for the semidilute and concentrated range. In both sections, first, flexible, uncharged macromolecules (the so-called *random coil polymers*) will be discussed, followed by a discussion of the effect of branching, chain stiffness, and charge. The rheology of charged globular proteins will not be discussed specifically, in view of their analogies with charged particles discussed in Section 10.3.2. Below, we will use "polymers" to denote linear, flexible, essentially uncharged macromolecules, and "macromolecules" for all differently shaped ones.

11.1 Very dilute and dilute macromolecular solutions

The rheological behavior of macromolecules with a random coil structure (polymers) is very different from that of dispersed particles, although there are also several analogies. The main difference stems from the fact that macromolecular coils are not hard, impermeable entities, but allow for fluid flow through them to an extent that is determined by the molar mass M and the quality of the solvent.

The simplest model of a *random coil polymer* is a linear chain of n segments with length l, where each segment is free to adopt any orientation with respect to its neighbors, and where all orientations have an equal probability. The total length of the molecule is nl. The conformation of the molecule as a whole can then be described by a random walk through space; all steps have the same length l, but they can be in any direction. The end-to-end distance of the chain r (**Figure 11.3**) is, on average, zero. Theory shows that, for large n, the *root-mean-square end-to-end distance* r_m of a chain, consisting of n freely jointed segments of length l, is given by

$$r_m \equiv \langle r^2 \rangle^{0.5} = l\sqrt{n} \qquad (11.1)$$

The quantity r_m can be considered as a mean coil diameter. Another measure of the coil diameter is the *radius of gyration* r_g, which is the root-mean-square distance of the segments from the center of mass. For the same conditions as those in Equation 11.1, r_g is given by

$$r_g = \frac{r_m}{\sqrt{6}} = l\sqrt{\frac{n}{6}} \qquad (11.2)$$

For real polymers, the chain will be less flexible because the bonds between the segments cannot assume arbitrary directions. The bond angles between the segments are fixed. In addition, rotation about the bonds is not entirely free, for example, because of the bulky side

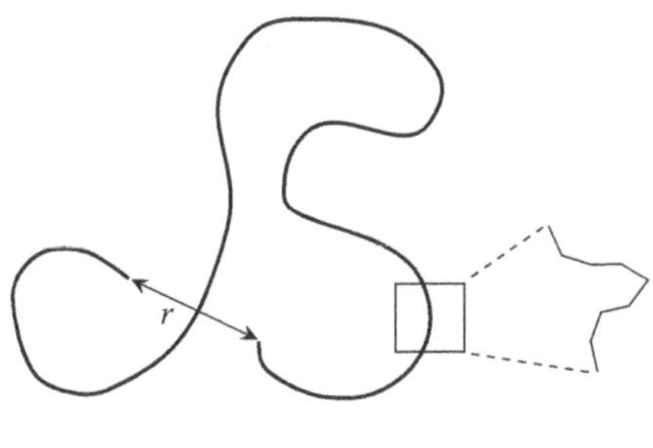

Figure 11.3 *Conformation of a long macromolecule. Indicated is the end-to-end distance r.*

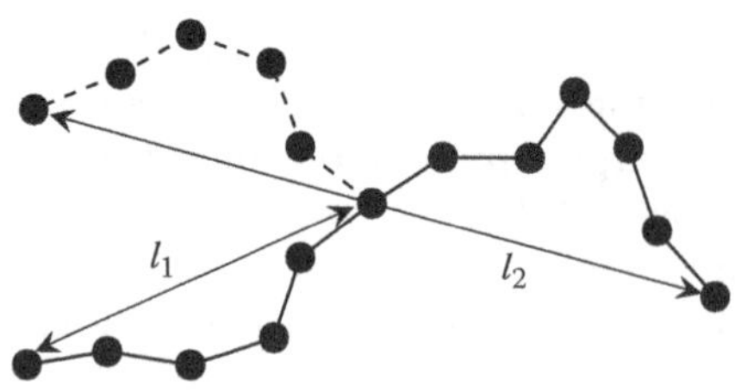

***Figure 11.4** Effect of stiffness of a polymer chain on its conformational freedom. Conformation over a few segments is not determined by chance. However, this is the case over longer distances, for example, l_2.*

chains. Nevertheless, the same laws may hold for the polymer conformation by taking into account that the position of a bond with respect to another one that is several segments away can still be random. This implies that a long chain can still be described as a random chain of n', the so-called "statistical chain elements" of an average length b, $n'b = nl$. This is illustrated in **Figure 11.4**. As a consequence, a statistical chain element contains b/l segments. The ratio b/l is a measure of the chain stiffness. Equation 11.2 reads now as

$$r_g = b\sqrt{\frac{n'}{6}} \tag{11.3}$$

The volume that a random coil occupies in solution, including entrapped solvent, will in any way be proportional to r_g^3. Because r_g is proportional to $\sqrt{n'}$ and n' is proportional to the molar mass M of the macromolecule, the said volume will be proportional to $M^{1.5}$. Because the mass of each macromolecule is proportional to M, the specific volume, that is, the volume occupied per unit mass of macromolecule, is proportional to $M^{0.5}$. This implies that the volume occupied by a certain mass of macromolecules increases with the molar mass or molecular length, other things being equal. The coils are more expanded for larger M or n.

Equation 11.3 implies that the specific volume occupied by a polymer molecule is larger than that predicted by Equation 11.2 by a factor $(b/l)^{1.5}$. Experimental values for b/l vary greatly, for example, ~5 for amylase in 0.33 molar KCl and ~100 for xanthan at neutral pH and ionic strength ~0.01 molar. If n' is smaller than about 25, Equation 11.3 no longer holds, because the average distribution of segments is no longer Gaussian. Then, the molecule will on average have an elongated shape, and r_g will be larger than predicted by Equation 11.3. The extreme form will be a rod-like molecule with a length nl.

11.1.1 Intrinsic viscosity

Starting from the definition given in **Table 10.1** for the intrinsic viscosity, it follows directly from Equation 10.2 that for a dispersion of hard spherical particles, $[\eta]_\varphi = 2.5$. On concentration basis, this gives for the intrinsic viscosity of hard particles $[\eta]_c$ (dimension m^3 kg^{-1}) (Equation 10.9)

$$[\eta]_c = \frac{\varphi}{c}[\eta]_\varphi = \lim_{c \to 0} \frac{2.5\varphi}{c}, \tag{11.4}$$

where φ is the volume fraction dispersed particles or macromolecules and c is the concentration. The ratio $\varphi/c = 1/\rho_{pc}$, where ρ_{pc} is the density of the polymer coil (kg m^{-3}). The hydrodynamic volume V_h and with that, φ, will be proportional to the volume that a molecule occupies and thus $V_h \propto r_g^3$. Taking this into account, one obtains for the intrinsic viscosity of a polymer

$$[\eta] = \lim_{c \to 0} \frac{2.5\varphi}{c} = K'M^{1/2}, \tag{11.5}$$

where K' is a proportionality constant.

Equation 11.5 can also be written in the form as it was originally derived by Flory (1953)

$$[\eta] = \varphi_o \frac{6^{3/2} r_g^3}{M}, \tag{11.6}$$

where φ_o is the so-called Flory–Fox constant (2.5×10^{23} for $[\eta]$ in m^3 kg^{-1}, r_g in m and M in kg mol^{-1}).

In the derivation of Equation 11.5, several implicit assumptions were made, which implies that this equation only holds for special conditions. The main ones are that the volume taken up by the polymer segments and the net interactions between the polymer segments can be neglected. The first leads to the so-called *excluded volume effect* and the second to the *solvent quality effect*.

Regarding the excluded volume effect, in reality polymer segments occupy a certain volume that cannot be occupied by another segment at the same time. Therefore, instead of describing a polymer molecule by a random walk, one should use a so-called self-avoiding random walk. This implies that the molecule will be more expanded than an ideal random coil and that r_m is proportional to n' to the power 0.6 instead of 0.5.

When dissolved, a polymer segment will interact both with the surrounding solvent molecules and with other segments belonging to the same or other polymer molecules. Depending on the balance between segment–segment, solvent–solvent, and segment–solvent interactions, the net mutual interaction between the polymer segments can be attractive, repulsive, or in between when they approach each other closely. The net Helmholtz interaction energy U_{net} (= interaction energy at constant volume) between two polymer segments is given by

$$U_{net} = 2U_{12} - U_{11} - U_{22}, \tag{11.7}$$

where U_{net} (J) stands for the attraction energy between solvent molecules (subscript 1) and polymer segments subscript (2), etc. All U are attractive. When $U_{net} < 0$, solvation of the segments will occur. In polymer literature, the tendency to interaction between polymer segments and solvent molecules is mostly expressed in Flory's solvent–segment interaction parameter χ, which is a dimensionless number proportional to U_{net}. When $U_{net} = 0$ and with that $\chi = 0$, there is no net solvent–segment interaction, and the conformation of the macromolecule will follow the self-avoiding random walk mentioned above. In this case, the quality of the solvent for dissolving the polymer is considered good. For larger χ (i.e., U_{net} is positive), solvent quality is less good, and for χ = 0.5 the net attraction between the polymer segments and the excluded volume effect just compensate each other. The polymer molecule behaves as if it has no volume. Then the dependency of the intrinsic viscosity on the molecular weight of the polymer can be described by Equation 11.5. Such a solvent is called an *ideal* or *theta solvent*.

Another way to deal with the excluded volume and solvent quality effect is by introducing a *linear expansion coefficient* α, which depends on χ and n. Equation 11.2 is then modified to

$$r_g = \alpha l \sqrt{n/6} \tag{11.8}$$

For a good solvent ($\chi = 0$), α is larger than unity because of the excluded volume effect. As the solvent quality decreases, the mutual attraction between the segments partly compensates the excluded volume and α will be closer to unity. Because α depends on n, the relation between r_g and n will also be affected or, in other words, those between r_g and M. A complication is that the dependence of α on M is also a function of the solvent quality. In general,

$$\alpha^5 - \alpha^3 \approx (1 - 2\chi)\sqrt{n} \tag{11.9}$$

For $\chi = 0$, α is large, so $\alpha^5 \gg \alpha^3$ and $\alpha \propto n^{0.1}$. Although this is a small exponent, it does lead to large expansions when n is large. It leads to r_g being proportional to $n^{0.6}$. For $\chi = 0.5$, that is, under theta conditions, $\alpha = 1$ and $r_g \propto n^{0.5}$. For solvents characterized by an intermediate χ, $\alpha \propto n^x$ and $r_g \propto n^{0.5+x}$. As a consequence, one obtains for the intrinsic viscosity

$$[\eta] = KM^a, \tag{11.10a}$$

where in most cases $a = 0.5 + 3x$, that is, in the range 0.5–0.8. Equation 11.10a is known as the Mark Houwink equation, originally introduced as an empirical equation.

For heterodisperse polymers, one can consider $[\eta]$ as being due to a polymer solution additively composed of various fractions with different molecular weight, M. In this case, M has to be replaced by the *viscosity-averaged molecular mass* given by

$$M_v = \left(\frac{\sum c_i M_i^a}{\sum c_i}\right)^{1/a} = \left(\frac{\sum n_i M_i^{1+a}}{n_i M_i}\right)^{1/a}, \tag{11.11}$$

where a is the Mark Houwink exponent. This results in

$$[\eta] = KM_{\mathrm{v}}^{a} \tag{11.10b}$$

The equations given above assume a fixed relation between the radius of gyration r_g and the effective hydrodynamic radius under the applied measuring conditions. In polymer science, the hydrodynamic radius r_h of a molecule is usually defined as the radius that can be calculated from the Stokes–Einstein equation, which reads as

$$D_0 = \frac{k_B T}{6\pi\eta_c r_h}, \tag{11.12}$$

where D_0 is the diffusion coefficient at zero concentration, k_B is Boltzmann's constant, and T is the absolute temperature. The diffusion coefficient is normally measured by dynamic light scattering. However, the hydrodynamic radius calculated in this way does not have to be similar to the hydrodynamic radius calculated from the viscometric data. According to theory, r_h of an unbranched polymer is, to a good approximation, given by

$$r_h = (2/3)r_g \tag{11.13}$$

Because r_h, calculated from dynamic light scattering data, and r_g, calculated from viscometry data, are measured under different conditions, they will depend in a different way on the molecular weight of a polymer, and therefore for heterodisperse polymers the relation between both will depend on the polymer size distribution. Often, values slightly higher than 0.67 are observed for r_h/r_g, for example, ~0.75. For branched macromolecules, measured values of r_h/r_g are clearly higher than 0.67.

11.1.2 Nonideal macromolecules

11.1.2.1 Nonrandom coil macromolecules

Several macromolecules used in food for thickening are highly branched. Amylopectin is a good example. The volume of a branched molecule will be less than that of a linear polymer molecule with the same number of segments and a random coil conformation. The exponent in the relation between r_g and n' will be generally smaller than 0.5; for example, for amylopectin, exponents of 0.41 and 0.43 have been reported. As a result, the dependence of the intrinsic viscosity on M of branched macromolecules is characterized by a much lower Mark Houwink exponent a. If the conformation of the branched molecule approaches that of a perfect sphere, $a \rightarrow 0$. This implies that $[\eta]$ does not depend on the molecular weight of the macromolecule and is equal to $2.5/\rho_{pc}$ (see discussion below, Equation 11.4). For rod-shaped macromolecules, a becomes much larger, reaching up to 1.8. This is the main reason why high molecular weight polysaccharides that have a stiff or flexible rod shape

(**Figure 11.1**), such as xanthan gum, give such a high apparent viscosity even at low weight concentrations (**Figure 11.2**).

11.1.2.2 Hetero-macromolecules

Many macromolecules are built of different segments. Clear examples are proteins, but many polysaccharides are also hetero-macromolecules. As a result, the stiffness and the bulkiness of the chain may vary along the chain. The same applies for the solvent quality. This implies that the excluded volume parameter of the macromolecule has a kind of average value. In addition, specific interactions may occur between the different side groups.

11.1.2.3 Polyelectrolytes

Many macromolecules used in foods contain electrically charged groups. The most common are polyacids with carboxyl groups ($-COO^-$) (e.g., pectin, alginate, and xanthan gum) or $-SO_3^-$ groups (carrageenans) and polyampholytes, which contain both negatively and positively charged groups. Proteins are the most prominent example of the last group. The number of charges depends, of course, on the number of potentially charged groups, but in addition on the degree of ionization, and hence on the pH. Another factor of importance is the ionic strength I, which is determined by the concentration of ions present and their valency. One should be aware that the polyelectrolyte also contributes to the ionic strength. The electric charge on the molecules will strongly affect their conformation in solution, because equal charges will repel each other and unequal charges will attract each other. Its extent strongly depends on the charge density and on the ionic strength. A repulsive force between two electrically charged particles is proportional to the square of the electric potential ψ (V) they generate. In solution, shielding by ions cause the surface potential ψ_0 to decrease with distance h from the other charge according to

$$\psi = \psi_0 e^{-\kappa h}, \tag{11.14}$$

where κ is a shielding parameter given by Equation 10.16. In water at room temperature, κ is given by

$$\kappa \approx 3.2\sqrt{(1/2)\sum_i m_i z_i^2} \approx 3.2\sqrt{I} \tag{11.15}$$

if κ is expressed in nm^{-1} and the ionic strength I in mol L^{-1}. It follows from Equation 11.14 that ψ is reduced to ψ_0/e at a distance h equal to $1/\kappa$. $1/\kappa$ is a measure of the distance over which electrostatic interactions are significant. In food, values of I are usually in between 0.05 and 0.8 mol L^{-1}, implying values for $1/\kappa$ of about 0.35–1.4 nm at high and low I, respectively. At high I, the value of $1/\kappa$ is lower than the size of the building blocks (0.5–0.7 nm) for most polysaccharides. Charges on the molecule will only sense each other when they approach each other quite closely, implying that the conformation of the macromolecule will only be affected to a relatively small extent. On the other hand, at low I, charges on adjacent segments of highly charged polysaccharides will repel each other, leading to a clear expansion of the macromolecule. Some polysaccharides, such as xanthan gum, may even adapt a (flexible) rod-like conformation. The extent to which this occurs strongly depends on the linear charge density. This parameter is inversely proportional to the (average) mutual distance between the charged groups given by $nl/|z|$, where $|z|$ is the valency (number of charged group times their valency) of the polyelectrolyte.

A precise calculation of the expansion of polyelectrolyte molecules as a function of the ionic strength of the solution, pH, and linear charge density of the molecule is not possible for the macromolecules used in food. For instance, in proteins and most charged polysaccharides, the charges are not evenly distributed along the chain. In any case, the expansion of a polyelectrolyte molecule increases with decreasing $\kappa \cdot nl/|z|$. Its effect on viscosity may be large (**Figure 11.5**). At very high salt concentrations, charged polyelectrolytes will behave similarly to uncharged polymers.

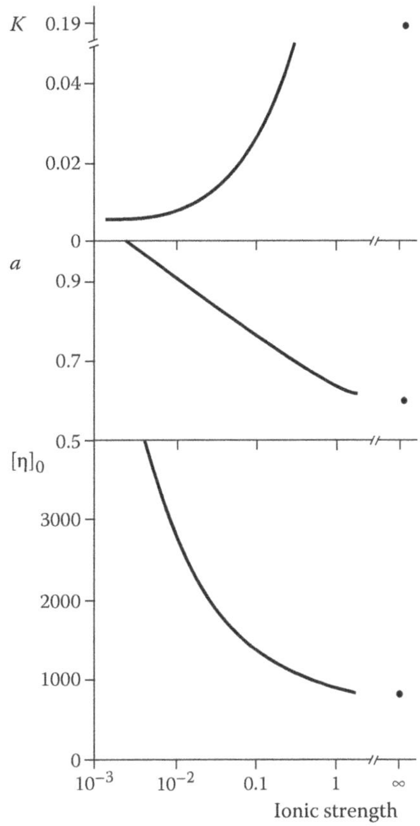

***Figure 11.5** Mark Houwink parameters K (ml g^{-1}) and a and intrinsic viscosity at low shear rate [η]$_0$ (ml g^{-1}) of solutions of carboxymethylcellulose (Na salt, M 10^6 Da) as a function of the ionic strength (mol L^{-1}). (Reprinted from Walstra, P.,* Physical Chemistry of Foods, *Marcel Dekker, New York, 2003. With permission.)*

11.1.3 Concentration effects

11.1.3.1 Dilute solutions

At macromolecular concentrations that are so high that the flow patterns around the molecules start to affect each other, an extra rise in viscosity is observed because of these interactions. The solutions become semidilute. A common method of expressing the effect of a finite, but still low, concentration is by treating viscosity as a function of the concentration by a series expansion of c similar to Equation 10.5

$$\eta = \eta_c(1 + [\eta]c + k_H[\eta]^2c^2), \qquad (11.16)$$

where k_H is the so-called *Huggins constant* (dimensionless). It accounts for the effect of pair interactions between the macromolecules. The trend is that k_H is larger for poorer solvents. Equation 11.16 can be used to obtain [η] by plotting $(\eta - \eta_c)/\eta_c c$ as a function of c and extrapolation to $c \rightarrow 0$.

For polyelectrolytes, on top of the value that k_H would have for uncharged polymers, there will be an extra effect resulting from electrostatic interaction between the polyelectrolyte molecules. The latter is very difficult to quantify and will depend on factors such as polyelectrolyte conformation, its charge density, and ionic strength.

11.1.3.2 Transition to semidilute solutions

The intrinsic viscosity of a macromolecular solution can be considered a measure of the hydrodynamic size of the macromolecule; hence, it allows users to calculate an overlap concentration c_o^*, which demarcates the boundary between dilute and semidilute solutions. A way to estimate c_o^* is as follows. A homodisperse sol of fully packed spheres has a volume fraction of 0.74. For a macromolecular solution, the same is valid for the volume fraction swollen macromolecules φ^*. Assuming that, for the radius of such a macromolecule, we can take r_g, φ^* is given by

$$\varphi^* = N_{AV} \frac{c_o^*}{M} \frac{4\pi}{3} r_g^3, \qquad (11.17)$$

where N_{AV} is Avogadro's number (6.022×10^{23} mol^{-1}). Combining this equation with Equation 11.6, which only applies for random coil polymer solutions, to eliminate r_g gives:

$$\varphi^* = \frac{4\pi}{3 \times 6^{3/2}} N_{AV} c_o^* \frac{[\eta]}{\varphi_o} \qquad (11.18)$$

For $\phi^* = 0.74$, Equation 11.18 leads to

$$c_o^* = \frac{1.08}{[\eta]} \quad (c_o^* \text{ in kg m}^{-3}) \qquad (11.19)$$

Often, one can take, as a good approximation of c_o^*, simply 1/[η]. Because the hydrodynamic volume of a dissolved macromolecule depends on its expansion and, with that, on its molecular weight, chain stiffness, extent of branching, etc., the overlap concentration will likewise do so. These factors are accounted for in [η].

Dilute and semidilute polymer solutions exhibit shear rate thinning behavior. The main reason for this can be understood qualitatively as follows. In the absence of flow, a dissolved polymer molecule will, on average, have a random coil structure. In a flow field, liquid may flow through the coil, whereby each segment will experience a frictional drag, causing a small disturbance of the conformation of the polymer chain. The chain will try to rearrange itself by diffusion to regain its equilibrium conformation. This elastic response can be interpreted as being caused by an elastic spring with a certain response time. Moreover, in a shear flow field, a polymer molecule is continuously rotated and elongated in one direction and compressed in the other (Section 3.3). As a result, the solvent is locally expelled from it and locally imbibed. At low shear rates, the molecule is fully deformed twice during every completed rotation. This leads to additional energy dissipation and hence a relatively large apparent viscosity. With increasing shear rate, the time scale of the deformation (roughly $1/\dot{\gamma}$) becomes shorter than the time needed for the polymer to restore its conformation, implying that the molecule no longer deforms fully during a rotation, and the transiently elongated coils align in the direction of the flow. This results in a lower energy dissipation and, consequently, in a lower viscosity.

Theory has been developed to describe this behavior based on the molecular properties of the macromolecule. However, since in foods, macromolecular solutions are normally not diluted and the shear rate thinning of these solutions is relatively minor (typically less than a factor of 2 over the entire range of experimentally accessible shear rates), these theories are considered to be outside the scope of this book.

11.2 Semidilute and concentrated macromolecular solutions

Macromolecular solutions with concentrations above the overlap concentration c_o^* are considered semidilute and those with concentration far above c_o^* are considered concentrated. The boundary between semidilute and concentrated solutions is often taken at $c[\eta] \approx 4$–10. The marked difference between the dispersions of hard particles and random coil macromolecular solutions is that, in the latter case, the coils can interpenetrate. A schematic layout of the difference between dilute, semidilute, and concentrated solutions of macromolecules is given in **Figure 11.6**. In the concentrated region, extensive overlap is seen between the macromolecular chains. In three dimensions, the molecules interpenetrate each other. Another effect that takes place going from dilute to concentrated solutions, is that when coil overlap starts, the expanding effect of a good solvent becomes more and more screened by the segments from neighboring molecules and the chains begin to collapse back to the dimension they have in a theta solvent. The environment of the molecules changes from pure solvent to a mixed one, which consists of solvent molecules and chain segments of neighboring macromolecules.

Going from a dilute macromolecular solution to a concentrated one, the so-called *correlation length* ξ decreases. The correlation length is a parameter that quantifies the distance over which fluctuations in a system are correlated. In a dilute solution, ξ is roughly equal to the radius of gyration. In a

Figure 11.6 *Schematic figure of a dilute (left panel), semidilute (middle panel), and concentrated (right panel) macromolecular solution. (Reprinted from Walstra, P.,* Physical Chemistry of Foods, *Marcel Dekker, New York, 2003. With permission.)*

concentrated solution, the concentration of the macromolecules is fairly even. For such solutions, ξ is usually defined as the diameter of the "blobs" illustrated in **Figure 11.7**. Its value will decrease with increasing overlap of the polymer coils, and also with increasing stiffness, solvent quality, and concentration of the macromolecule. The molecular weight does affect the transition from dilute to semidilute and next to concentrated, but not ξ in a concentrated solution. For most food macromolecules, the concentrated region is only reached at fairly high polymer concentrations, which is roughly above 3%.

This mutual interpenetration of the coils has a clear effect on the concentration and molecular weight dependence of the viscosity of polymer solutions

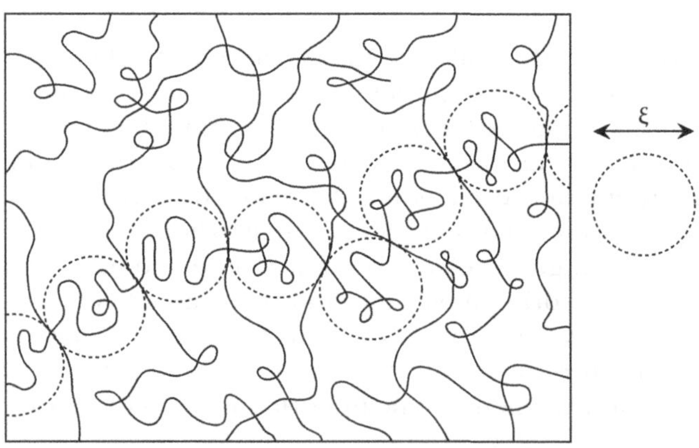

Figure 11.7 *Explanation of the concept of correlation length ξ in a concentrated polymer solution. (Reprinted from Walstra, P.,* Physical Chemistry of Foods, *Marcel Dekker, New York, 2003. With permission.)*

and on properties such as shear thinning behavior. Experimental results show that, for many polysaccharides at very low shear rates, the viscosity as a function of c shows a pronounced increase in viscosity above a specific critical concentration c^* (**Figure 11.8**). In this figure, the logarithm of the zero shear (rate) specific viscosity is plotted as function of log c. The polysaccharides involved are usually assumed to have roughly a random coil conformation in solution under the indicated conditions. The critical concentration c^* is considered to mark the onset of significant coil overlap and interpenetration.

As discussed above, the value of $c[\eta]$ can be considered a measure of the extent of overlap. This variable can be used to describe the dependence of viscosity on concentration and molecular weight. When the data shown in **Figure 11.8** and many other ones are replotted as function of log $c[\eta]_0$, a general relation between the relative or specific viscosity at very low shear rate and $c[\eta]_0$ is obtained (**Figure 11.9**). The subscript 0 indicates that the intrinsic viscosity has been determined at a shear rate extrapolated to zero—the so-called zero shear intrinsic viscosity. The critical concentration c^* is found to be about $4/[\eta]_0$ for a range of different disordered polysaccharides. At this concentration, $(\eta_{sp})_0 \approx 10$. This experimentally observed critical concentration does not correspond with c_0^* defined by Equation 11.19 as the concentration at which the

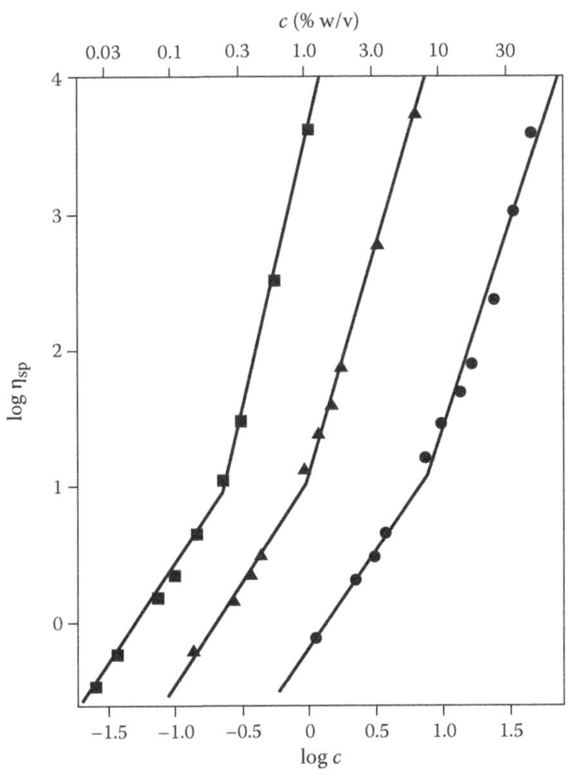

Figure 11.8 *Specific viscosity η_{sp} as function of concentration c for guar gum (■), dextran (▲), and high guluronate alginate (•), equilibrated against 0.2 molar NaCl. (After Morris, E.R. et al.,* Carbohydr. Polym., *1, 5–21, 1981.)*

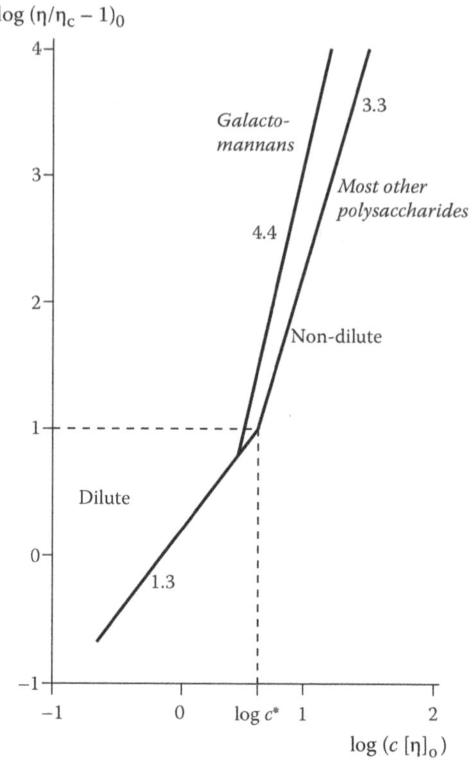

Figure 11.9 *Specific viscosity of polysaccharide solutions at very low shear rates versus dimensionless concentration ($c[\eta]_0$); $[\eta]_0$ is the intrinsic viscosity extrapolated to zero shear rate. (Reprinted from Walstra, P.,* Physical Chemistry of Foods, *Marcel Dekker, New York, 2003. With permission.)*

molecules start to overlap. It is likely that some degree of overlap is required before the clear change in slope is observed (see further below). In addition, as mentioned above, expansion of the polymers will decrease as they start to overlap. For several polymers, the transition is more gradual than those shown in **Figures 11.8 and 11.9**. It occurs over a limited concentration range.

For concentrations below c^*, the slope of the curves is between 1.2 and 1.4. This is higher than a slope of unity, which would be expected for dilute solutions. A reason for this may be that most polysaccharides are not perfect random coils, but are so stiff that parts of the molecules are somewhat rod-like—then, they may already hinder each other in shear flow at concentrations well below the overlap concentration. At concentrations above c^*, the slope of the plot of $\log(\eta_{sp})_0$ as a function of $\log c[\eta]_0$ becomes much higher (3.3 for most polysaccharides). This increase is generally explained by the formation of the so-called *entanglements* between the molecules (**Figure 11.10**). These are formed at concentrations far exceeding c_0^*. The extent of entanglement formation increases with the extent of interpenetration of the polymer coil, and thus with the concentration and molecular mass.

For galactomannans, a higher slope of about 4.4 is observed in **Figure 11.9**. This is supposed to be attributable to specific attractive interactions between side groups on the polymer chain. One should be aware that the above relations between viscosity and $c[\eta]_0$ only hold for very low shear rates. At higher shear rates, c^* will be about the same, but the slope of $\log(\eta_{sp})_0$ versus $\log c$ will be much lower (e.g., about 2). This is attributable to the strong shear rate thinning character of concentrated polymer solutions, the extent of which increases with the polymer concentration.

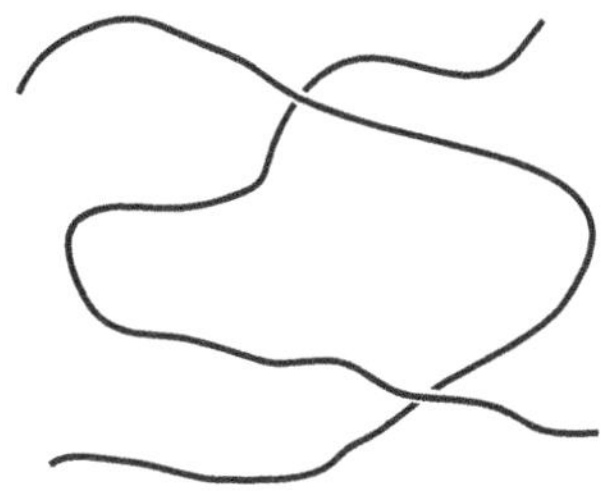

Figure 11.10 *Schematic layout of an entanglement.*

Because the degree of overlap between polymer molecules increases with their molecular mass M at a constant concentration, a very strong dependence of viscosity on M is also observed for concentrated polymer solutions. Above the overlap concentration, $(\eta_{sp})_0$ is approximately proportional to $M^{3.4}$ at constant c. This relation is far from exact, but the much stronger dependence of η on M than in the dilute regime is unmistakable. Finally, as discussed in Section 11.1, $[\eta]_0$ increases with chain stiffness, implying a decrease in c^*.

It is generally assumed that, at concentrations and/or molar masses below the sharp change in slope of curves of $\log\eta$ versus $\log c$ or $\log M$, entanglements do not play a role, whereas above this point, they do. The critical concentrations c^* and molecular weight M^* at which the kink is found are related in the sense that c^*M^* is a constant. As noted above, at concentrations above the kink, the concentration of polymer in the solution is fairly even, down to distances between interchain crossings, the so-called correlation length ξ (**Figure 11.7**). In concentrated solutions, this concerns length scales clearly far below the size of the polymer coil (rather, this length is of the order of several segment lengths); it acts as a characteristic length scale. In this concentration regime, simple mean field arguments can be invoked to describe the rheological properties of the system. We will limit ourselves to a qualitative discussion of the effect of entanglements on the rheology of polymer solutions, considering a more profound discussion to be beyond the scope of this book.

At rest in concentrated solutions, interpenetration of the polymer coils will give rise to a dynamic entangled network structure. These entanglements will restrict the motion of the polymer, and therefore increase their relaxation time. For flow to occur, the entanglements have to disentangle. This requires a relatively large amount of energy, which is measured macroscopically as a high shear stress σ at a set shear rate $\dot{\gamma}$ or a low $\dot{\gamma}$ at a set σ or, in other words, a high viscosity (remember that energy dissipation is equal to $\eta\dot{\gamma}^2$). As the number of entanglements increases strongly with polymer concentration, and at constant weight concentration with polymer dimensions, this qualitatively explains the strong dependence of viscosity on concentration and molecular weight at concentrations above c^*.

At low shear rates and steady-state conditions, as many entanglements will be (re-)formed by Brownian motion as are disentangled, per unit time, and Newtonian viscosity ensues, that is, η is independent of the shear rate (first Newtonian plateau) (**Figure 11.11**). With increasing shear rates, the extent of disentangling increases, and above a certain shear rate this can no longer be compensated for by entanglement formation. The time available for the formation of new entanglements by Brownian motion during shearing is roughly the reciprocal of the shear rate $\dot{\gamma}$, whereas the time needed for disentanglement is less dependent on $\dot{\gamma}$. The latter time is determined by the relaxation time of the polymer molecules, which is roughly proportional to the viscosity at the shear rate involved. A new steady state develops in which the number of entanglements between the polymer molecules is lower. This implies that, for further shearing to occur, fewer entanglements have to be loosened, resulting in lower energy dissipation and thus, in a lower apparent viscosity. Moreover, the energy dissipated during disentangling of an entanglement will also become lower; it is roughly proportional to the prevailing apparent viscosity. The end result is that both the actual number of entanglements present per unit volume during shearing, and the amount of energy dissipated in the disentangling process, decrease with increasing shear rate. The polymer solution exhibits a shear rate thinning behavior.

At high shear rates, the time available for formation of entanglements becomes too short, and the polymer molecules remain fully disentangled. Then, the viscosity will be much smaller than that at low shear rates, and again independent of shear rate—implying a Newtonian behavior (second Newtonian plateau; **Figure 11.11**). In many cases, the second Newtonian plateau cannot be observed experimentally, because at the shear rates required departures from shear flow occurs.

In contrast to what is observed for dilute solutions, the apparent viscosity of a concentrated polysaccharide solution may decrease by several orders of magnitude with increasing shear rate. The extent of shear rate thinning increases with concentration and at constant weight concentration with molecular weight (**Figure 11.12**). Because the relaxation time for the formation of new entanglements increases with increasing viscosity, for higher viscosities entanglement formation can only keep up with the rate of disentanglement due to

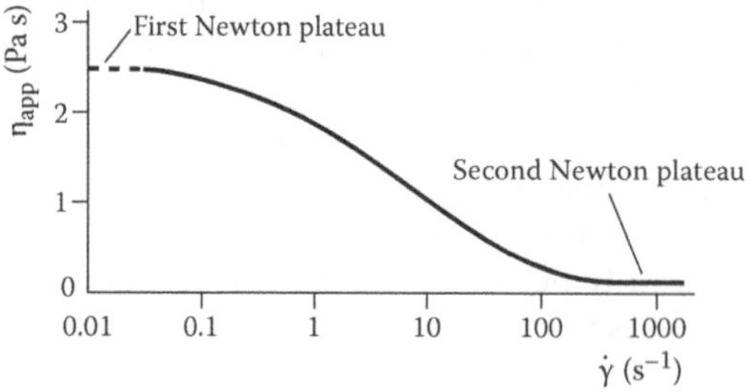

Figure 11.11 *Apparent viscosity of a 1% guar solution in water as a function of rate of shear, exhibiting two Newtonian (often also called Newton) plateaus.*

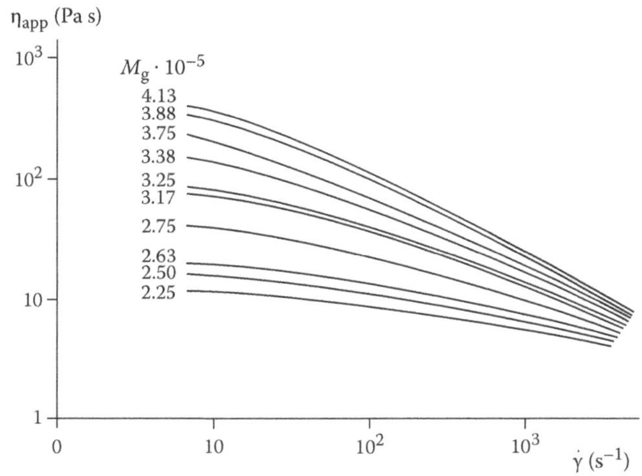

Figure 11.12 *Apparent viscosity of a 2% Na carboxymethyl cellulose in water solutions as a function of rate of shear. M_g, weight-averaged molar mass. (Courtesy of H. Beltman.)*

applied shearing over longer time scales (lower shear rates). Because viscosity increases with increasing polymer concentration and molecular weight, the shear rate at which the shear thinning region sets in will decrease (longer time scale) with increasing polymer concentration or molecular weight (**Figure 11.12**).

The time needed for the formation and disentanglement of entanglements depends on the molecular weight of the polymer molecules. The variation in this time will be larger for solutions containing a polymer characterized by a wide molecular mass distribution than that for a solution containing the same polymer, but characterized by a narrow M distribution. As a result, the shear thinning region will proceed over a narrower range of $\dot{\gamma}$ for the latter solution.

Another aspect is that disentangling and re-formation of the entanglements will take some time. This may cause concentrated polymer solutions to become thixotropic.

11.2.1 Gel-like properties

The picture given above described a concentrated polymer solution as being a temporary network that is destroyed during flow. This concept implies that, upon fast deformations too small to cause disentangling, the system will have solid-like (gel-like) properties whereby the entanglements behave as the temporary cross-links (relaxation time of the entanglements > observation time, i.e., $De > 1$) (Doi and Edwards 1986). Various theories have been developed to estimate the relaxation time of these cross-links. One rather successful theory is the *reptation* approach. The central idea is that a polymer molecule in a concentrated polymer solution cannot move freely in all directions, because of the topological constraints imposed by its neighbors, but that it can move along the contour of its own chain in the "tube" formed by its neighbors. The chain is folded within the tube (**Figure 11.13**), and the motion of these "folds" allows the molecule to diffuse through the tube in a twisting, snake-like motion. De Gennes (1979) gave this process the nickname "reptation."

Related to the reptation process, two dominant relaxation processes can be distinguished: a short-time process due to fluctuations of the polymer within the tube and a longer-time process related to the disengagement of the polymer from the tube. Adaptation of the tube to an applied strain or stress takes much longer than the fluctuations of the polymer in the tube. The tube disengagement time τ_d is given by

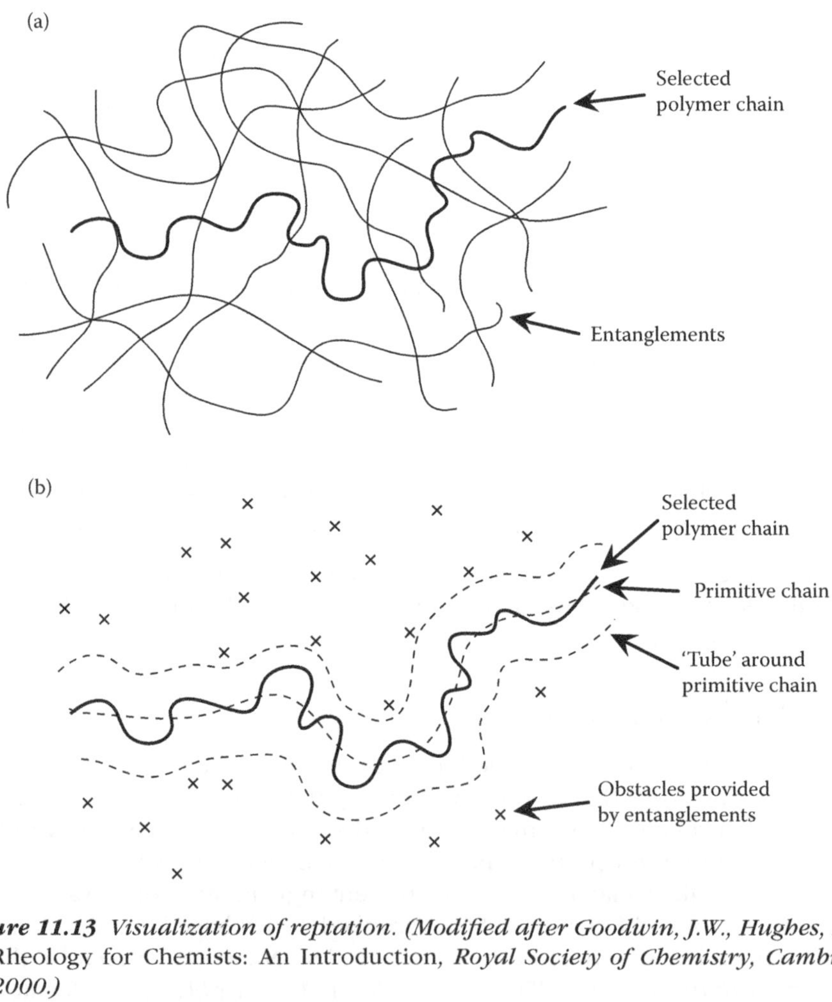

Figure 11.13 *Visualization of reptation. (Modified after Goodwin, J.W., Hughes, R.W.,* Rheology for Chemists: An Introduction, *Royal Society of Chemistry, Cambridge 2000.)*

$$\tau_d = \frac{L^2 n\zeta}{\pi^2 k_B T}, \tag{11.20}$$

where L is the length of the tube and $n\zeta$ is the friction coefficient (N s m^{-1}) of the whole chain in the tube. Because both L and $n\zeta$ are proportional to M, Equation 11.20 predicts τ_d to be proportional to M^3. The time scale (relaxation time) of the fluctuation of the polymer within the tube will depend on the length or the part of the chain involved and its stiffness. According to the so-called Rouse model, this results for the longest relation time τ_f

$$\tau_f = \frac{r_g^2 n\zeta}{\pi^2 k_B T} \tag{11.21}$$

Because n as well as r_g^2 are proportional to M, τ_f is proportional to M^2. So, for longer chains, the relaxation process due to chain fluctuations is much shorter than the tube disengagement time. The difference between the two characteristic relaxation times increases with M. At times, between the two relaxation processes, the polymer behavior is independent of time and shows solid-like (or better gel-like) behavior, implying that the reaction of the material on a stress or strain will be predominantly elastic. Over time scales longer than τ_d, the polymer solutions will react in a (predominantly) viscous manner. The gel-like behavior in this time-scale region will be discussed further in **Chapter 12**.

Another consequence of the behavior of concentrated polymer solutions as a temporary network is that normal stresses may develop on deformation. The size of these normal stresses will depend on polymer characteristics, the relaxation time of the temporary bonds (entanglements), the time scale of deformation, and the way deformation occurred. For non-Newtonian liquids, the first normal stress difference N_1 may become substantial compared to the shear stress when their reaction on a stress contains an important elastic component, as will be the case for concentrated polymer solutions. It may lead to phenomena such as rod climbing (or Weissenberg effect) and substantial die swell (**Figure 11.14**).

Rod climbing may occur when a rotating rod is placed in a concentrated solution of polymers. The shear flow due to the rotating rod will deform the large polymer molecules (compared to water molecules) in the direction of the curved streamlines and with that, the temporary network. This strong deformation of the different polymer molecules and of the network as a whole gives rise to an opposing force in the molecules and on the temporary bonds between them. This results in an effective force (stress) directed to the rod. This stress will be higher when there is less time for stress relaxation in the network. It causes the polymer molecules to move toward the rod and to climb upwards along the rod, the higher the shear rate becomes, instead of being thrown outward as is the case for Newtonian liquids.

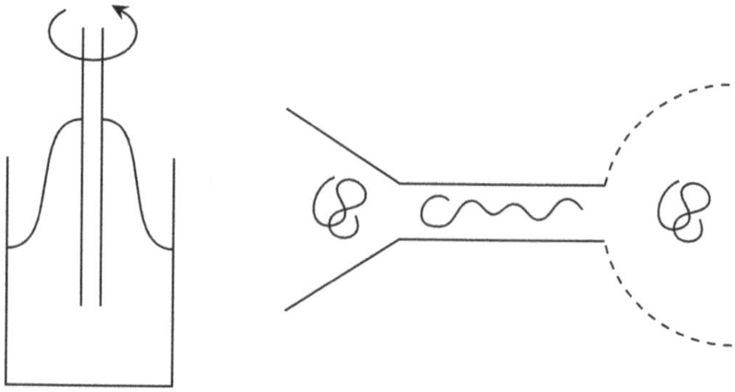

Figure 11.14 Phenomena of rod climbing in a rotating rod (left panel) and die swell (right panel).

In uniaxial elongational flow, as occurs in a constriction in a tube or before the die in an extruder, polymer molecules will be substantially stretched out in the direction of the flow. The polymer molecules will try to restore their average equilibrium random coil conformation. In a tube or die, this can only proceed through relaxation of the conformation of parts of the polymer chain and, more slowly, by relaxation of the entanglements between the molecules. After leaving the die, the normal stresses can also relax by contraction of the volume elements in the liquid jet in the flow direction, leading to an increase in its diameter. As discussed above, the time needed for the disentanglement of polymer chains increases with molecular weight and concentration of the polymers. The discussion above implies that die swell will be more pronounced for more concentrated systems, higher M of the polymers, shorter dies, and higher flow rates.

To summarize, the formation of entanglements in concentrated macromolecular solutions has three effects:

1. Shear rate thinning behavior because (re-)formation of entanglements takes time.
2. Thixotropic behavior as a result of the finite time needed for disentanglement on increasing $\dot{\gamma}$ and for their reformation on decreasing $\dot{\gamma}$.
3. Elastic behavior for deformations shorter than the tube disengagement time, $\tau_d(\tau_d/\dot{\gamma} > 1)$.

References

De Gennes, P.G. 1979. *Scaling Concepts in Polymer Physics*. Ithaca, NY: Cornell University Press.

Doi, M., and S.F. Edwards. 1986. *The Theory of Polymer Dynamics*. Oxford: Oxford University Press.

Flory, P.F. 1953. *Principles of Polymer Chemistry*. Ithaca, NY: Cornell University Press.

Goodwin, J.W., and R.W. Hughes. 2000. *Rheology for Chemists: An Introduction*. Cambridge: Royal Society of Chemistry.

Morris, E.R., A.N. Cutler, S.B. Ross-Murphy, and D.A. Rees. 1981. Concentration and shear rate dependence of viscosity in random coil polysaccharide solutions. *Carbohydr. Polym*. 1: 5–21.

Walstra, P. 2003. *Physical Chemistry of Foods*. New York: Marcel Dekker.

12

Solids and Solid-Like Materials

For consumers, solid products are usually associated with a high modulus and visible fracture or yielding on deformation. They encompass a large variety of products such as hard-boiled sweets, crispy products, vegetables, fruits, potatoes, nuts, hard and semihard cheeses, bread, cake, pastry, pies, and various dairy desserts. All these products fracture on deformation. Products that exhibit yielding on deformation include butter, margarines, various marmalades, and mayonnaises. Food products such as tomato ketchup, apple sauce, and many other sauces are considered to be at the border between solids and liquids. In rheological terms, solid materials are those that show elastic behavior on deformation, and yield or fracture on large deformations.

At first sight, the descriptions described above overlap to a large extent. Both are primarily based on mechanical properties. However, there are also several clear differences, which are two-sided. First, products containing a very weak network such as buttermilk, tomato juice, and orange juice are, strictly speaking, solids in rheological terms, whereas consumers will classify them as liquids. Second, viscoelastic products characterized by a relatively high tan δ (e.g., 0.3–1) such as several types of cheese (e.g., the center of ripened Camembert and Brie), bread dough, and relatively concentrated solutions of polysaccharides are in rheology classified as being in between solids and liquids, but not products such as tomato ketchup, as long as the tan δ of these products is low. A classification purely based on rheological arguments is given in **Figure 4.3**. A similar classification based on fracture characteristics would result in a distinction between brittle, plastic, and viscoelastic materials. This classification overlaps only partly with the one exclusively based on rheological characteristics. In addition to the classification mentioned above, a distinction is often made between hard and soft solids primarily based on the modulus or stiffness.

Another way to classify solid (and semisolid) materials is based on their structure, whereby structure is defined as the distribution in space of the components in the system. This is a purely geometrical concept, as it just concerns angles and distances between the components and shapes and their sizes. The physical building blocks of the structure may be called structural elements (see footnote in Section 4.2). Structural elements can be particles, such as emulsion droplets, air bubbles, crystals, starch granules, cells, and macromolecules. If these particles are separated from each other, the system is called a dispersion. The relation between their rheological behavior and the structure of these systems is discussed in **Chapters 10 and 11** for liquid dispersions. In the next chapters, the relation between mechanical properties and structure will be discussed for solid and solid-like materials.

The distinction, as will be discussed in the following chapters, between solids and solid-like materials will be based on a combination of structural and mechanical (rheological and fracture) characteristics of the systems. It is done in this way to combine in the most feasible manner the differences seen by, primarily, researchers and, to a lesser extent, consumers. The order of the chapters is primarily based on the degree of difficulty of the relation between structure and mechanical properties. The discussion will be limited to relatively simple systems and to the more general principles. Composite materials with a highly complicated structure are considered to be outside the scope of this book. In those cases, the relations are often quite product-specific.

As an introduction to the following chapters, first some general characteristics will be discussed regarding the mechanical behavior of solid materials at small deformations as a function of time scale/water content. In **Chapter 13**, the relation between mechanical properties and structure will be discussed for gels, whereby a distinction will be made between the so-called polymer and particle networks. A special section (Section 13.7) is devoted to weak networks. In **Chapter 14**, simple composite materials will be discussed. This will be followed by chapters on gel-like close packed materials (**Chapter 15**), dry and wet cellular materials (**Chapter 16**), and hard solids (**Chapter 17**).

12.1 Rheological behavior of elastic materials at small deformations

As discussed in Section 4.2, the rheological behavior of materials at small deformations depends on the time scale of the applied deformation or stress. For elastic materials, this behavior can be best illustrated by considering the change in storage (G') and loss (G'') modulus as a function of frequency for a pure viscoelastic polymer system (**Figure 12.1**).

At (very) high oscillation frequencies of the applied stress (or strain) (region A), the system will behave like a glassy material, that is, as an amorphous solid. Here, solid is defined as a material having an apparent viscosity larger than a specified value, usually 10^{12} Pa s. An amorphous solid differs from a crystalline solid by not showing a regular periodicity in atom or molecule density. In the glassy state, the frequency of deformation is so high that the polymer

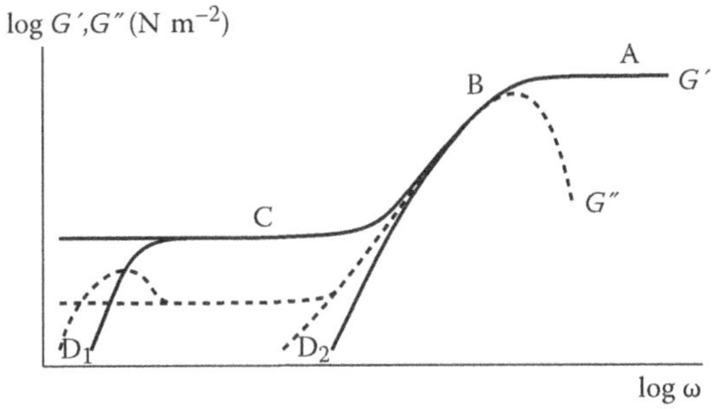

***Figure 12.1** Schematic layout of storage modulus G′ and loss modulus G″ of polymer systems as a function of oscillation frequency ω in dynamic measurements. For further explanation, see main text.*

molecules have no time to adapt their conformation to the applied change in shape of the material. In principle, even no rotation around or adaptation of the angle between covalent and other bonds will occur. As a result, no energy dissipation takes place during the periodic application of the stress. The storage modulus G' is high, typically about 10^9–10^{10} Pa, and G'' is much lower. The material behaves as a brittle solid.

At lower frequencies (region B), adaptation to the change in shape can take place, at high frequencies side, by rotation around bonds, adaptation of bond angles, movement of side chains, etc. With decreasing frequencies, larger segments of the molecules may adapt their conformation. These processes lead to higher energy dissipation and, with that, to lower energy storage, where tan δ (G''/G') increases and G' decreases with decreasing ω. At still longer time scales (lower ω), the polymer molecules can adapt their conformation over their whole length or when they are mutually cross-linked over the length between the cross-links. In the latter case, G' will not decrease further on further lowering of ω. The material behaves as a rubber (region C). The transition from the glassy state to the so-called rubbery state is aptly called the glass–rubber transition.

In the rubbery state (region C), the time scale of the deformation is so long that the part of the macromolecules between the cross-links can fully adapt their conformation to changes in the shape of the material. A prerequisite for the occurrence of this region is that the life span of the cross-links between the polymer molecules is longer than the time scale of the deformation. In region C, G' is independent of the frequency. The energy dissipation per second is low and with that, so is G''. The term "rubbery state" is attributed to historical reasons because the type of behavior shown in **Figure 12.1** was first studied for rubber-like materials.

When the time scale of the deformation becomes of the order of the relaxation time τ_{rel} of the bonds, G' starts to decrease again and G'' first increases and at still lower ω also decreases. When ω becomes equal to $1/\tau_{rel}$, a crossover between G' and G'' takes place, and the material starts to behave like a liquid. In these so-called terminal zones (regions D_1 and D_2), G' is theoretically proportional to ω^2 and G'' to ω. The latter implies that $\omega G'' = \eta$ is independent of ω (Equation 7.23). Terminal zone D_1 applies for materials with physical cross-links (including entanglements) between the molecules.

For covalent cross-linked materials, there will be no terminal zone because under usual conditions, the relaxation time of these bonds is too low to be observed experimentally. When there are no chemical or physical cross-links between the polymer molecules, there will be no rubbery region, but a direct transition from glassy behavior to the terminal zone D_2. The size of the rubbery region is determined by the distance in time scale between the glass–rubber transition and the onset of the terminal zone. For concentrated polymer solutions, the onset of the terminal zone is determined by the disentangle time of the entanglements, that is, roughly the relaxation time of the whole molecule. This time is longer when the molecular weight is higher (Section 11.2.1).

Since, in most materials (including all food products) that show rubber-like behavior the cross-links are characterized by a large spectrum of relaxation times, G' is not constant in region C, but decreases slowly with decreasing ω. Because these relaxation processes lead to energy dissipation, G'' and tan δ will be higher than those for ideal rubber behavior. An important class of food products showing rubber-like behavior is gels, even if the (molecular) basis of it is very different from that in the originally studied polymer systems. For instance, gels formed by aggregated particles as emulsion droplets and fat crystals usually show rubber-like behavior at small deformations.

Another very important aspect is that the transition from the glassy state to a rubber (gel-like) state is not only related to the time scale of the deformation, but may also be due to other factors that favor the mobility of the molecules forming the glass. Important factors are the temperature and the presence of the so-called plasticizers. A higher temperature leads to a higher mobility of the molecules (proportional to $k_B T$, where k_B is the Boltzmann constant), inducing a transition from glassy behavior to that found in region B, C, or D at constant ω in oscillatory tests. Plasticizers are additives that increase the fluidity of the material in which they become incorporated. The most important plasticizer for food products is water, but other small molecules such as sugar molecules and oil may also act as plasticizers. A well-known effect of the uptake of plasticizer is the loss of crispiness of crispy products because of the uptake of water during storage at high relative humidity, which induces a transition in fracture behavior from brittle fracture to elastic plastic fracture.

Finally, many foods are composite products. The different components in such products may be characterized by a glass–rubber transition at different temperatures or plasticizer content. Similarly, they may exhibit different terminal zones. Together, this may make the figures characterizing log G' and log G'' as a function of log ω (much) more complicated than the idealized picture shown in **Figure 12.1**. However, this does not mean that the main message does not hold—that the mechanical behavior of solid and solid-like materials depends on the time scale of their deformation and on factors that affect the mobility of molecules such as temperature, presence of other molecules, and extent of chemical and physical interactions between the molecules.

13

Gels

13.1 Introduction

It is difficult to provide a clear-cut definition of a gel. In general, *gels* are characterized by a preponderance of solvent and the presence of a three-dimensional network of connected molecules or particles, at least over the time scale considered. Rheologically, they are characterized by a predominantly elastic behavior over the time scale considered and a modulus that is relatively small (generally $<10^7$ Pa) compared with real solids (generally 10^9–10^{11} Pa). Various types of gels can be distinguished, where the precise division may depend on the criteria used.

A division based on gel structure is as follows (partly after Flory 1974);

1. *Polymer networks*; a further subdivision can be made between covalent (e.g., cured rubbers) and physically cross-linked networks (e.g., entangled networks and gelatin gels). Another subdivision is between networks with long flexible polymer chains between the cross-links, (e.g., gelatin gels) and those with stiff polymer chains between the cross-links (e.g., gels of many polysaccharides). There are no food gels containing only covalent cross-links. Covalent cross-links may contribute to protein gels, but generally the majority of the bonds are physical cross-links.
2. *Particle networks*; a further subdivision can be made between networks of hard particles (e.g., gels of emulsion droplets or fat crystals) and of deformable particles (e.g., milk gels such as set yogurt, which are made up of physically cross-linked protein aggregates).
3. *Well-ordered lamellar structures*, including gel mesophases. This type of gels may occasionally occur in foods, but the needed concentration of small molecule surfactants is generally too high to be acceptable for food. Therefore, the rheology of this last type of gels will not be discussed.

Gels of types 1 and 2 are illustrated in **Figure 13.1**.

The formation of gels proceeds in different ways, depending on the system being considered, and parallel to it, theories will differ that describe gel formation. A theory describing gel formation in a concentrated solution of long flexible polymers by random formation of cross-links will, for understandable reasons, differ from one describing gel formation by aggregation of particles as protein particles and emulsion droplets. A common feature of all theories is that a space spanning network has to be formed that is able to transfer stress from one side of the gel to the other. An extensive discussion of these theories is considered to be outside the scope of this book, except insofar as this is useful for understanding the relation between the structure of gels and their rheological and fracture behavior.

With respect to the rheological and fracture behavior of gels, the extremes are gels with long flexible polymers between the cross-links and gels of hard particles. These will be discussed in Sections 13.2 and 13.3, respectively. In addition, some intermediate types will be discussed as gels of stiff polymer

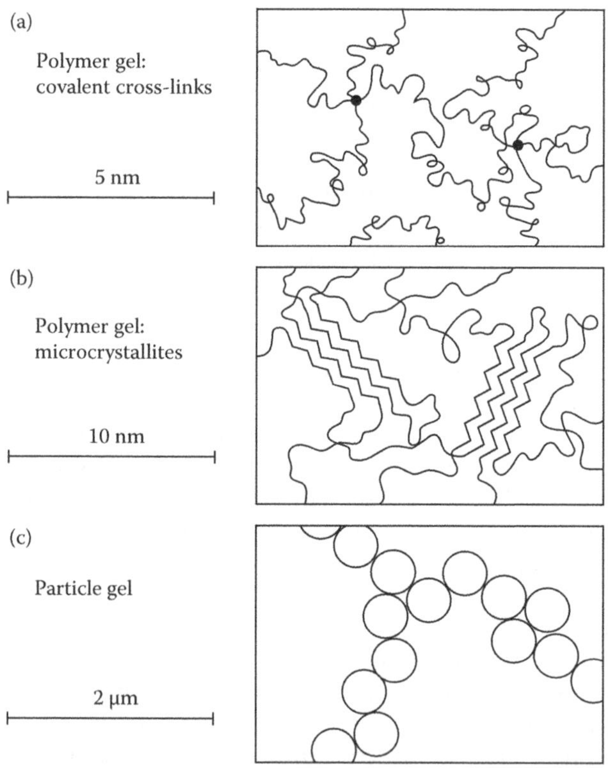

Figure 13.1 *Highly schematic illustrations of three types of gel structure. Dots in panel (a) denote covalent cross-links. Note that scale differs between different panels. (Reprinted from Walstra, P.,* Physical Chemistry of Foods, *Marcel Dekker, New York, 2003. With permission.)*

chains. One should be aware that macromolecules such as globular proteins (e.g., soy and whey proteins) behave in this respect more as deformable particles than as long, flexible polymers. Therefore, we will use the term polymer gel, and not macromolecular gels, for gels characterized by long flexible or stiff molecules between the cross-links.

Often, gels are characterized by their small deformation properties, using a well-established rheological technique and/or by their large deformation and fracture properties in a (semi-)empirical way. However, several methods are available for determining the fundamental characteristics of the large deformation and fracture behavior of gels, and there are qualitative models that describe how these can be related to the gel structure. In general, the large deformation and fracture properties of gels primarily determine their physical stability and their handling, usage, and eating characteristics. **Table 13.1** gives an overview of the desired mechanical properties in relation to some important gel properties. Sagging of dispersed particles is especially of interest for products that are, on further consideration, very weak gels. Examples include various drinks such as buttermilk, tomato juice, and orange juice, and some deserts such as custards. In these drinks, the gel network is mostly so weak that during drinking consumers will not notice its presence. Their yield stress is of the order of 0.05–0.5 Pa (corresponding to a water pressure of 5–50 μm). This type of gels will be discussed further in Section 13.7.

Below, we will first discuss several general aspects of the relation between mechanical properties and gel structure. Next, in the following sections, various types of gels relevant for food will be discussed.

At small deformations, gels are characterized by a modulus. First, a general expression will be derived for the modulus, based on a simplified picture of a gel. In this picture, the gel is assumed to be built of strands, which are mutually cross-linked. A strand can consist of a polymer chain or a linear chain of aggregated particles. A force f_s applied to such a chain results in a reaction

Table 13.1 Required Mechanical Characteristics of Gels Needed to Obtain a Desired Property

Desired Property	Essential Parameters	Relevant Conditions
"Stand-up"	Yield stress (or modulus if quite low)	Time scale
Firmness	Modulus or yield stress or fracture stress	Time scale, strain, and strain rate
Shaping	Yield stress + restoration time on setting	Several
Immobilization of solvent/no expulsion of liquid	Permeability, yield stress	Stress, time
Sagging of dispersed particles	Yield stress + short restoration time	Exerted stress, time
Handling, slicing	Fracture stress and work of fracture	Strain rate
Eating characteristics	Yield and/or fracture properties, modulus	Strain rate, strain (for modulus)
Strength	Fracture properties and/or strain hardening	Stress + time scale, strain rate

Source: Partly after Walstra, P., *Physical Chemistry of Foods*, Marcel Dekker, New York, 2003.

force in the chain, which is proportional to the deformation, Δx, times the change in the interaction force due to a change in distance between the cross-links/particles, $\mathrm{d}f_s/\mathrm{d}x$. The parameter $\mathrm{d}f_s/\mathrm{d}x$ is the so-called *spring constant* of the strand (N m^{-1}). For a polymer gel, the resistance to deformation of a chain is due to a change in distance between the cross-links, and for particle gels it is due to a change in distance between the particles (**Figure 13.2**). By multiplying both terms by the number, N, of elastically effective strands per cross section of the gel, one obtains the following equation:

$$\sigma = -N\frac{\mathrm{d}f_s}{\mathrm{d}x}\Delta x \tag{13.1}$$

The local change in distance can be recalculated to a macroscopic strain γ or ε by dividing Δx by a characteristic length C determined by the geometry of the network. In formal rheology, C (= $\Delta x/\gamma$) is a tensor. Because f_s can generally be expressed as $-\mathrm{d}F/\mathrm{d}x$, where $\mathrm{d}F$ is the change in Gibbs energy, one obtains

$$\sigma = CN\frac{\mathrm{d}^2F}{\mathrm{d}x^2}\gamma \tag{13.2}$$

Because $G = \sigma/\gamma$, and $\mathrm{d}F = \mathrm{d}H - T\mathrm{d}S$, where $\mathrm{d}H$ is the change in enthalpy and $\mathrm{d}S$ the change in entropy, one gets for G

$$G = CN\frac{\mathrm{d}^2F}{\mathrm{d}x^2} = CN\frac{\mathrm{d}(\mathrm{d}H - T\mathrm{d}S)}{\mathrm{d}x^2} \tag{13.3}$$

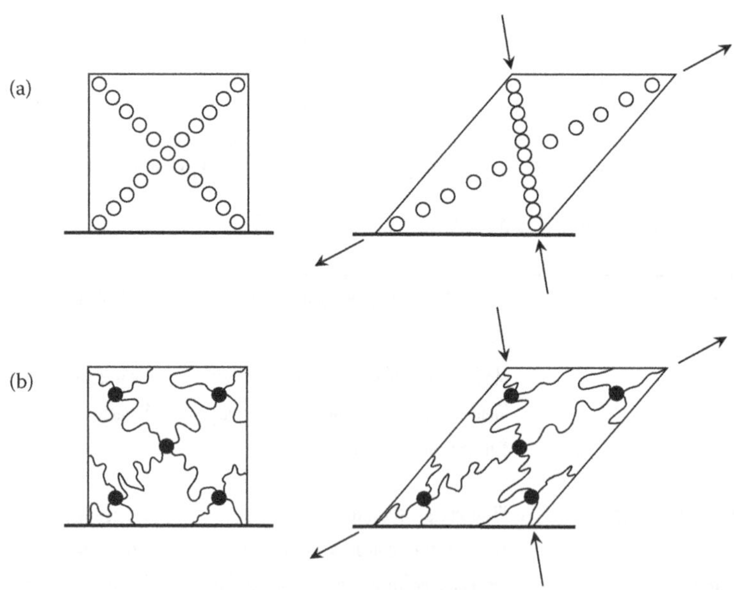

***Figure 13.2** Schematic representation of deformation of a network due to an applied shear force: (a) particle network, (b) polymer network.*

For gels with long flexible polymer chains between the cross-links, the increase in Gibbs energy on deformation is predominantly due to a decrease in conformation entropy of the chains between the cross-links. These networks are dominantly disordered with the exception of the (physical) cross-links where the chains are locally ordered. Because of this, the enthalpy term in Equation 13.3 may be neglected compared with the entropy term. On the other hand, for particle networks, the entropy term may be neglected. For systems consisting of hard particles, resistance to deformation stems primarily from a change in the colloidal interaction forces between the particles, whereas for gels of globular proteins, deformation involves distortion of the tertiary structure of the molecules and of their aggregates. For gels with "stiff" polymers between the cross-links, it will depend on the bending stiffness of the chains and their lengths, which term in the right-hand side of Equation 13.3 prevails, and whether one of the two may be neglected. Be aware that, strictly speaking, Equation 13.3 is only valid for identical strands. In most real (food) systems, this condition is not satisfied. For example, the bonds between particles in a particle gel may be size-dependent.

At larger deformations, the quantity df_s/dx is no longer independent of Δx, leading to a decrease or increase of the measured modulus (nonlinear behavior) with increasing deformation. Another complication is that the constant C may become dependent on γ and/or ε. For most food systems, G will, at least initially, decrease. For several systems at larger deformation, an increase in σ/γ and/or σ/ε is observed, which is the phenomenon of strain hardening (Section 4.3.2). In spite of these objections, at large deformations the resistance to deformation will also consist of an entropic and an enthalpic contribution, whereby the ratio between both components may depend on the deformation.

13.2 Polymer networks

The simplest structure that a polymer network can have is depicted in **Figure 13.1a**: long linear polymer molecules that are cross-linked at various places along the chains. Gel formation can be imaged to result from a random formation of cross-links between overlapping polymer molecules until a space spanning (percolating) network is formed. To obtain such a network, a minimum concentration is required of polymer and/or cross-links between them, the so-called *percolation threshold.*

As mentioned above, for networks of long flexible polymers, the enthalpy term may be neglected compared to the entropy term in Equation 13.3. Based on statistical thermodynamic arguments, Flory (1953) derived a relationship between the modulus of gels consisting of ideal cross-linked long flexible polymers, and the number of chains between the cross-links per unit volume, ν. It is based on the idea that, in the undisturbed state of the gel, the chains (be aware that it concerns, in fact, parts of the whole polymer chain) between cross-links have such a conformation that their entropy is at a maximum. The distribution in space of the statistical chain elements in a chain between two cross-links is assumed to be Gaussian, which will be true if their number n' is large. The entropy of a polymer chain between two cross-links is given by

$$S = k_B \ln N_c, \tag{13.4}$$

where N_c is the number of conformations available to the chain (number of possible realizations) and k_B is the Boltzmann constant.

Each deformation then leads to a change in distance between the cross-links and, with that, to lower N_c, that is, to a lower entropy of the chain. Moreover, it is assumed that deformation is affine, that is, the strain is the same everywhere down to the change in distances between the cross-links and that the enthalpy contribution to the resistance to deformation may be neglected. Further assumptions are that the number of cross-links per original polymer molecule is large, that all chains between two cross-links have about the same contour length, and finally that the network contains negligible numbers of entanglements (**Figure 11.10**), close loops, and dangling ends (**Figure 13.3**). Next, the number of chains with a specified end-to-end distance can be calculated for the undisturbed and the deformed state ν^0 and ν^d, respectively. From such statistical considerations, the change in entropy ΔS can be calculated

$$\Delta S = k_B \sum_i \left(\nu_i^d \ln \frac{\nu_i^0}{\nu_i^d} \right) \tag{13.5}$$

Elaboration leads to the simple equation

$$G = \nu k_B T, \tag{13.6}$$

where ν is the number of chains between two cross-links per unit volume (or two times the number of cross-links, assuming that each cross-link connects four chains each leading to another cross-link; this will be the case when there are much more than two cross-links per polymer molecule). This equation can also be written as

$$G = \nu \frac{M_c}{N_{AV}} \frac{N_{AV}}{M_c} k_B T = \frac{c}{M_c} R_g T, \tag{13.7}$$

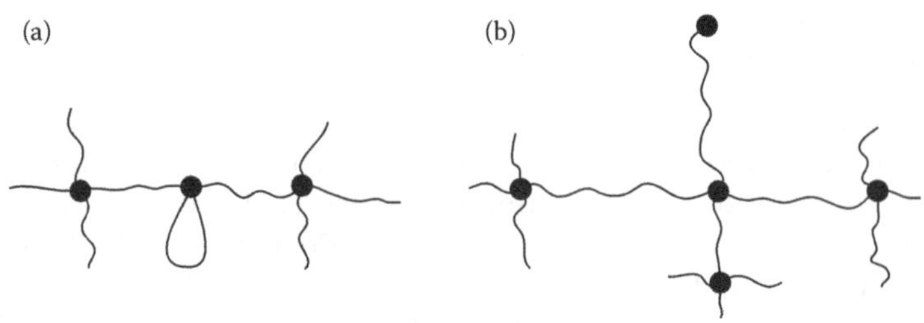

***Figure 13.3** Examples of elastically inactive cross-links in polymer networks: (a) close loops, (b) dangling ends.*

where M_c is the average molecular weight of the chain between two cross-links and R_g is the gas constant. The theory on which Equations 13.6 and 13.7 are based is known as the rubber theory of elasticity. Both equations predict that the modulus is proportional to the concentration; in practice, however, stronger dependencies are found, for example, $G \propto c^2$ for gelatin gels. Moreover, for example, te Nijenhuis (1981) found for 1.95 w/w% gelatin gels a modulus of 500 N m^{-2}, implying M_c to be 94 kg mol^{-1}, although the weight average M of the whole molecule was only 70 kg mol^{-1}. The conclusion must be that part of the molecules does not contribute to the elasticity of the network, and/or that part of the cross-links is elastically ineffective, for example, because they are parts of dangling ends or closed loops. A correction for dangling ends derived by Flory (1953) reads as

$$G = \frac{cR_gT}{M_c}\left(1 - 2\frac{M_c}{M_N}\right), \tag{13.8}$$

where M_N is the number-average molecular weight before cross-linking. Other complications are the presence of a sol fraction consisting of molecules not connected to the main network (which occur especially at low concentrations and for short molecules), and the formation of entanglements (which occur especially at higher concentrations). The latter will also cause the modulus to depend on the time scale of the deformation and, hence, on the frequency in oscillatory measurements.

Another characteristic of Equations 13.6 and 13.7 is that they predict that the modulus is proportional to the absolute temperature T. This is a direct consequence of the assumption that the resistance to deformation is attributable to a change in entropy of chains between the cross-links (Equation 13.3). In foods, at first approximation gelatin gels can be described by the theory on which Equations 13.6 and 13.7 are based. However, it is well known that on heating these gels become weaker and melt at about 30–35°C. The reason is that the semicrystalline cross-links (**Figure 13.4b**) responsible for the gel formation melt, resulting in a gelatin solution.

In Equations 13.6 and 13.7, there is no term that accounts for the nature of the cross-links between the polymer chains. The theory was originally developed for covalent bonds, implying no relaxation of the cross-links under common conditions. This indicates G' to be independent of time scale of the stress application and tan δ to be low (typically <0.03). For gels with physical cross-links, the same holds as long as the cross-links do not relax during the time scale of the stress application. When they do, both equations can still be used as long as the resistance to deformation is primarily due to a change in entropy of the polymer chains. However, then G' will depend on the time scale of the measurement and tan δ may become larger.

At high concentration, polymer solutions may also form a gel as a result of the formation of entanglements. Various theories have been developed to

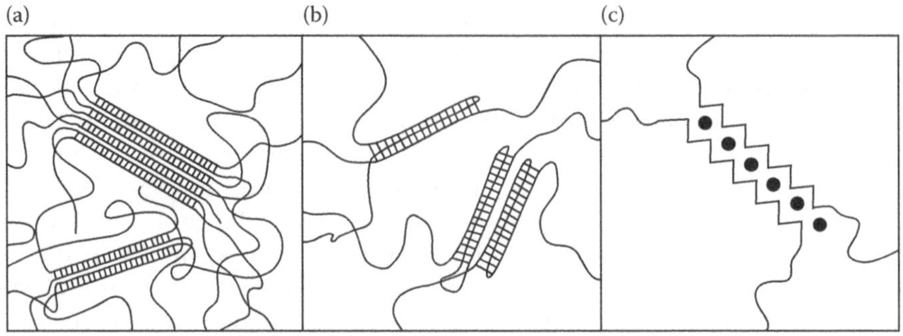

Figure 13.4 *Various types of junctions in polymer gels. (a) Stacked double helices, e.g., in carageenans, (b) partly stacked triple helices in gelatin, (c) "egg box" junction, e.g., in alginate; dots denote calcium ions. Highly schematic; helices are indicated by hatching. (Reprinted from Walstra, P.,* Physical Chemistry of Foods, *Marcel Dekker, New York, 2003. With permission.)*

estimate the shear modulus and its frequency dependence based on entangled concentrated polymer solutions and the assumption that enthalpy terms may be neglected compared to entropy terms. Because the latter is rarely (if ever) the case for macromolecules added to food as thickening and gelling agents, these theories are considered to be outside the scope of this book.

13.2.1 Large deformation behavior of polymer gels

On deformation, the distances between the cross-links will be changed, implying a distortion of the Gaussian distribution of the statistical chain elements. This will happen more readily for a smaller number of statistical chain elements n' in the chain between two cross-links. Moreover, at still larger deformation, the finite extensibility of the chain will start to play a role. It implies that at large deformation the chemical bonds in the primary chains become distorted. Consequently, the increase in Gibbs energy is no longer due to a decrease in conformational entropy solely, but is also due to an increase in bond enthalpy. This will already be the case at smaller strains when the value of n' is lower. (In a crystalline solid, only the increase in bond enthalpy contributes to the elastic modulus.)

A theory has been developed that takes these aspects into account. Some results of it are shown in **Figure 13.5**, which shows the relation between stress and strain for polymer gels subjected to large deformations in elongation. The calculations leading to the results shown involved some assumptions about the network structure, but these do not detract from the trends observed. It is clear that, for short chains between the cross-links (small n'), the curve relating the stress to the strain turns upward at low strains, implying that the chains become fully stretched. Further elongation of the test piece would then lead to breaking of the chains. For larger n', gels can be deformed further before the curve becomes approximately vertical. One should be aware that

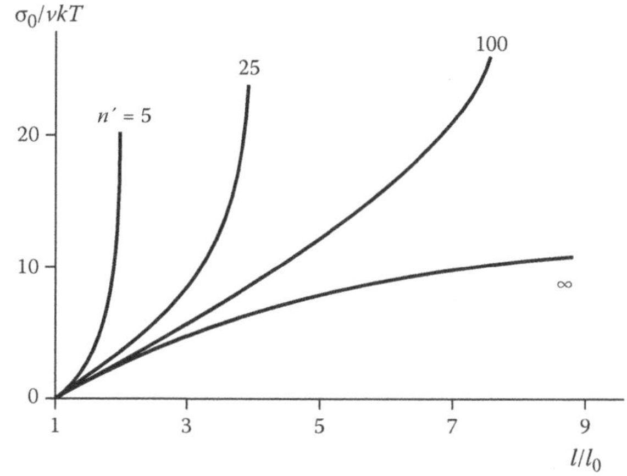

Figure 13.5 *Effect of the number of statistical chain elements in a chain between two cross-links n' (indicated) on relation between stress σ_0 and relative elongation l/l_0 of a polymer gel. σ_0, measured force divided by original cross section area of a cylindrical test piece; ν, number of chains between two cross-links per unit volume; l, length; l_0, original length of test piece. (Based on calculations by Treloar, L.R.G.,* The Physics of Rubber Elasticity, *Clarendon Press, Oxford, 1975.)*

for low n' (roughly $n' \leq 5$), the distribution of the statistical chain elements between cross-links will not be Gaussian at all, implying that for such gels, Equations 13.6 and 13.7 do not hold.

Another effect that may play a role in the stiffening of polymer gels at large deformation is the so-called strain induced crystallization implying the formation of extra cross-links due to the stretching of the chains. It is likely that its effect is usually much smaller than that due to the finite extensibility of the chains.

For polymer gels that are the result of the formation of physical bonds, usually the cross-links will break (e.g., unzipping of the semicrystalline regions forming the bonds in gelatin gels) before rupture occurs in the covalent bonds in the main polymer chain. For all polymer gels, fracture stress and strain will depend on the stochastic nature of the structure and on the fracture force of the bonds. For physically cross-linked gels, the latter parameter will also vary in a stochastic manner, resulting in a relatively large scatter of experimental results.

13.2.2 Polysaccharide gels

In the first instance, researchers tried to describe polysaccharides (and protein) gels by using the theory for flexible macromolecules. However, polysaccharide chains are much stiffer than the polymers for which the rubber theory was developed. For poly-oxyethylene, length b of the statistical chain elements is 1.6 nm, whereas for locust bean gum and carboxymethyl cellulose (CMC) it is about 8 nm, for alginate and pectin ~15 nm, and for xanthan even about 60 nm. This high stiffness is, for several polysaccharides, at least partly due to the presence of bulky side groups. As a result, in general, for polysaccharide gels n' will be low (≤5), implying that the rubber theory cannot be applied. This will be further illustrated for a study on calcium alginate gels by Segeren et al. (1974).

Dynamic measurements on alginate gels with a concentration of 9.4 kg m^{-3} gave for G' 2×10^3 Pa and for tan δ 0.01 independent of the oscillation frequency. G' was proportional to T. By using Equation 13.7, M_c could be calculated resulting in 12 kg $kmol^{-1}$, whereas M of the used alginate was 130 kg $kmol^{-1}$. This would indicate ~11 cross-links per alginate molecule. However, the observed value for M_c implies ~74 sugar units per chain section between two cross-links (M sugar unit 0.162 kg $kmol^{-1}$). Each cross-link contains about 20 sugar units, leaving 54 sugar units for the chains in between. For alginate, the length of a statistical chain element is about 15 nm. The length of a sugar unit is about 0.5 nm, which indicates that a statistical chain element contains about 30 sugar units or, in other words, n' of the chains between two cross-links is about 2. This low value implies that the rubber theory may not be applied to these systems. In accordance with these data, large deformation experiments showed a linear region to a shear strain γ of 0.13, which would also fit with $n' \leq 2$. At larger shear strain, the gels exhibited strain hardening and finally fractured at a shear strain of about 0.8. The strain hardening is assumed to be related to the deformation of the alginate junction zones.

Similar data have been obtained for other polysaccharide gels. For instance, for pectate and agar (or agarose) gels, linear regions of the order of 0.1–0.2 were also found. So, in general, polysaccharide gels cannot be described as networks built up of flexible polymer chains. In contrast, they resemble much more a network of stiff rods. Because n' will be lower when the extent of cross-linking is higher, this will all the more be the case for more highly cross-linked gels.

The temperature dependence of polysaccharide gels is primarily determined by the temperature dependency of the cross-links. For polysaccharides such as carrageenan and agar, where gelation is attributed to helix formation (**Figure 13.4a**), gel formation will take place below a critical temperature T_g and the gel will melt again above its melting temperature T_m. There may be a pronounced hysteresis between T_g and T_m; for example, for agar, T_g and T_m are about 38°C and 90°C, respectively. For κ- and ι-carrageenan, the hysteresis are about 5–10°C. For these polysaccharides, T_g depends on the cations present and may vary between 40°C and 70°C. Below T_m, the modulus of the gels increases with decreasing T. Polysaccharides as alginate form thermo-irreversible gels, which do not melt below 100°C. For these gels, gelation is due to formation of so-called "egg box" junctions (**Figure 13.4c**). Some chemical modified polysaccharides form gels above a certain temperature. This mainly concerns some cellulose esters such as methylcellose, which contain

$-OCH_3$ groups and hydoxypropyl methylcellulose. They form thermoreversible gels at temperatures above 30–90°C depending on concentration, degree of methylation, and other structural details such as uniformity of the substitution. Presumably, gel formation is mainly due to hydrophobic bonding.

Often, polysaccharide gels are subdivided between "strong" and "weak" gels. Strong gels such as alginate and agar gels are characterized by a relatively high modulus and a low tan δ (generally <0.1). The linear region on deformation is up to a strain of about 0.1–0.25 and at larger strains the gels will fracture. The gels exhibit relatively brittle behavior. Weak gels exhibit gel-like viscoelasticity under small oscillatory strains. For example, xanthan gels and κ-carrageenan gels in the presence of aggregation impeding anions may show weak gel behavior at high enough concentrations. These gels have a lower modulus and a higher tan δ compared with strong (or true) gels. The linear region is even smaller than that for strong gels, for example, ~0.05 for a 2% w/w/xanthan gel (Na^+ ion form). At larger strains, the gels yield and flow subsequently. The stress versus strain behavior exhibits clear stress overshoot at intermediate strains, its extent increasing with increasing shear rate. For xanthan gels, stress overshoot was at maximum at a strain of about 1. Weak gel systems recover their gel-like character afterward at rest, whereas strong gels never "heal" after fracture without melting and resetting. This behavior makes weak gels very suitable for dressings, etc. The gel character under low stresses prevents sedimentation of dispersed particles and flow out over the food, like a low viscosity liquid, after application.

Pectin may form both weak and strong gels depending on the conditions and the degree of substitution of the carboxyl groups with methyl groups. Low methoxyl pectin (i.e., with many nonsubstituted carboxyl groups) may form a gel in the presence of calcium ions because of the formation of so-called "egg box" junctions. Strong gels are formed that are relatively brittle. On the other hand, highly substituted high-molar-mass pectin may form a weak gel in a highly concentrated sugar solution at low pH as a result of the formation of junctions of stacked helices. The low pH neutralized most carboxyl groups and the sugar lowers the solvent quality. Weak pectin gels confer jelly-like consistency to traditional jams and marmalades.

Gels formed by swelling of starch granules as a result of heating under quiescent conditions behave in a different way. They can be described as a system fully packed with deformable particles. This type of gels will be described in Section 15.1.

13.3 Particle networks

Particle networks may be formed by various types of aggregation processes. In the first primitive models for particle networks, it was assumed that the spherical particles aggregate to strands that are mutually linked to each other at regular distances. In the so-called ideal network model, it was assumed that all particles contributed to the same extent to the stiffness of the network. The network was assumed to consist of straight strands that formed a regular cubic network with strands in x, y, and z directions in a Cartesian coordinate system. For hard particles of a diameter d, the total length of the stress carrying chains per cubic meter is then equal to the number of particles times d, resulting in $(6\varphi/\pi d^3)d$. The number of stress carrying strands, each 1 m long, in one direction per unit cross section (m^{-2}), is then given by $(1/3)(6\varphi/\pi d^2) = 2\varphi/\pi d^2$. For shearing deformation in the linear region, $\gamma \sim 2\Delta h/d$, where Δh is the extension of the distance between the particles on application of a force. This gives for the shear modulus G of the network:

$$G = \frac{\sigma}{\gamma} = \frac{(df_s/dh)(2\varphi/d^2)}{2\Delta h/d} = \frac{\varphi}{d\Delta h}\frac{df_s}{dh}, \qquad (13.9)$$

where f_s is the interaction force between the particles. Equation 13.9 predicts that the modulus is proportional to the particle volume fraction, φ. This prediction turned out to be in conflict with experimental results (see below). This objection could be circumvented by assuming that particle networks are built up of aggregates of particles and, in addition, that the ratio of the size of the aggregates (number of particles in an aggregate) over the number of particle–particle bonds between the aggregates depends on φ.

This last assumption turned out to be also better in agreement with newer particle aggregation theories. Considering only the most relevant ones there is, on one hand, random cluster–cluster aggregation of spherical particles as described by fractal aggregation and, on the other hand, there are aggregation processes closely controlled by the structure of the particles involved. An example of the latter is gel formation by fat crystals. Fat crystals are normally strongly anisodimensional, typically plate- or needle-like. Below, we will first discuss network formation by spherical particles and the mechanical properties of the gels formed, and in Section 13.5, we will focus on gel formation by fat crystals.

As discussed in Section 10.3.3, aggregation of spherical particles at rest usually leads to aggregates (flocs) with a fractal structure. In such aggregates, the number of particles in it, N_{agg}, scales with the aggregate radius R_{agg} as $(R_{agg}/R)^{D_f}$, where D_f is the fractal dimensionality (<3) and R_p is the particle radius (Equation 10.21). This scaling implies that the volume fraction of particles in the aggregate, φ_{agg}, decreases with increasing R_{agg} (Equation 10.23). As φ_{agg} approaches the original volume fraction φ of particles in the system, the aggregates may touch each other and may even interpenetrate to some extent, and a gel will be formed. This implies that the fractal dimensionality of the aggregating flocs is retained in the gel (albeit with a slight modification owing to the interpenetration), whereas at a scale $\gg R_{agg}$, the gel is homogeneous and has, of course, a dimensionality of 3 (**Figure 13.6**). A reasonable

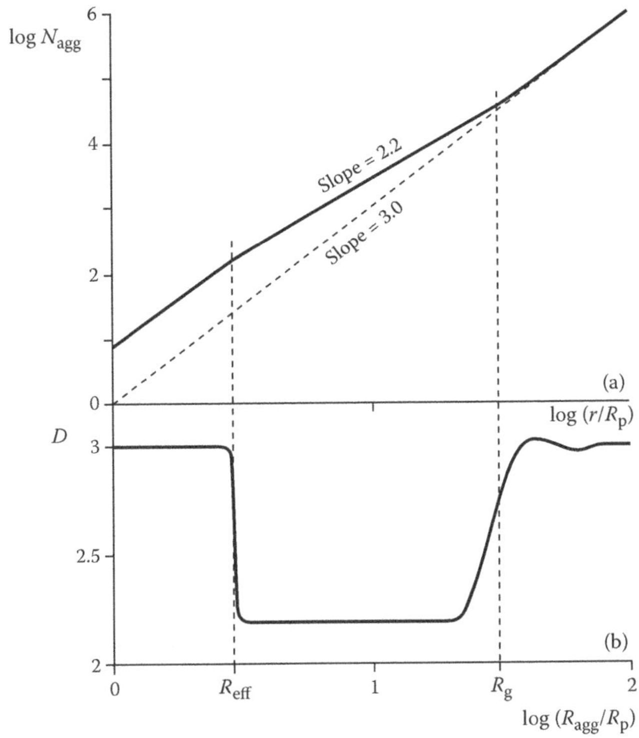

Figure 13.6 *Effects of length scale (distance r from a "central" particle of radius R_p) on fractal properties of a particle gel. (a) Log of particle number N_{agg} in a sphere of radius (r/R_p). (b) Fractal dimensionality; D_f, as a function of log (R_{agg}/R_p). Schematic. (Reprinted with some modifications from Walstra, P.,* Physical Chemistry of Foods, *Marcel Dekker, New York, 2003.)*

approximation for the radius of the aggregates/clusters in the gels can be obtained from Equation 10.23 by equating φ_{agg} and φ. This results in

$$R_g \propto R_p \varphi^{1/(D_f-3)}, \tag{13.10}$$

where R_g is the average radius of the fractal clusters forming the gel. The average number of particles in an aggregate of radius R_g is given by

$$N_{agg,g} \propto (R_g/R_p)^{D_f} = \varphi^{D_f/(D_f-3)} \tag{13.11}$$

Both R_g and $N_{agg,g}$ are smaller for larger φ and smaller D_f. For example, the values $\varphi = 0.1$ and $D_f = 2$ yield $R_g = 10R_p$ and $N_{agg,g} = 100$. This implies that a gel will be formed soon after aggregation starts.

Readers should be aware that Equations 13.10 and 13.11 as well as Equations 10.21 through 10.23 are scaling relations. They should contain numerical constants to be precise. However, these constants generally do not differ greatly from unity.

According to Equations 13.10 and 13.11, there is no lower limit for the concentration at which a gel will be formed. However, in practice, there will be one for various reasons. First, the vessel should be larger than $2R_g$. For the formation of macroscopic gels, this will usually not be an important restriction but, for aggregation processes in very small "vessels" such as emulsion droplets, it will. Second, the aggregates formed should not show significant sedimentation before a gel is formed, because otherwise demixing may occur. Third, aggregation should take place under quiescent conditions because velocity gradients can deform and may even break up the aggregates that form. The result is that, usually dense aggregates are formed that do not fill the whole space as soon as open ones and those may sediment quickly, resulting in a precipitate. Finally, at low φ, the network formed may be so weak that it is not noticed visually and is easily broken up by small velocity gradients due to, e.g., transport or temperature gradients. For these reasons, in practice there is a minimum particle volume fraction $\varphi \approx 0.01$.

Another complication is that during the aggregation process, rearrangements in the configuration of the aggregating particles may occur. It concerns a change in the mutual position of particles directly after bonding. This is illustrated in **Figure 13.7a**. The particles "roll" over each other until they have obtained a higher coordination number (number of particle–particle bonds), which implies a more stable configuration. Because the time required for this process increases strongly with the size of the aggregating particles/"aggregates," above a certain size of these "aggregates" network formation proceeds faster

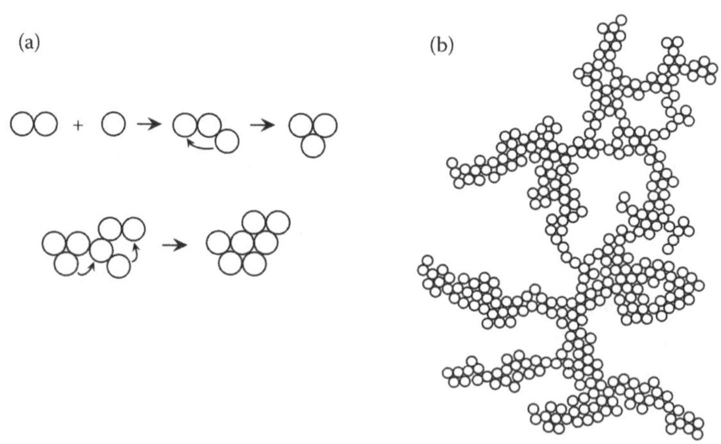

***Figure 13.7** Rearrangement processes that may occur during particle aggregation. (a) Examples of particles rolling over each other so that a higher coordination number is obtained. (b) Example of a resulting fractal cluster in two dimensions, when rearrangements have occurred during aggregation. (Redrawn from Walstra, P.,* Physical Chemistry of Foods, *Marcel Dekker, New York, 2003.)*

than the rearrangement process and fractal clusters may still be formed, as illustrated in **Figure 13.7b**, although the building blocks of the aggregates now are larger. The size of the primary particles forming the fractal cluster then has to be replaced by the size of the small compact "aggregates" formed as a result of the rearrangement process. Equation 10.21 should be replaced by

$$N_{\text{agg}}/n_{\text{p}} = (R_{\text{agg}}/R_{\text{eff}})^{D_{\text{f}}}, \tag{13.12}$$

where n_{p} is the number of particles forming the effective primary particles with radius R_{eff}.

These so-called short-term rearrangements are very common during aggregation processes. For acid-induced milk gels, as present in yogurt, it has been shown that by increasing the acidification temperature from 20°C to 40°C, the size of the effective primary building blocks increases from that of the protein particles present originally in the milk to building blocks containing about 40–50 primary protein particles (Lucey et al. 1997). If rearrangements proceed very extensively, precipitation may occur instead of gel formation, depending on the primary particle concentration. This may lead to a measurable critical volume fraction.

Readers should be aware that the gels formed are not fractal; the aggregates/clusters forming the gels are. Both at length scales smaller than R_{eff} and longer than R_{g}, the dimensionality is 3 (**Figure 13.6**). Fractal behavior is only seen at length scales between R_{eff} and R_{g}. This means that the fractal region decreases with increasing value of φ. This implies that for high φ the aggregates forming the gel will no longer be fractal, roughly for $\varphi > 0.3$.

The contact area between two aggregates in a gel will scale as R_{g}^2. If we assume the structure of the clusters building the gel to be scale-invariant, the number of links/bonds between two adjacent clusters will be independent of their size and, hence, independent of the original particle concentration. Links will primarily be formed via strands at the periphery of the clusters. This implies that the number of links N_{lc} between adjacent clusters per cross section of the gel will scale with R_{g}^{-2}. It results in the next relationship between N_{lc} and system properties;

$$N_{\text{lc}} \propto R_{\text{g}}^{-2} \propto R_{\text{eff}}^{-2}\varphi^{2/(3-D_{\text{f}})} \tag{13.13}$$

The number N_{lc} in Equation 13.13 (common SI unit m^{-2}) may be equated to N in Equation 13.3, which relates G' to some structure parameters of a gel.

Depending on the structure of the strands, an expression relating C to R_{g} can also be deduced. Below, this is done in a rather intuitive way. For more rigorous mathematical derivations, readers are referred to the literature (e.g., Bremer and van Vliet 1991; Mellema et al. 2002a). As mentioned above, C is a characteristic length relating the local extension of bonds to the macroscopic strain. For the case where a gel is built up of fractal aggregates, we assume that the local strains in the gel will be uniform and equal to the macroscopic strain (affine in rheological terms) at length scales larger than the size of the aggregates R_g. Then C relates the deformation of the bonds between the particles (or of the particles if they are deformable) in the stress carrying strands to the macroscopic strain γ or ε (which is equal to $\Delta R_g/R_g$). The number of deformable bonds equals the length l_a of the stress carrying strands in an aggregate divided by the diameter (d) of the "rigid" units. The size of the rigid units will depend on the structure of the gel forming "fractal" aggregates. Depending on the model, the rigid units are the particles forming the "fractal" aggregates or the aggregates themselves (see below). Working this out, it becomes evident that C will scale with $\Delta R_g/(\gamma \cdot l_a/d) = R_g/(l_a/d)$. The term dF^2/dx^2 in Equations 13.2 and 13.3 will strongly depend on how the strands are deformed, by bending or by extension. In the latter case, dF^2/dx^2 will be independent of the length of the strand. For bending, the force needed to obtain a given deformation scales per strand, like R_g^{-1}, so that dF^2/dx^2, will be proportional to $\propto R_g^{-2}$.

Four different situations will be distinguished depending on the structure of the gel-forming aggregates and their mutual links (**Table 13.2**). It should be noted that in practice the structure of a gel may also fall in between two of these models.

1. *Fractal strands.* In this model, the rigid units are the particles forming the fractal aggregates. The strands in the aggregates have a fractal structure, implying that their length is proportional to R_g^x, where x is the so-called chemical dimensionality of the strand that may vary between 1 and 1.3. The length l of the strands in a gel is then proportional to R_g^x. Deformation of the strands will be by bending. This

Table 13.2 Scaling Relations between Parameters in Equation 13.3 and the Radius of the Aggregates, R_g, Making Up the Network for the Various Structural Models Discussed in the Text

Structural Model	Structure	$l_a/2R$	$C \propto R_g 2R/l_a$	d^2F/dx^2	$NC(d^2F/dx^2)$
Fractal strands	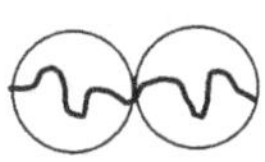	R_g^x	R_g^{1-x}	R_g^{-2}	$R_g^{-(3+x)}$
Hinged strands		–	R_g	R_g^{-2}	R_g^{-3}
Stretched strands	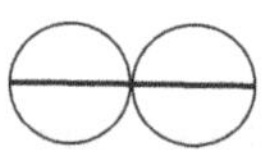	R_g	–	–	R_g^{-2}
Weak links	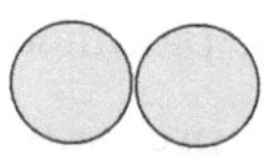	–	R_g	–	R_g^{-1}

Note: The circles schematically denote the aggregates forming the gel; inside only the stress carrying strands (of length l_a) are shown. It is assumed that the stress on the gel acts only in the horizontal direction. R is the radius of the undeformed particles, $l_a/2R$ gives the number of bonds that will be deformed *in an aggregate* on deformation of the gel, and x is the chemical dimensionality. C is elastic constant of the stress carrying strands. For all models, the number N of stress carrying strands per cross section of the gel scales with $\propto R_g^{-2}$.

model implies that the aggregates must be highly rarefied, and that the strands are long and slender. Such a gel will be very weak and prone to restructuring, for example, due to the formation of extra bonds between the strands. This will invalidate the model.

2. *Hinged strands.* This category of gels may be formed as a result of rigidifying of curved strands, for example, due to rearrangements after gel formation. Effectively, the aggregates behave as the rigid units. Furthermore, it is assumed that the aggregates are still so loosely interconnected that bending of the bonds between the aggregates dominates the deformation behavior. The length l of the strands will be independent of R_g. Because some rearrangements will nearly always occur in the fractal aggregates after gel formation, this model is generally more realistic for food products than that based on fractal strands.
3. *Stretched strands.* This will be the case when a series of particles of constant size are linked in a perfectly straight line or when the stress carrying structures are highly interconnected (**Figure 13.8**). In the latter case, part of the strands will deform in bending and the others by stretching. Because stretching a strand requires a much larger force than that used for bending, the bending force can be neglected in the calculation of the overall force needed for deforming the gel. The particles forming the aggregates are the rigid units. This implies that the number of deformable bonds in an aggregate is proportional to R_g. When stretching straight strands that are affine deformable over length scales larger than the diameter of the particles, the relationship between the local extension and the overall strain will be independent of their length, so C is independent of R_g.
4. *Weak link.* In this model, similar to situation 2, the aggregates forming the gel are assumed to be rigid as a result of ongoing cross-linking between the particles after the gel has been formed. On deformation of the gel, only the relatively weak links between the aggregates are

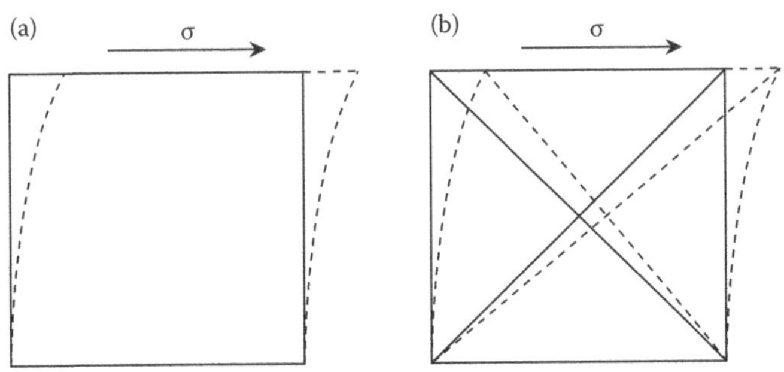

Figure 13.8 *Highly schematic picture illustrating the way in which strands will deform under an applied shear stress for structures built of straight strands. (a) Deformation material primarily involves bending of vertical strands; (b) deformation involves both stretching/elongation of diagonal strands and bending of vertical strands.*

deformed by stretching. This generally means that both φ and D_f will be high, and that the primary particles are quite rigid. Such systems have rarely been observed in foods. The number of deformable links in the gel as a whole is proportional to $1/R_g$ and therefore C scales with R_g.

In **Table 13.2**, an overview is presented of how the various models for the structure of the particle gels work out on the scaling relation between the modulus of the gels and the size of the fractal clusters.

Based on the data shown in the right column in **Table 13.2** for the various structural models and by using Equation 13.13, it is possible to write the next relation between the shear modulus G of a gel built up of aggregates with a fractal structure and parameters characterizing gel structure

$$G = NC\frac{dF^2}{dx^2} \propto (R_{\text{eff}})^{-\alpha}\varphi^{\alpha/(3-D_f)} \tag{13.14}$$

The value of α is shown in **Table 13.3** for the different models.

As is clear from **Table 13.3**, the dependence of G on the volume fraction particles may vary greatly depending on the structure of the network formed on gelation. As an example, in **Figure 13.9**, the storage and loss modulus are shown as a function of the casein concentration for two types of acid-induced casein gels. The slope of the log–log plots is 2.6 in the case of HCL-induced gels (type 1 gels) and 4.6 for the gels formed by glucono-δ-lactone (GDL) addition (type 2 gels). The moduli of the type 1 gels are, especially for the lower casein concentrations, much higher than those of type 2 gels. Both gels are formed under quiescent conditions. However, for type 1 gels, gelation starts at a lower T (~10°C). The voluminosity (mL g^{-1}) of the casein particles is then relatively high. During further heating, the voluminosity will decrease, which leads to a straightening of the particle strands making up the network. The fractal dimensionality that can be calculated from these curves is 2.24 for type 1 gels by applying the stretched strand model and 2.35 for type 2 gels in combination with the hinged strands model. These fractal dimensionalities are in agreement with those obtained from other data (Bremer et al. 1990).

Table 13.3 Model-Dependent Exponents α, υ, and β in Scaling Relations between Storage Modulus, G', Fracture Stress, σ_{fr}, and Fracture Strain γ_{fr} and Volume Fraction φ of Particles Making Up the Gel Network, Respectively, for Nonbrittle Strands

Structural Model	α (Equation 13.14)	υ (Equation 13.19)	β (Equation 13.20)
Fractal strands	$3 + x$	2	$-x$
Hinged strands	3	2	0
Stretched strands	2	2	0
Weak links	1	2	1

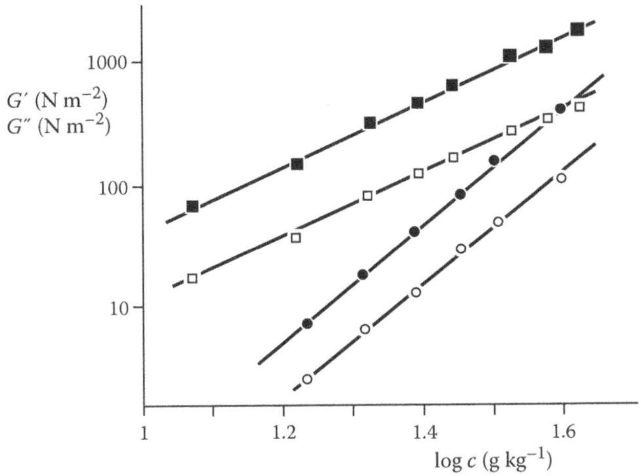

Figure 13.9 *Storage G′ (filled symbols) and loss modulus G″ (open symbols) as a function of casein concentration of acid-induced gels (pH 4.6) made of Na caseinate. Gels made by acidification at 2°C followed by slowly heating to 30°C (■, □) or at 30°C by glucono-δ-lactone addition (•, ○). Aging time about 16 h. Angular frequency 1 rad s^{-1}. (Redrawn from Bremer, L.G.B. et al.,* Colloids Surf., *51, 159–170, 1990.)*

The large difference in moduli can also be explained based on the two models. Moduli will be much higher for gels in which the resistance to deformation is based on tensile deformation of straight strands than for bending deformation of relatively long thin strands. In agreement with the difference in network structure, the fracture strain for type 2 gels was a factor of 2 higher than for the type 1 gels (1.1 and 0.5, respectively). The fracture stress was roughly the same for both types of gels, about 100 Pa.

Effect of rearrangements. As mentioned above, various types of rearrangements may occur both during gel formation and after a gel is formed. In general, rearrangements may occur at different length scales. An overview is given in **Figure 13.10**. Particle fusion will mainly occur during gel aging and causes an increase in the stiffness of the chains. It may occur between particles that are built up of smaller units, for example, protein molecules in protein aggregates or triglyceride molecules in fat crystals. Interparticle rearrangements will mainly occur during aggregation. It has already been discussed above and will lead to an increase in the effective particle radius, R_{eff} in Equations 13.12 and 13.14. Rearrangements at strand level may involve, among others, ongoing cross-linking between dangling ends and strands and mutual cross-linking between flexible strands. This and particle fusion may lead to tensile stresses in strands nearby. This causes stretching (straightening) of strands and eventually even fracture of strands. Then the fractal character will be lost in the clusters forming the gel. If many strands break and the gel is, moreover, not constrained by vessel walls, macroscopic contraction

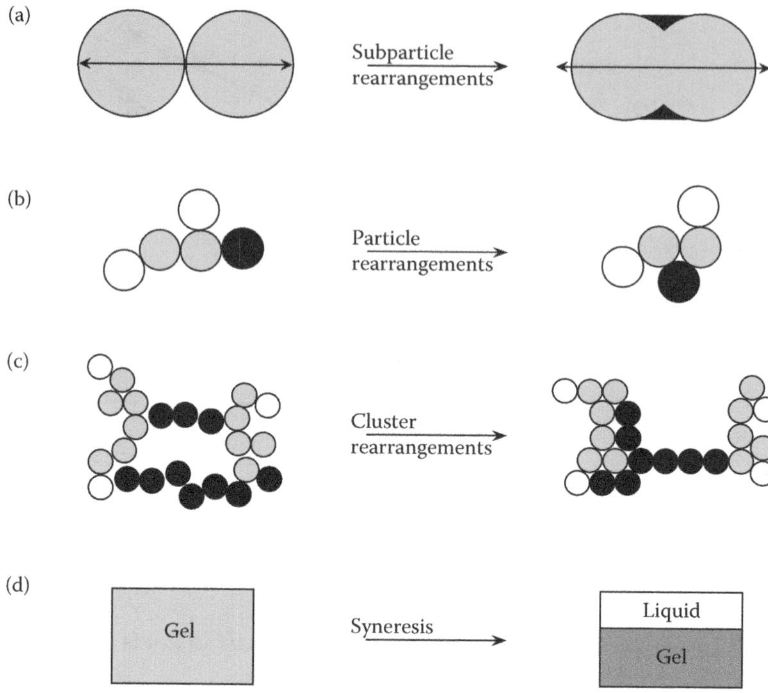

Figure 13.10 *Four types of rearrangements that can be distinguished in particle gels at various levels: (a) molecular or subparticle level leading to particle fusion; (b) particle level leading to interparticle rearrangements; (c) involving strands of particles; and (d) macroscopic level giving syneresis. (Reprinted from Mellema, M. et al.,* Adv. Colloid Interface Sci., *98, 25–50, 2002b. With permission.)*

may set in and liquid is forced out of the (shrinking) gel. This process is called syneresis. During rheological measurements, it often causes slip between the gel and the measuring body. Depending on the stage in which rearrangements occur, their effect on the modulus of the gels may differ. An overview of possible rearrangements in particle gels and their main effects on the moduli is given in **Table 13.4**.

Fusion of particles may eventually result in a loss of identity of the original particles. The interaction between the original particles becomes just as stiff as the rest of the particles. If this is the case, it is inappropriate to speak further of interaction energy between particles. An alternative is to assign a certain modulus to the strand. For a cylindrical chain of length l_c and radius R_c, where the stiffness of the particles is the same as that of the bonds in between, the Young's modulus E is given by

$$E = \frac{\sigma}{\Delta l_c / l_c} = \frac{f}{\pi R_c^2} \frac{l_c}{\Delta l_c}, \tag{13.15}$$

Table 13.4 Main Effects of Rearrangements in Particle Gels on Moduli

Stage	Effect on Geometric Structure	Effect on Moduli (Equation 13.3)
During aggregation	Increase in R_{eff}, hence also of R_g	Lower N; may (partly) be compensated by larger d^2F/dx^2
	Increase in D_f, hence also of R_g	Stronger dependence N (and C) on φ
After a gel has been formed	Fusion of particles	Larger d^2F/dx^2 term or often, in fact, E of strands
	Straightening of strands	Smaller dependence of C on φ; increase in effective E of strands
	Displacement of particles to center of fractal clusters[a]	Larger holes; fractal description will be lost
	Fracture of strands	Lower N, fractal description will be lost

Note: For explanation of symbols, see text.

[a] Importance in practice probably small.

where f is the tensile force in the strand. In the linear region, it holds that $\Delta l_c/l_c = \Delta x/x = \Delta R_c/R_c$. The parameter f is equal to $(d^2F/dx^2)\Delta x$. This ultimately leads to

$$E = \frac{1}{\pi R_c}\frac{d^2F}{dx^2} \tag{13.16}$$

And for the shear modulus of the gel to

$$G = NC\pi R_c E \tag{13.17}$$

One should realize that parameter C, and with that G, will strongly depend on the way in which the strands are deformed, being in tension or by bending.

13.3.1 Large deformation behavior of particle gels

The main structure parameters characterizing large deformation behavior of particle gels will be the resistance to deformation of the aggregates building up the network and of the bonds between them, the curvature of the strands, and the deformation of the particles. When the deformation of the gel is affine down to the aggregate level, the stress on these aggregates will be the same as on the gel, and so will be their yielding or fracture stress, σ_{fr}. The latter is given by

$$\sigma_{fr} \propto f_{fr}/R_g^2, \tag{13.18}$$

where f_{fr} is the tension force within the strand at fracture, which is a constant independent of the size of R_g. Note that fracture of the strands in most gels will be determined by their fracture force in tension even when the gel at rest contains curved strands. Expect that, for the case where strands are brittle, curved strands will be straightened on deformation before they fracture. The latter will be the case for gels built up of protein particles. Thus, σ_{fr} scales with R_g^{-2}

independently of the volume fraction particles and the structural model, so using Equation 13.13 gives:

$$\sigma_{fr} \propto (R_{eff})^{-\upsilon}\varphi^{\upsilon/(3-D_f)} \propto (R_{eff})^{-2}\varphi^{2/(3-D_f)} \qquad (13.19)$$

For the fracture strain, it is different. It is easy to imagine that for straight strands of hard particles that are deformed affine in tension, the Cauchy fracture strain, γ_{fr} will be proportional to $\Delta h/h$, where h is the equilibrium distance between the particles and Δh is the increase in distance up to bond rupture. Since, for this model, both Δh and h do not depend on φ, fracture strain will also be independent of φ. If Δh does not differ too much between gels that contain similar particles, but of different size, γ_{fr} will be larger for the gel consisting of smaller particles.

For the weak link model h should be replaced by $2R_g$ giving $\gamma_{fr} \propto R_g^{-1}$. Because R_g decreases with increasing φ (Equation 13.13), γ_{fr} will be larger when φ is higher.

For the hinged and fractal strands models, the curvature of the strand is an important factor determining γ_{fr}. The stronger it becomes, the larger is the fracture strain. For the fractal strands the curvature of the strands will increase with decreasing φ, whereas for hinged strands this will not be the case. For the latter case, a strand with two small bends (higher φ) will have the same length as a strand with one similar curved bend (lower φ). Then if the strands fracture after straightening, γ_{fr} will be independent of φ. For fractal strands, the number of bonds that deform under an applied stress increases with R_g^x. Working this out, it can be seen that γ_{fr} is proportional to R_g^x. If the network strands are brittle, fracture may occur before the strands are stretched, causing a somewhat stronger dependence of γ_{fr} on R_g for the hinged and the fractal strands models. Next, we can write the following general scaling equation

$$\gamma_{fr} \propto R_{eff}^{-\beta}\varphi^{\beta/(3-D_f)}, \qquad (13.20)$$

where β depends on strand curvature, their brittleness, and the ratio between the stiffness of the fractal cluster over that of the bonds in-between. For flexible strands, β varies from $-x$ for gels with curved fractal strands in the clusters, to 1 for gels where deformation is located in the bonds between the fractal clusters (weak link model) (**Table 13.3**).

Another point not discussed explicitly so far, is that particles forming the network in a gel may be deformable. This will especially be the case if these particles are aggregates of protein molecules as is the case, for example, for casein (milk) gels formed by acidification or enzyme action. For particle volume fraction approaching φ at maximum packing, the modulus will approach the modulus of the protein particles and does not go to infinity. Another effect is that the linear region and especially the strain at fracture will be higher than one would expect for a gel of nondeformable particles (see data for casein gels and polystyrene latex gels in **Table 13.5**).

Table 13.5 Linear Region and Fracture Strain in Shear for Various Gels

Material	Type of Gel	Linear Region	Fracture Strain
Vulcanized rubber	Gel with flexible chains between cross-links	1–3	2–5
Gelatin	Gel with flexible chains between cross-links	0.5–1	0.8–1.5
Alginate	Gel with stiff chains between cross-links	~0.1	1–1.5
Soy protein isolate	Heat set protein gel	~0.05	0.3–1
Casein (milk) gels	Gel from deformable particles.	~0.03	0.5–1.1[a]
Polystyrene latex gels	Gel from hard spherical particles; $R \sim 50$ nm	~0.01	0.05–0.2
Margarine	Gel from hard anisometric fat crystals	~0.0003	0.1–0.3

Source: Various sources.
Note: Values will depend on, e.g., concentration, deformation speed, salt concentration, pH, and batch of polymer.
[a] Acid-induced casein gels.

Summarizing, the discussion above illustrates that particle gels may exhibit a huge variety in the dependency of their mechanical properties on the concentration particles. This makes it impossible to predict the quantitative effect of a change in concentration of gel properties, for example, the modulus (stiffness or rigidity) and fracture stress. For the fracture strain, it is even more difficult since, depending on the structure of the gels, an increase in concentration may result in no effect, an increase, or a decrease in the fracture strain.

13.4 Comparison of polymer and particle gels

Although the basis, in terms of a change in Gibbs energy, for the resistance against a small deformation is completely different between polymer networks and particle networks, a simple plot of the dynamic moduli against angular frequency often looks very similar (compare **Figures 13.11a and b**). In contrast to this similarity, the size of the linear region and the large deformation behavior, including the fracture strain, are very different between these two types of gels (**Figure 13.12** and **Table 13.5**). The latter is caused by the very different structure of the networks, although there are some similarities.

Flexible polymer chains can be deformed strongly before they will rupture, the more so the longer they are. Fracture strain is large, and related directly to the ratio of the (contour) chain length between cross-links and the distance between them at rest. As discussed in Section 13.2.1, at large strain an increase in the resistance against deformation may occur. For particle gels, the fracture strain depends strongly on the structure of the gel and on the deformability of the particles.

As particles are normally much less deformable than polymer molecules and the strands of aggregated particles are much less curved and flexible, the fracture strain for particle gels will be much smaller than that for flexible polymer gels (**Figure 13.12** and **Table 13.5**). The fracture strain of gels consisting of

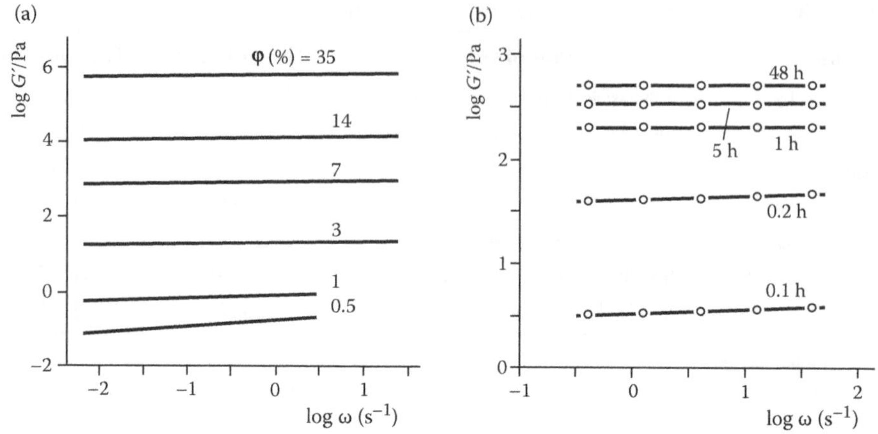

Figure 13.11 *Comparison between frequency dependence of storage modulus of a particle gel (a) and a polymer network (b). Particles are latex sols at different volume fractions φ (indicated) and polymer is gelatin, 1.95% (w/w) at various aging times. (Reprinted from van Vliet, T., Lyklema, J., in* Fundamentals of Interface and Colloid Science, Vol. IV, Particulate Colloids, *ed. Lyklema, J., Academic Press, Amsterdam, 2005. With permission.)*

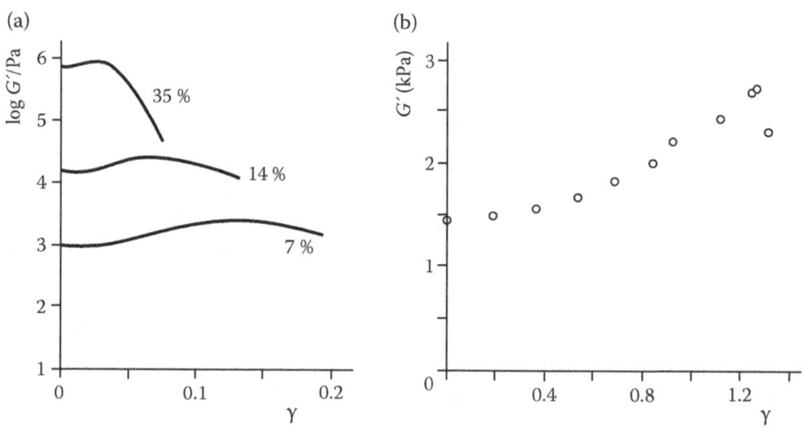

Figure 13.12 *Comparison between storage modulus as a function of shear strain for: (a) a particle gel and (b) a polymer network. At the ends of curves, gels were fractured. (a) Particles are lattices at different volume fraction (indicated); (b) polymer is gelatin, 4% w/w, after aging for 4 h at 15°C. (Reprinted from van Vliet, T., Lyklema J., in* Fundamentals of Interface and Colloid Science, Vol. IV, Particulate Colloids, *ed. Lyklema, J., Academic Press, Amsterdam, 2005. With permission.)*

stiff macromolecules and of deformable particles will assume an intermediate position.

Also, regarding the dependence of the modulus on the volume fraction or concentration of the material that forms the network, similarities can be seen. The exponent α in Equation 13.14 is for $\alpha = 2$, the same as that obtained in percolation models for the case of isotropic interaction forces between the interacting macromolecules. In that model, a network of connected macromolecules is considered in which, at random, connections are formed and the modulus is calculated, or vice versa. Below a certain number of connections, the modulus will be zero. The concentration of bonds at which the modulus increases from zero to a finite number is called the percolation threshold. By replacing the number concentration of bonds by the weight concentration of molecules (or the volume fraction of particles), a percolation concentration c_{tr} (or volume fraction) can be defined. It results in an equation of the form

$$G \propto (c - c_{tr})^2 \qquad (13.21)$$

The occurrence of the same power as in Equation 13.14 for stretched strands (**Table 13.3**) can be understood if one realizes that this percolation problem is only related to the number of elastically effective strands, just as in the stretched strand model and not to details regarding how it is deformed. In contrast to the percolation models, the fractal model does not anticipate a general critical volume fraction for gelation to occur. The only result of a lower volume fraction will be a coarser gel (Equation 13.10). However (as discussed above), in practice, because of disturbances by flow or sedimentation, etc., a critical volume fraction for gelation will be observed.

13.5 Heat-set protein gels

When a solution of globular proteins is heated above the denaturation temperature of the protein, a gel may be formed if the concentration is above a critical value c_{tr}. Gels are only formed if at least a part of the protein molecules has been heat denatured and their formation is irreversible on cooling. Generally, gel formation is a relatively slow process, lasting at least several minutes and sometimes even hours. The modulus of the gels strongly increases on cooling (**Figure 13.13**). The magnitude of c_{tr} depends on the nature of the proteins, the physicochemical conditions, and the rate of heating. Protein gels may also be formed by changing the conditions that affect molecular or colloidal interactions, such as pH, ionic strength, addition of specific salts (e.g., Ca^{2+}), or enzyme action. Examples are acid-induced and rennet-induced milk gels, and cold gelation by pH change or specific ions addition of thermally induced globular-protein aggregates (e.g., of β-lactoglobulin or ovalbumin). These gels will not be discussed in this section. The main principles of the properties of these gels are discussed in Section 13.3 and those of gelatin gels in Section 13.2.

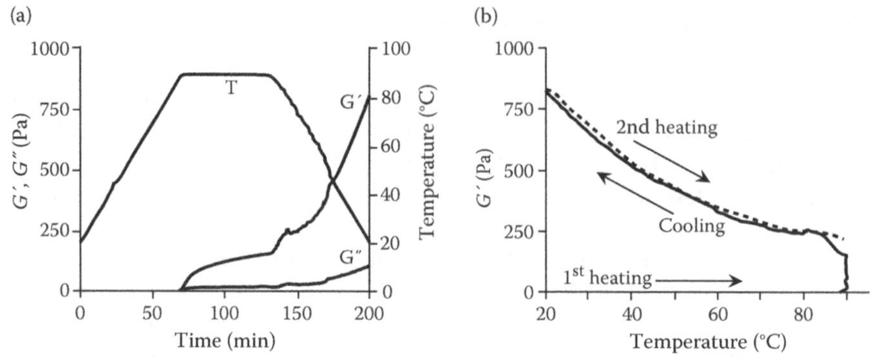

Figure 13.13 *Storage modulus G′ and loss modulus G″ of an 11 w/w% soy protein isolate dispersion at pH = 7 as a function of time during a heating and cooling cycle (temperature T indicated) (a) and G′ as a function of temperature (b). In b, full line represents the first heating and cooling cycle, and dotted line denotes the second heating curve. (Reprinted from Renkema, J.M.S., van Vliet, T.,* J. Agric. Food Chem., *50, 1559–1573, 2002. With permission)*

Details of gel formation and properties of heat-set gels vary widely among the proteins involved. This is because globular proteins differ greatly in molecular structure and conformation stability and, in addition, in food preparation heat-set gels are usually made up of protein mixtures such as isolates (or concentrates) of soy proteins or whey proteins. In these mixtures, the various protein molecules react in a different way during heat treatment and often mutually as well [e.g., β-conglycinin and glycinin in soy protein isolate (SPI) at pH 3.8]. In general, gel formation will involve a number of consecutive reactions: (1) protein molecules become denatured; (2) denatured molecules aggregate to form (roughly spherical or elongated) particles: and (3) these particles aggregate to form a space-filling network. There are indications that in several cases, step (3) may initially proceed as fractal aggregation as described in Section 13.3. However, usually a gel is already formed before all protein molecules are denatured, as a result of which, reactions (1)–(3) proceed at the same time for a large part of the gel formation process (**Figure 13.14**). Thus,

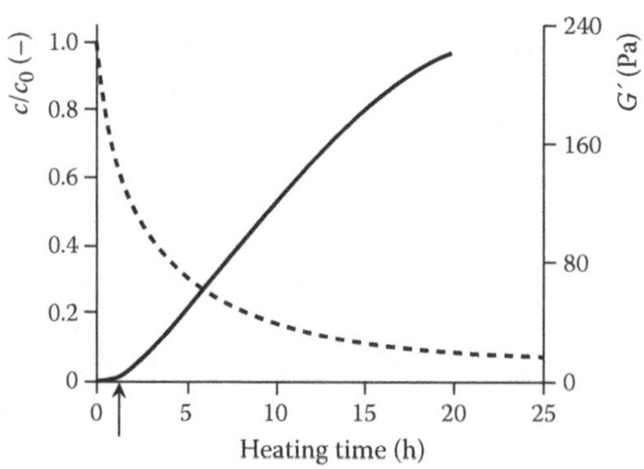

Figure 13.14 *Fraction "native" whey isolate (WPI) (c/c_0) (dashed curve) and storage modulus G′ (full curve) as a function of heating time at 68.5°C for an initial WPI concentration 44.5 kg m^{-3} and 0.4 M NaCl; arrow indicates the gel point. (After Verheul, M., Roefs, S.P.F.M.,* Food Hydrocolloids, *12, 17–24, 1998.)*

after a space filling network is formed, denatured protein molecules or protein particles are deposited onto the network. The network will keep about the same overall geometric structure, although pores will become somewhat smaller, whereas the strands and especially the bonds between the particles will become stiffer. This will upset the fractal properties of the gels at least to a large extent. When reaction (3) proceeds slowly compared to reactions (1) and (2), a high minimum concentration for gel formation will be observed and/or precipitation of protein aggregates may occur.

Based on appearance and network structure as seen by microscopy, the network structure of heat-set gels can be divided roughly into fine-stranded and coarse aggregated networks. Fine-stranded gels are generally clear gels. They may be completely transparent and are built up of relatively thin strands with a thickness several times the size of a single protein molecule (thickness generally between 15 and 50 nm). These gels typically form at a pH far away from the isoelectric point at low ionic strength. Coarse gels are nontransparent turbid gels composed of roughly spherical particles with diameters in the range of 100–1000 times that of a single molecule. Intermediate structures also exist that contain fine-stranded and coarse structures. In general, gel structure is coarser as the pH approaches the isoelectric point or when the ionic strength is higher.

In addition to coarseness of the strands, their curvature is expected to affect the mechanical properties of the gels. When the curvature of the strands is higher, the gels are expected to have a larger fracture strain and a lower modulus (Section 13.3) regardless of whether or not they are built up of fractal aggregates (clusters). Another likely secondary factor affecting the modulus is the thickness of the curved strands. For cylindrical strands of equal length, the bending force scales with their diameter to the power four, which implies a lower modulus for networks built up of thinner strands. Furthermore, the modulus is expected to depend on the properties of the network on molecular scale as the stiffness of the denatured molecules and of their mutual interactions.

No clear effect of curvature of the strands on the fracture stress is expected since, during large deformations, all stressed curved strands will be straightened, implying that they all will fracture in tension. On the other hand, coarseness may affect fracture stress because higher coarseness generally implies larger defects in the network structure, and this will give a lower fracture stress.

Because it is difficult to quantify strand curvature based on light microscopy (including confocal laser microscopy) pictures (and of other factors involved), it is difficult to predict the mechanical properties of heat-set gels based on such pictures alone. However, as indicated implicitly in the paragraphs above, such pictures can be of great help in understanding the effect of conditions such as pH, ionic strength, heating rate, and time, on mechanical properties and so allow a more focused product development.

As an example of heat-set gels and the influence of pH and ionic strength, some results for SPI gels will be discussed. Gel formation of the SPI dispersions (12 w/w%) was induced by heating from 20°C to 95°C at a heating rate of 1 K min^{-1}, keeping T at 95°C for 60 min, and cooling down at a rate of 1 K min^{-1}. **Table 13.6** gives a summary of some small and large deformation characteristics. Young's moduli showed the same trends as the shear moduli shown.

Gels formed at pH 7.6 and at pH 3.8 without added NaCl have likely a more fine-stranded structure and the other ones a coarser network structure. From **Table 13.6**, it follows that there is no unequivocal relation between the modulus and fracture strain of the gel and its coarseness. For example, at pH 3.8 (0 M), the structure is not coarse and has a high modulus and low fracture strain, whereas for gels at pH 7.6 the structure is also not coarse but the modulus is low and the fracture strain is intermediate. Taking into account both coarseness and strand curvature can explain the difference in modulus between pH 3.8 (0 M) and pH 7.6. At pH 3.8, fracture strain is low, implying on average more stretched strands and with that a high modulus, whereas at pH 7.6 strands are curved, leading to a low modulus. Also at pH 5.2, fracture strains are large, so strands are more curved and therefore moduli are lower than at pH 3.8. The fracture strain of the gels at pH 5.2 (0.5 M) is about the same as that at pH 7.6, whereas the modulus is much higher. For both gels, deformation at small strain is likely, mainly by bending the protein strands. However, the strands of the coarse gel at pH 5.2 will be much thicker and will therefore require a higher bending force than those in the fine-stranded gels at pH 7.6. Moreover, at pH 7.6 less protein is incorporated in the gel network.

Table 13.6 Mean Values of Storage Modulus, Fracture Stress and Fracture Strain of 12 w/w% Heat-Set Soy Protein Isolate Gels

pH	Added NaCl	Storage Modulus (kPa)	Fracture Stress (kPa)	Fracture Strain (–)	Microstructure and Appearance
3.8	0 M	20.5	14	0.28	Fine, white
	0.2 M	14.0	12	0.38	Coarse, white
	0.5 M	15.2	13	0.40	Coarse, white
5.2	0 M	[a]	31	1.1	White
	0.2 M	7.3	25	1.05	Coarse, white
	0.5 M	11.2	21	0.68	White
7.6[c]	0 M	1.1	10	0.79	Fine,[b] light brown, jelly-like
	0.2 M	1.9	8	0.61	The same as 0 M
	0.7 M	1.4	10	0.69	The same as 0 M

Source: Data from Renkema, J.M.S., *Food Hydrocolloids* 18: 39–47, 2004. With permission.
Note: pH and concentration of added NaCl indicated.
[a] No data because of protein precipitation, Young's modulus was about the same as for the pH 5.2, 0.2 M NaCl gel.
[b] No clear micrographs could be obtained.
[c] At pH 7.6 less protein is incorporated in the network than at pH 3.8 and 5.2.

These two factors and a possible difference in interaction energy between the protein molecules may explain the observed results.

For the studied SPI gels, fracture stress σ_{fr} was observed to depend much less on the conditions than the modulus (**Table 13.6**). This is understandable, since this parameter is not expected to depend on strand curvature. It will depend on the amount of protein incorporated in the gel network. This factor is likely, at least partly, responsible for the lower σ_{fr} observed for the gels made at pH 7.6. In addition, gel coarseness via its effects on defect size and differences at molecular scale as a function of pH and ionic strength will play a role. Because the dependence of fracture stress on conditions is much smaller that it is for the modulus, it can be concluded that the effects of differences in gel structure at molecular scale plays a smaller role in determining the influence of gel forming conditions on the modulus than differences at mesoscopic scale.

The above discussion for soy protein gels suggests that a direct relation between network structure and elastic modulus of heat-set gels can be obtained if both coarseness and curvature of the strands are taken into account—provided that other factors such as interaction forces do not change substantially. A rheological characterization of fracture strain and storage modulus as a function of curvature of the strands and coarseness of the network structure is shown schematically in **Figure 13.15**.

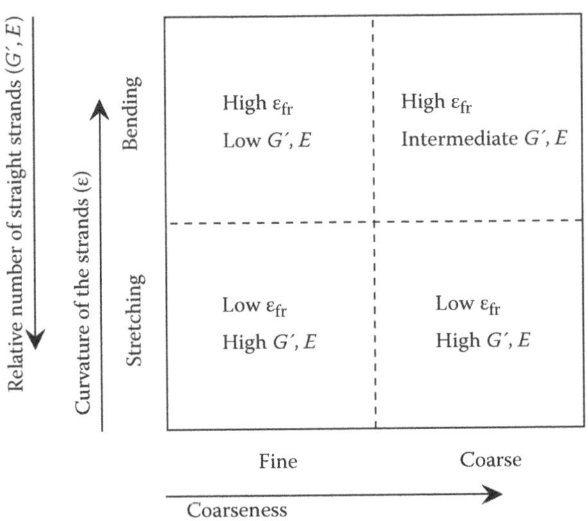

Figure 13.15 *Rheological characterization of protein gels as a function of structural parameters, i.e., curvature of the strands, relative number of strands deformed in stretching deformation, and coarseness of gel (thickness of strands). Dotted lines between categories indicate a gradual transition. Only main trends are indicated. ε_{fr} fracture strain; G′, storage modulus; E, Young's modulus. (After Renkema, J.M.S.,* Food Hydrocolloids, *18, 39–47, 2004.)*

13.6 Plastic fats

A plastic fat consists of a space-filling network of triglyceride crystals, where the continuous phase is triglyceride oil. Products such as butter and margarine contain, in addition, emulsified moisture droplets (water with salt, flavor components, etc.). Butter and standard margarine contain about 80% fat. At room temperature, the volume fraction of crystals in the fat is generally below 0.5. Therefore, a plastic fat can be considered a particle gel. However, their properties are clearly different from gels formed from hard spherical particles as discussed in Section 13.3 for various reasons. The main ones are

- Plastic fats are formed by cooling a liquid triglyceride mixture to a temperature in which at least part of the triglyceride molecules starts to crystallize. Network formation encompasses three processes that proceed initially in the following order: formation of nuclei, growth of crystals, and aggregation of crystals. However, after a little while, all three processes happen simultaneously (**Figure 13.16**). Network

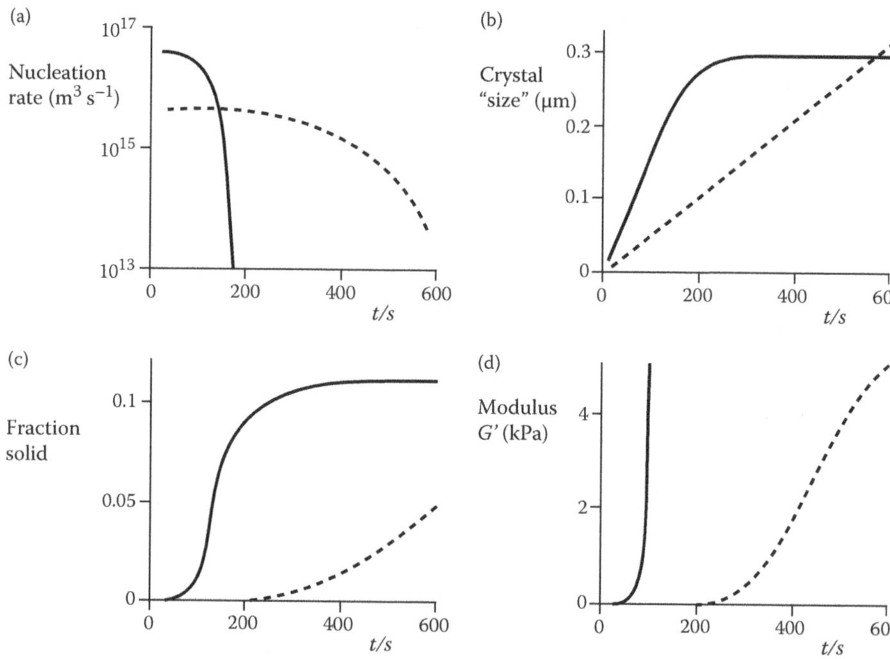

***Figure 13.16** Course of main processes involved in formation of a simple crystal network as a function of time t after cooling of a simple mixture of 12% hydrogenated palm oil in sunflower oil. (a) Nucleation rate; (b) crystal size; (c) fraction solid fat; (d) shear modulus, G. Crystal size is given as diameter of a sphere of equal volume. Full curves, a relatively high initial supersaturation of crystallizing triglycerides; dashed curves, for a relatively low supersaturation. (Reprinted from Walstra, P.,* Physical Chemistry of Foods, *Marcel Dekker, New York, 2003. With permission.)*

formation likely proceeds as schematically illustrated in **Figure 13.17**. Initially, fat crystal nuclei are formed that grow out to small crystals (frame 1) and next, as a result of colloidal interaction forces, they start to aggregate to form small (fractal) aggregates (frame 2). Fat crystal growth is a relatively slow process, taking a relatively long time before the supersaturation of the crystallizing triglycerides has

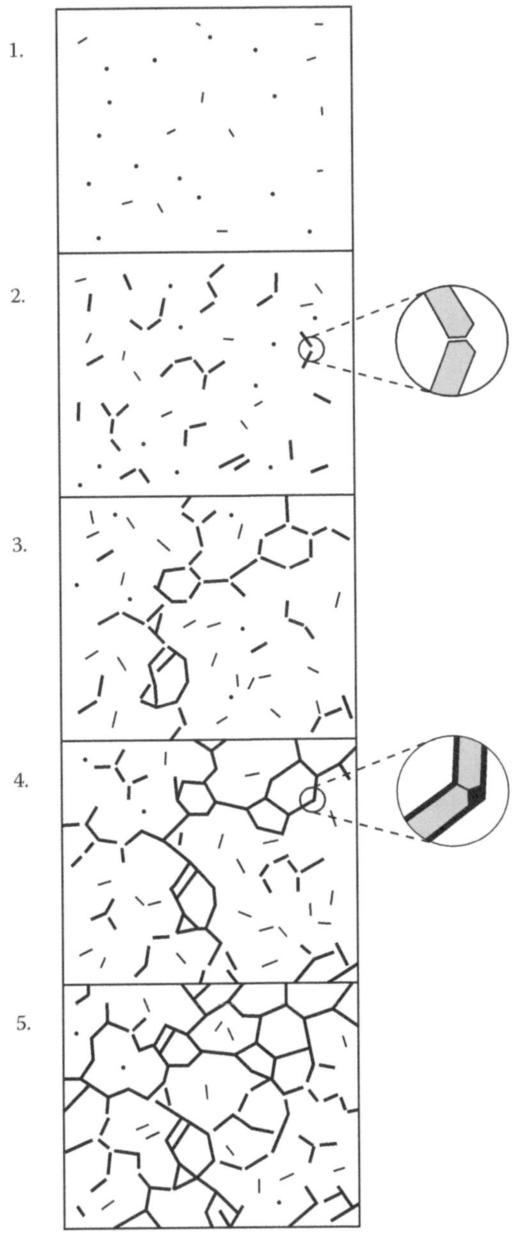

Figure 13.17 *Various stages in formation of a fat crystal network. For further explanation, see text. Approximate scale; several μm for the edge of a frame. (Reprinted from Walstra, P.,* Physical Chemistry of Foods, *Marcel Dekker, New York, 2003. With permission.)*

decreased to a point that no new nuclei are formed. So the formation of new crystals, their growth and aggregation, both mutually and with the network formed, will go on (frames 3 and 4).

- As result of the ongoing depositing of crystallizing triglycerides on the crystals, they will grow in size and, in addition, this may lead to sintering of adjacent crystals (compare detailed pictures of frames 2 and 4 in **Figure 13.17**). The latter process leads to tremendous stiffening of the crystal–crystal bonds and, with that, to a large increase in the stiffness/modulus of the fat network. Moreover, other processes may start to take place as recrystallization, which may, among others, also lead to sintering of crystal–crystal bonds.
- Fat crystals are never spherical; mostly, they have a platelet or needle-like structure. They are strongly anisometric. Fats with a wider compositional range tend to give smaller crystals that are more anisometric (more slender).

13.6.1 Small deformation properties of plastic fats

The rheological properties will strongly depend on the structure of the fat crystal network. Important factors are the spatial structure of the network, the morphology (size and shape) of the fat crystals, and the bonds between them. First, these factors will be discussed more extensively in relation to their properties in the linear region. Linear regions are very small, as may be expected for networks composed of hard particles, although the particles may be subject to bending because of their strong anisometric shape. Typical values measured in shear are 10^{-4}, 3×10^{-4}, and 10^{-3} for hydrogenated palm oil (HP) crystals in sunflower oil, a margarine fat, and butter, respectively. The difference in the linear region between butter and margarine has been ascribed to a difference in crystal shape, being more slender and hence more bendable for butter.

Spatial network structure. As can be deduced from **Figure 13.16**, by comparing frame c and d, a network is already formed at a solid concentration of the order of 0.01. At this stage, the network may still have some characteristics typical for a network built up of fractal aggregates. However, if this is the case it will soon cease to be so, primarily because of incorporation in the network of aggregates formed at a later stage and, in addition, as a result of the ongoing deposition on the network of crystallizing triglyceride molecules and small crystals. In a later stage, the structure will also change because of recrystallization, which will affect the structure of the crystals. In addition, data on the permeability of fat crystal networks show that these do not have a fractal structure (Walstra et al. 2001).

As discussed in the sections above, a method to obtain information on the network structure is by determining the concentration dependence of the system involved. For fat systems, this is complicated because, by changing the concentration fat that will crystallize on cooling, one will also change the crystallization conditions resulting in crystals with a (largely) different size (and shape). The solubility x_s of a single triglyceride expressed as a mole fraction at absolute temperature T is given by

$$\ln x_s = \frac{\Delta H_m}{R_g}\left(\frac{1}{T_m} - \frac{1}{T}\right), \tag{13.22}$$

where ΔH_m is the molar enthalpy of melting of the solute, R_g is the gas constant, and T_m is the melting temperature of the pure solute. For ideal solutions, the driving force for crystallization is given by the difference in chemical potential μ between a supersaturated solution (mole fraction dissolved x) and a saturated solution (mole fraction dissolved x_s):

$$\Delta\mu = R_g T \ln\left(\frac{x}{x_s}\right) = R_g T \ln\beta, \tag{13.23}$$

where ln β is called the supersaturation. It increases with decreasing temperature and increasing x. The supersaturation affects the rate of nuclei formation and, with that, crystal size as well as the form in which the triglyceride will crystallize. Together, they may strongly affect the rheological properties of a fat (**Figure 13.18**). This figure shows that, for the studied fat system, the final value of shear modulus G' is higher for a larger initial supersaturation despite the fact that the final volume fraction solid was the same for the three systems. The decrease in G' in time for the highest supersaturation is likely due to a transition in the form in which the triglyceride was crystallized or due to growth of large crystals at the expense of the smaller ones (the so-called Oswald ripening).

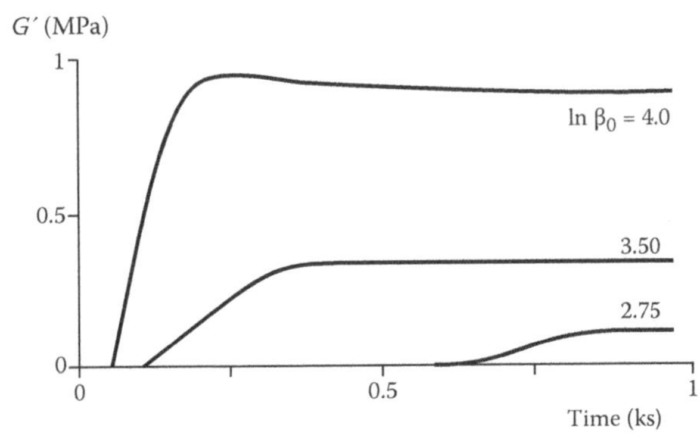

Figure 13.18 *Storage modulus G′ as a function of time of hydrogenated palm oil in sunflower oil systems at various initial supersaturation ln β_0 (indicated near the curves). Oscillation frequency 0.1 Hz. (Reprinted from Walstra, P. et al., in* Crystallization Processes in Fats and Lipid Systems, *ed. Garti, N., Sato, K., Marcel Dekker, New York, 2001. With permission.)*

The effect of the concentration of solid fat can be investigated by studying systems containing different concentrations of solid fat in which crystallization is induced at constant initial supersaturation $\ln \beta_0$ by adjusting the crystallization temperature. Examples of measured data for G' as function of the volume fraction solid fat φ, where $\ln \beta_0$ was kept constant, are shown in **Figure 13.19**. In agreement with the data shown in **Figure 13.18** for a higher $\ln \beta_0$, a higher G' is obtained at the same φ (although differences were not significant for $\ln \beta_0 = 2.75–3.25$). The slopes of G' versus φ are very steep (4.1 for $\ln \beta_0 = 4.0$ and about 7 for the lower $\ln \beta_0$). Similar slopes were observed for other fat systems.

In view of the considerable importance of the mechanical properties of margarine and butter, in the literature various models have been discussed to explain the elastic modulus of fats. In the earliest model, the crystal particle network was considered to consist of aggregated single, fairly isometric crystals in a regular cubic network of (straight or bent) strands of aggregated fat crystals in x, y, and z direction in a Cartesian coordinate system. Aggregation was supposed to be due to van der Waals attraction and hard-core repulsion. Such a model predicts that G' is proportional to φ (Equation 13.9), which is clearly not the case. An alternative model is to assume that the fat crystal clusters formed on aggregation have a fractal structure. For crystals aggregated due to van der Waals attraction $(\mathrm{d}f_s/\mathrm{d}x) \propto (A/h_0^3)$, Equation 13.14 would become

$$G \propto \frac{A}{h_0^3}\varphi^{\alpha/(3-D)}, \tag{13.24}$$

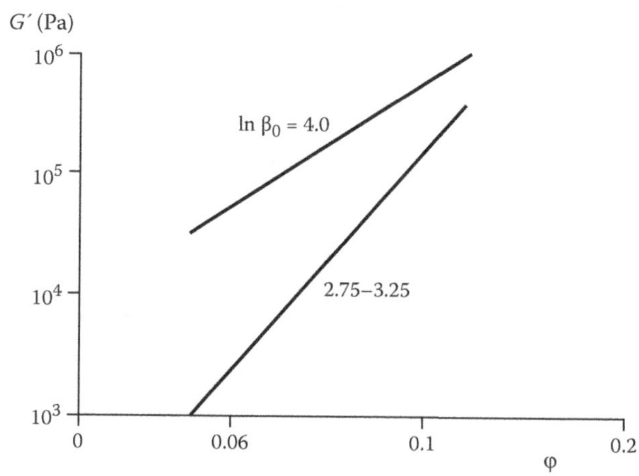

***Figure 13.19** Storage modulus G' measured 2 h after crystallization was started as a function of volume fraction solid fat φ for hydrogenated palm oil in sunflower oil systems at various initial supersaturation $\ln \beta_0$, indicated near the curves. (Reprinted from Walstra, P. et al., in* Crystallization Processes in Fats and Lipid Systems, *ed. Garti, N., Sato, K. Marcel Dekker, New York, 2001. With permission.)*

where A is Hamaker constant, h_0 is the hardcore distance between the particles, and parameter α depends on the flexibility of the stress carrying strands of particles in the network (Section 13.3). Generally, α can vary between 1 for particle networks built up of straight strands of stiff aggregates linked by deformable bonds and 4 for networks containing fully flexible strands. Calculated fractal dimensionalities D_f are then in the range of 2.0, for $\alpha = 4$ and a slope of 4 for G versus φ, to 2.86, for $\alpha = 1$ and a slope of 7.0. However, all these values are clearly higher than the value of 1.7 for D_f obtained by light scattering for 0.25% HP in sunflower oil systems (Kloek et al. 1997) and 0.5 wt.% glycerol tristearate in olive oil or in paraffin oil (Vreeker et al. 1992). For the latter system, D_f was found to increase slowly to 2.0 on storage lasting about 1 week due to recrystallization. In addition, already during an early stage in network formation, sintering of crystal–crystal bonds will occur because of the ongoing deposition of triglyceride molecules on the aggregated crystals, which will strongly limit the flexibility of the strands and make higher values of α unlikely. In conclusion, networks of solid fat in oil at higher volume fractions can no longer be considered as built up of aggregates with a fractal structure. This is in spite of the fact that during the initial stages of aggregation, such clusters may be formed.

Another aspect of importance for the stiffness of fat networks is the ongoing recrystallization and Ostwald ripening causing a transition of particle–particle bonds formed by van der Waals attraction and Born repulsion to solid crystalline bonds (the so-called sintering process). Such bonds can be broken again by extensive deformation (working) of the fat network. Its effect can, for example, be noticed by comparing the modulus G' and tan δ of gels formed by crystallization of 10% HP and by a pure tripalmitine (PPP) in sunflower oil (**Table 13.7**). A part of the fat was strongly deformed by spreading it with a spatula over a flat plate. For the HP fat, this resulted in a drastic reduction

Table 13.7 Storage Modulus G' and tan δ of Fats Containing 10% Solid Hydrogenated Palm Oil (HP) or Pure Tripalmitine (PPP) in Sunflower Oil Measured at Rest, 1 h after Working the Sample and 20 h after Working at 10°C

Solid Fat	Treatment	G' (kPa)	tan δ
HP	Rest	703	0.02
HP	1 h after working[a]	39	0.09
HP	20 h after working[a]	110	0.06
PPP	Rest	35	0.11
PPP	1 h after working[a]	30	0.09
PPP	20 h after working[a]	32	0.08

Source: Based on data from Kloek, W., Mechanical properties of fats in relation to their crystallization, Thesis, Wageningen University, Wageningen, the Netherlands, 1998. With permission.

Note: Frequency 0.1 Hz.

[a] Working was performed by spreading the fat after crystallization with a spatula over a flat plate.

of the modulus and an increase of tan δ, whereas for the PPP fat no changes were observed. During the spreading action, part of the sintered bonds will be broken and replaced by bonds due to van der Waals attraction. It is likely that these bonds are mainly bonds between fat crystal aggregates. The van der Waals bonds formed will be much weaker and more prone to relaxation, which explains the strong decrease in G' and the increase in tan δ. During storage of the fat after working, G' increases again due to sintering, albeit much more slowly than it did after initial crystallization. The pure PPP in sunflower oil fat will not show so much recrystallization, and with that sintering, as HP fat does. In HP fat, mainly compound fat crystals will be formed during initial crystallization that are prone to recrystallization. The huge effect of applying intensive large deformations on the structure of fat is further illustrated in **Figure 13.20**, showing G' and tan δ of the HP fat as a function of the measuring frequency in Hz (= $\omega/2\pi$) for the same treatments as in **Table 13.7**. The increase in tan δ due to working is seen over the whole time-scale range

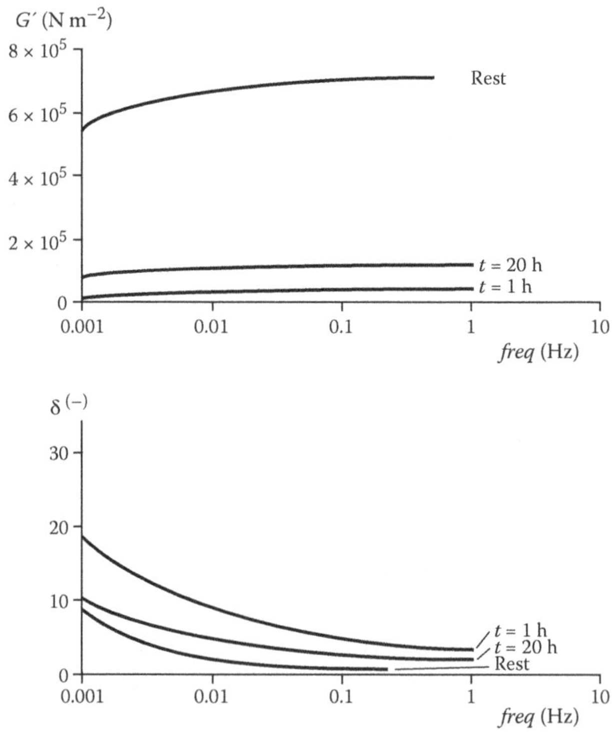

Figure 13.20 *Storage modulus G′ (upper panel) and loss tangent δ (lower panel) as a function of oscillation frequency for 10% hydrogenated palm oil in sunflower oil dispersions crystallized at 10°C, directly after crystallization in the rheometer (at rest), 1 h after working and 20 h after working (indicated). (After Kloek, W., Mechanical properties of fats in relation to their crystallization, Thesis, Wageningen University, Wageningen, the Netherlands, 1998.)*

of the applied deformation of 0.16–160 s (1/2π frequency). The nonworked HP fat was very elastic at the higher frequencies; however, at lower frequencies, some relaxation of bonds (likely those due to van der Waals attraction that have a lower energy content) can be seen.

13.6.2 Large deformation properties of plastic fats

As noted above, fat systems are characterized by a very small linear region. On application of larger strains, the modulus decreases and tan δ increases, with the extent and the reversibility of changes depending on the fat system. These properties and the resistance to deformation at large strains as a function of strain rate and deformation time are much more important for determining the usage properties of plastic fats compared to small deformation properties.

Large deformation properties of fats are usually determined by uniaxial compression of a cylindrical testing piece or by a penetration test using a cone or a rod. Uniaxial compression testing allows the determination of stress versus strain curves and therefore is normally preferable for fundamental research studies. Penetration tests are well suited as quality control tests for margarine production. Below, uniaxial compression tests will be discussed first.

An idealized stress–strain curve and some literature data for margarine and butter are shown in **Figure 13.21** (left and right panels, respectively). Near the origin, the relation between the stress σ and the strain ε should be, in the ideal

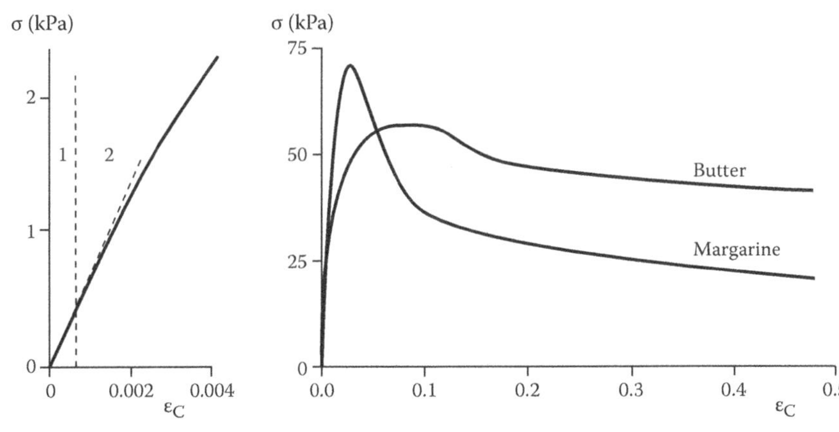

Figure 13.21 *Idealized stress versus Cauchy strain curve for a plastic fat (left panel) and stress versus Cauchy strain curves for butter and margarine determined by a well-lubricated uniaxial compression test (right panel). (Left panel reprinted from Walstra, P. et al., in* Crystallization Processes in Fats and Lipid Systems, *ed. Garti, N., Sato, K., Marcel Dekker, New York, 2001. With permission. Right panel reprinted from De Bruijne, D.W., Bot, A., in* Food Texture, Measurement and Perception, *ed. Rosenthal, A.J., Aspen Publishers, Gaithersburg, MD, 1999. With permission.)*

case, linear. However, because of the small linear region of 0.01–0.1%, this implies that for observing linear behavior the test specimen must have perfect flat and parallel upper and lower end planes. For test pieces with a height of 2 cm, deviations must be smaller than 2–20 μm, which is (almost) impossible to achieve in practice. Therefore, in region 1, σ will increase in a nonlinear manner with ε, whereby initially most of the deformation will be reversible, but irreversibility will increase with ε. Presumably, at smaller strains, mainly van der Waals bonds will be broken and re-form, but at larger strains weaker sintered bonds will also be broken that will not re-form in a short time span.

In region 2, yielding and a decrease in σ with increasing ε is observed to be indicative of the phenomenon *stress overshoot* (Section 7.3). In this region, extensive breakdown of bonds between the aggregates occurs resulting in a decrease in the stress needed to deform the fat. Normally, plastic fat test specimens will not fall apart into pieces on deformation although there may be a visual change in structure and formation of cracks. They do not fracture but yield. As discussed in Section 5.2.6, yielding occurs when on deformation cracks are formed at various places in the material, but no fracture propagation occurs because of the large energy dissipation. This may be the case in materials containing a relatively homogeneous network interspaced with a continuous liquid phase at the relevant length scale. Energy dissipation will primarily be attributable to the flow of the liquid phase by itself and relative to remaining parts of the solid network and secondary to friction between the fat crystal aggregates. The pores between the fat crystal aggregates will act as crack stoppers. Cracks formed will immediately fill with liquid. Energy dissipation will become less when the amount of oil present in the fat is lower. This favors the occurrence of fracture above yielding. So fats will fracture when they contain a high amount of solids at a set temperature or as a result of extensive crystallization at low temperature. Chocolate bars, which contain nearly 100% solid fat at room temperature (~22°C), fracture on deformation, but at higher temperatures (e.g., ~30°C) they exhibit yielding.

Because local yielding will directly lead to a lower local resistance to deformation, fairly extended slipping planes are often formed in the fat. This implies that after yielding velocity gradients will vary greatly throughout the fat. Finally, a regime of plastic flow is reached (**Figure 13.21**, right panel). The material flows, but it has a high apparent viscosity, is shear thinning, is thixotropic, and exhibits some elasticity. The high viscosity is primarily attributable to the presence of the remains of the once continuous crystal network. These remains incorporate a large part of the liquid oil, resulting in a high volume fraction dispersed particles and, with that, in a high apparent viscosity. Moreover, because of the irregular shape and spiky surface of these aggregates they will regularly touch each other during flow, giving the system some elastic properties. When the deformation is stopped the material solidifies nearly immediately, although its stiffness and yield stress will, at least initially, be much lower than those in the undisturbed state.

The stress needed to cause yielding of a fat is very important for many practical applications because cutting, spreading, and shaping all involve yielding. On application of a stress, fat will first yield and next start to flow with a strain rate increasing with increasing σ. Often, at larger σ, the relation between σ and the strain rate becomes about linear. The fat will roughly exhibit a Bingham-type behavior. By extrapolation of the linear part, a Bingham yield stress σ_B can be obtained. It is considered to be a good measure of the firmness or hardness of a fat. Just like the modulus of a fat, the yield stress depends primarily on its solid content, the interaction forces between the crystals (van der Waal attraction and extent to which sintering occurs), geometric structure of the fat crystal aggregates and the network, and size and shape of the fat crystals (milk fat containing 20% crystals in the form of large spherulites is usually pourable, whereas milk fat containing 10% small crystals may be solid-like with a high σ_y).

In industry, fat is often crystallized by using surface-scraped heat exchangers (votators) that allow fast cooling. The oil is pumped through a cylinder that is deeply cooled from the outside, while the content is extensively agitated by means of a stirrer that scrapes the cylinder wall. All or most of the crystallizable fat is usually solidified at the outlet of the heat exchanger, implying that fat crystal aggregation occurs in an intensive flow field and that their formation and that of any fat crystal network will be strongly disturbed. The extent to which the fat is crystallized will depend on the residence time of the fat in the heat exchanger. This may strongly affect the mechanical properties of the resulting fat, as illustrated in **Figure 13.22**.

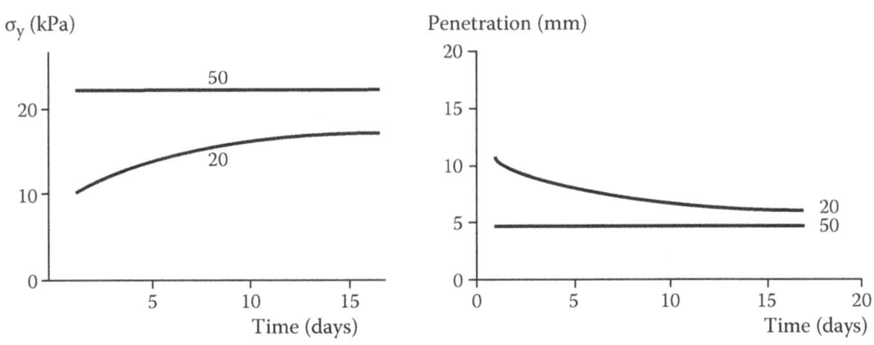

Figure 13.22 *Yield stress σ_y measured by uniaxial compression (left panel) and penetration depth of a 40° cone (right panel) in 15% hydrogenated palm oil in sunflower oil crystallized in a scraped surface heat exchanger at a throughput of 20 and 50 kg/h (indicated). Measuring temperature 20°C. (Left panel redrawn from Walstra, P. et al., in* Crystallization Processes in Fats and Lipid Systems, *ed. Garti, N., Sato, K., Marcel Dekker, New York, 2001. Right panel redrawn from Kloek, W., Mechanical properties of fats in relation to their crystallization, Thesis, Wageningen University, Wageningen, the Netherlands, 1998.)*

For the fat shown in **Figure 13.22**, nearly all crystallizable fat was solidified directly after leaving the scraped-surface heat exchanger for the low processing rate, whereas this was about 85% for the high processing rate. One day after processing, σ_y is higher and the penetration depth is smaller for the high processing rate and does not change any more during further storage. On the other hand, the yield stress of the product obtained at the lower processing rate increased in time. For cone penetration, the penetration is, among others, strongly determined by the yield stress of the system studied (Section 8.4.1.4). This relation has been shown to hold very well for plastic fats. For these systems, the yield stress is in general proportional to the penetration depth to the power $-n$, whereby n is about 1.6.

The fact that the fat in a scraped-surface heat exchanger is crystallized under high shear conditions will cause extensive rearrangements of the aggregates initially formed, resulting in the formation of compact aggregates. Moreover, sintered bonds formed between crystal or aggregates may be broken. Therefore, the product leaving the scraped-surface heat exchanger will obtain its mechanical properties from "compact" aggregates attracting each other by van der Waals forces. Despite the high shear rates, the crystals inside the aggregates may be sintered. Crystallization of additional solid fat (about 2% for the high-throughput system illustrated in **Figure 13.22**) at rest, which will certainly take place within 1 day, will cause sintering of the aggregates and, with that, a high σ_y and a low cone penetration depth. Once sintered bonds are formed, no further change in these parameters is observed, because sintered bonds are too strong to allow any rearrangement in the structure.

The fat processed at the low processing rate was completely crystallized when leaving the scraped-surface heat exchanger, as a result of which the aggregates will not become sintered directly afterward. They will attract each other by van der Waals forces, resulting in a relatively low σ_y and high cone penetration depth. Slow formation of sintered bonds between the aggregates because of recrystallization processes will result in a slow increase in firmness in time of these systems.

As discussed above, formation of a fat crystal network under higher shear conditions, which is common in industrial practice, leads to the formation of much more compact crystal aggregates than aggregation under quiescent conditions as in a rheometer at rest. This can be confirmed by polarized light microscopy. It is also found to greatly affect the dependence of the modulus and yield stress on solid fat concentration. Because, as argued above, Young's moduli cannot be obtained in the linear region, data for apparent Young's moduli E_{app} will be considered. The dependence of both σ_y and E_{app} was found to be proportional to φ^{μ}. Some literature data for products formed at rest and under high shear conditions are shown in **Table 13.8**. Apparently, intensive shear causes a decrease in the scaling exponent. This implies that at a higher level of solid fat, shear leads to a less effective use of the solid fat in giving the

Table 13.8 Slopes μ of Plots of Logarithm of Yield Stress σ_y and Apparent Young's Modulus as a Function of Logarithm of Concentration Solid Fat for Hydrogenated Palm Oil in Sunflower Oil Systems Crystallized under Rest and under High Shear Conditions

		High Shear	
Parameter	Rest	20 kg/h	50 kg/h
σ_y	3.9	1.1	1.9
E_{app}	3.8	1.2	2.4

Source: Based on data from Kloek, W., Mechanical properties of fats in relation to their crystallization, Thesis, Wageningen University, Wageningen, the Netherlands, 1998.

systems a certain stiffness and firmness. Systems will stay well spreadable up to a higher solid fat content. For the high shear low processing rate, the exponent measured for σ_y is close to that predicted by Equation 13.9, whereby the compacted aggregates have to be taken as the particles forming the network.

The data shown for the dependence of σ_y on the solid fat concentration for HP in sunflower oil at a processing rate of 50 kg/h correspond very well with literature data for a series of arbitrary margarines at temperatures between 10°C and 35°C, where on average an exponent of 2 is given.

13.7 Weak particle networks

Various food products may be considered liquid dispersions of particles that are too large to be stabilized by Brownian motion. On storage they may become inhomogeneous or demix due to sedimentation or creaming of the suspended particles. Examples of sedimenting or creaming particles in liquid dispersions are protein particles in buttermilk, cocoa particles and denatured protein particles in chocolate milk, fat globules in evaporated milk, oil droplets, and cellular fragments in fruit juices and nectars.

When Brownian motion and convection currents may be neglected, the steady-state rate of sedimentation or creaming of an isolated rigid particle in a homogeneous liquid can be obtained by equating the friction force $6\pi\eta Rv$ (Stokes law) the particle is subjected to and the sedimentation force f_s exhibited by it

$$f_s = \frac{1}{6}\pi d^3 g\Delta\rho \tag{13.25}$$

This gives for the sedimentation or creaming velocity v_s

$$v_s = \frac{g(\Delta\rho)d^2}{18\eta_c}, \tag{13.26}$$

where g denotes acceleration due to gravity (~9.8 m^2 s^{-1}), $\Delta\rho$ is the density difference between the suspended particle and the continuous phase, d is the particle diameter, and η_c is the viscosity of the continuous phase. The mean displacement x of a particle due to Brownian motion is given approximately by

$$x = \left(\frac{k_B T t}{3\pi\eta_c d} \right)^{1/2}, \tag{13.27}$$

where k_B is the Boltzmann constant, T is the absolute temperature, and t denotes time. Next, a Peclet number Pe_s can be defined as the ratio of the calculated time needed for a particle to move over its own diameter by Brownian motion to that by sedimentation, giving:

$$Pe_s = \frac{\pi d^4 g \Delta\rho}{6kT} \approx 10^{21} d^4 \Delta\rho \tag{13.28}$$

in SI units at room temperature. For $Pe_s > 1$, disturbance of sedimentation due to Brownian motion may be neglected. This implies for $d \geq 2$ μm for $\Delta\rho = 100$ kg m^{-3}. Convection currents may disturb the sedimentation of even larger particles. Particles that are smaller than 0.1 μm will remain in suspension almost indefinitely because of Brownian motion and convection currents.

In most food products, η_c depends on the shear rate $\dot{\gamma}$ applied. Therefore, the value of η_c to be inserted in Equation 13.26 depends on v_s, since v_s and the value of $\dot{\gamma}$ felt by the sedimenting or creaming particle are related. The tangential velocity component for creep flow around the particle should be considered. This gives, for the shear rate at the surface of the particle for a Newtonian liquid (van Vliet and Walstra 1989),

$$\dot{\gamma}_s = \frac{3v_{s,\infty}}{2R} = \frac{3v_{s,\infty}}{d}, \tag{13.29}$$

where $v_{s,\infty}$ is the velocity of the sphere relative to the fluid at such a distance that the fluid flow is not disturbed any more by the particle. For non-Newtonian liquids, the calculation of the effective $\dot{\gamma}_s$ is more complicated. Assuming that shear thinning behavior of the continuous phase can be described by a power law ($\sigma = k\dot{\gamma}^n$, Equation 4.10), the next relation can be deduced:

$$\dot{\gamma}_s = \left(\frac{\sigma}{k} \right)^{1/n} = (3x)^{1/n} \frac{v_{s,\infty}}{d}, \tag{13.30}$$

where x is an unknown factor that will probably not be far from unity and may depend on n. It should be stressed that Equation 13.30 applies only if the liquid exhibits power law behavior even at shear rates far lower than the $\dot{\gamma}_s$ calculated by this equation.

For particles with a diameter of 10 and 100 μm and $\Delta\rho = 500$ kg m^{-3} in water, the value of $\dot{\gamma}_s$ is about 3.5×10^{-3} and $3–5 \times 10^{-4}$ s^{-1}, respectively. These values of $\dot{\gamma}$ are much lower than those normally applied during usage (pouring, drinking, etc.) and also lower than the shear rates that can be applied by most shear rate-controlled viscometers. However, obtaining a sedimentation or creaming rate of less than 1 mm/day (which is still high for an acceptable shelf life of a beverage) requires a η_c of about 2.4 Pa s at the prevalent $\dot{\gamma}$ for a particle of 10 μm and $\Delta\rho = 500$ kg m^{-3}. For sedimentation to be less than 1 mm/week, that is, $v_s \leq 1.65 \times 10^{-9}$ m s^{-1} and $\dot{\gamma} = 5 \times 10^{-5}\,\mathrm{s}^{-1}$, η_c at this $\dot{\gamma}$ should be larger than 17 Pa s. On the other hand, many beverages should have a viscosity of about 10^{-2} Pa s, that is, 10 times that of water, at $\dot{\gamma} \sim 10^2$ s^{-1}. These numbers imply that the products must be extremely shear thinning. For larger particles, η_c at low $\dot{\gamma}$ has to be even larger. It is often very difficult to fulfill these requirements merely by adding a thickening agent (**Figure 13.23**, Remember that a yield stress implies that $\eta_{app}\dot{\gamma}$ is constant, giving a curve with a slope –1 in Figure 13.23 curve 3). Dispersions with a very high volume fraction of weakly aggregating particles and mixtures of polysaccharides that form a weak gel are mostly better suited. The first is, for example, the case in many turbid fruit drinks such as orange and tomato juices. The latter can be brought about by using, for example, a mixture of xanthan gum and a galactomannan.

In general, the difficulty in fulfilling the above requirements can be completely circumvented by the formation of a weak network with a yield stress high enough to prevent creaming or sedimentation. The sedimentation force f_s (Equation 13.25) acts roughly over the cross section $\pi d^2/4$ of the particle. This gives for the required yield stress σ_y

$$\sigma_y \geq \frac{2}{3} dg\Delta\rho \tag{13.31}$$

Equation 13.31 implies that for preventing sedimentation of particles of 10 and 100 μm and $\Delta\rho = 500$ kg m^{-3}, a yield stress of 0.034 and 0.34 Pa will be sufficient. Normally, a yield stress of 1 Pa is sufficient to prevent sedimentation. Such a yield stress is equivalent to the pressure exerted by a water column of 10^{-4} m or 0.1 mm. Consumers will not notice the presence of such a yield stress during pouring and drinking of a beverage. For instance, in chocolate milk stabilized by the addition of κ-carrageenan (mostly about 0.025 wt.%), a yield stress has been observed that is high enough to prevent sedimentation of the cocoa particles (**Figure 13.23**). Chocolate milk derives its stability from a network consisting of voluminous aggregates of milk protein and cocoa particles formed at elevated temperatures, which are cross-linked to each other after cooling by κ-carrageenan molecules adsorbed on these aggregates. In orange juice, the pulp particles are responsible for the formation of a yield stress and not the dissolved polysaccharides and proteins. Breaking these down with enzymes does not affect the flow curve significantly, whereas breakdown of the pulp particles leads to a loss of the yield stress. Moreover,

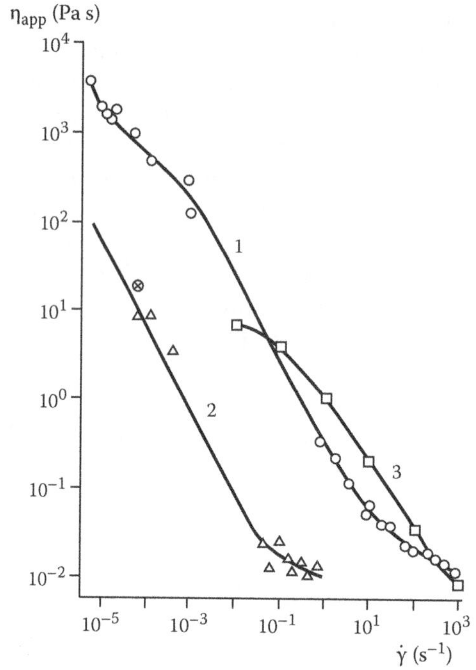

Figure 13.23 *Apparent viscosity η_{app} as a function of shear rate for (1) sterilized chocolate milk stabilized by 0.025% κ-carrageenan (stable against sedimentation), (2) orange juice (non-stable against sedimentation), and (3) 0.2% w/w xanthan gum solution. ⊗ indicates $\eta_{c,app}$ = 17 Pa s at $\dot{\gamma} = 5 \times 10^{-5}$ s^{-1} (see text). (After van Vliet, T., Walstra, P., in* Food Colloids, *ed. Bee, R.D. et al., Royal Society of Chemistry, Cambridge, UK, 1989.)*

the flow properties of various orange juices without pulp particles are virtually identical, whereas those of the original juices showed substantial variations, for example, σ_y varied between 10^{-5} and 10^{-3} Pa (van Vliet and Walstra 1989).

These observations support the statement that dispersions with a very high volume fraction of weakly aggregating particles are mostly better suited to get sufficient shear rate thinning systems. They also point to the remarkable fact that in these beverages, the suspended particles, which in the absence of a network would sediment directly, contribute to the formation of the network that retards or stops their sedimentation. Similar conclusions can be reached regarding, for instance, the stability of apricot nectar and tomato juice. However, to obtain a beverage that is physically stable on storage, some additional requirements must be fulfilled. These are: (1) after disturbing the network, it must restore itself in a relatively short time, generally within 10^3 s (this prerequisite excludes network forms by (mixtures of) polysaccharides involving bonds with a long relaxation time); (2) no syneresis (formation of a clear layer on top of the gel due to shrinkage of it as a result of ongoing cross-linking) may occur; and (3) the network may not settle under its own weight. In general, weak networks have a too low modulus (Young's moduli between 0.1 and 10 Pa) to prevent them from settling under their own weight except when the network is adhering to the wall of the container. Settling of a network under its own weight can be treated as uniaxial compression of it by gravitational forces until the latter are counterbalanced by the product of the Young's modulus E of the network and the strain ε. If there is no interaction between the network and the container, both stresses must be equal, which gives

$$E(\varepsilon, t)\varepsilon = \varphi_p \Delta\rho g h, \tag{13.32}$$

where φ_p is the volume fraction of suspended particles including those forming the network and h is the height of the network. Because most weak gels are, to some extent, viscoelastic and because large deformations may occur, E will depend on ε and on time. Moreover, during uniaxial compression, φ_p of the network compounds increases. Usually, E will increase more than proportional to the concentration of the network compounds, finally giving a smaller ε than calculated with Equation 13.32.

As an example, in chocolate milk the gravitational force due to the weight of the network is approximately $150 \cdot h$ Pa ($\varphi_{protein}$ and φ_{cocoa} are estimated to be 0.02 and 0.013, respectively, and $\Delta\rho$ is taken as 450 kg m^{-3}). For h = 0.1 m, this implies that $E(\varepsilon,t)\varepsilon$ must be 15 Pa to prevent the uniaxial compression of the network. Because, for chocolate milk, E is about 0.1–0.5 Pa, settling of the network under its own weight will occur unless h is small. The only way to obtain such a situation is to have the network adhere to the wall of the container.

References

Bremer, L.G.B., and T. van Vliet. 1991. The modulus of particle networks with stretched strands. *Rheol. Acta* 30: 98–101.

Bremer, L.G.B., B.H. Bijsterbosch, R. Schrijvers, T. van Vliet, and P. Walstra. 1990. On the fractal nature of the structure of acid casein gels. *Colloids Surf.* 51: 159–170.

De Bruijne, D.W., and A. Bot. 1999. Food texture. In *Food Texture, Measurement and Perception*, ed. A.J. Rosenthal. Gaithersburg, MD: Aspen Publishers.

Flory, P.J. 1953. *Principles of Polymer Chemistry*. Ithaca, NY: Cornell University Press.

Kloek, W., T. van Vliet, and P. Walstra. 1997. Structure of fat crystal aggregates. In *Food Colloids, Proteins, Lipids and Polysaccharides*, ed. E. Dickinson and B. Bergenståhl. *R. Soc. Chem. Spec. Publ.* 192: 168–181, Cambridge UK: Royal Society of Chemistry.

Flory, P.J. 1974. Introductory lecture. *Faraday Discuss. Chem. Soc.* 57: 7–18.

Kloek, W. 1998. Mechanical properties of fats in relation to their crystallization. Thesis, Wageningen University, Wageningen, the Netherlands.

Lucey, J.A., T. van Vliet, K. Grolle, T. Geurts, and P. Walstra. 1997. Properties of gels made by acidification with glucono-δ-lactone: 2. Syneresis, permeability and microstructural properties. *Int. Dairy J.* 7: 389–397.

Mellema, M., J.H.J. van Opheusden, and T. van Vliet. 2002a. Categorization of rheological scaling models for particle gels applied to casein gels. *J. Rheol.* 46: 11–29.

Mellema, M., P. Walstra, J.H.J. van Opheusden, and T. van Vliet. 2002b. Effects of structural rearrangements on the rheology of rennet-induced casein particle gels. *Adv. Colloid Interface Sci.* 98: 25–50.

Renkema, J.M.S. 2004. Relations between rheological properties and network structure of soy protein gels. *Food Hydrocolloids* 18: 39–47.

Renkema, J.M.S., and T. van Vliet. 2002. Heat-induced gel formation by soy proteins at neutral pH. *J. Agric. Food Chem.* 50: 1559–1573.

Segeren, A.J.M., J.V. Boskamp, and M. van den Tempel. 1974. Rheological and swelling properties of alginate gels. *Faraday Discuss. Chem. Soc.* 57: 255–262.

Te Nijenhuis, K. 1981. Investigation into the ageing process in gels of gelatin/water systems by the measurement of their dynamic moduli: Part 1. Phenomenology. *Colloid Polymer Sci.* 259: 522–535.

Treloar, L.R.G. 1975. *The Physics of Rubber Elasticity*. Oxford: Clarendon Press.

van Vliet, T., and J. Lyklema. 2005. Rheology. In *Fundamentals of Interface and Colloid Science*, vol IV. *Particulate Colloids*, ed. J. Lyklema, Chap. 6. Amsterdam: Academic Press.

van Vliet, T., and P. Walstra. 1989. Weak particle networks. In *Food Colloids*, ed. R.D. Bee, P. Richmond, and J. Mingins, 206–217. Cambridge, UK: Royal Society of Chemistry.

Verheul, M., and S.P.F.M. Roefs. 1998. Structure of whey protein gels, studied by permeability, scanning electron microscopy and rheology. *Food Hydrocolloids* 12: 17–24.

Vreeker, R., L.L. Hoekstra, D.C. den Boer, and W.G.M. Agterof. 1992. The fractal nature of fat crystal networks. *Colloid Surf.* 65: 185–189.

Walstra, P. 2003. *Physical Chemistry of Foods*. New York: Marcel Dekker.

Walstra, P., W. Kloek, and T. van Vliet. 2001. Fat crystal networks. In *Crystallization Processes in Fats and Lipid Systems*, ed. N. Garti and K. Sato, Chap. 8. New York: Marcel Dekker.

14

Composite Food Products

Most food products are composite materials when considered at mesoscopic and macroscopic length scales. Examples are dairy products (e.g., milk, yogurt, and cheese), bakery products (e.g., bread, cookies), meat and meat products, and candy bars. Various types of these products are composite solids such as cheese, margarines, butter, and bread dough. In this chapter, no extensive discussion will be given on all types of composite solids that can be found among food products. The emphasis will be on gelled systems. The discussion will focus on the main types of composite products that can be distinguished and limited to the relation between the structure of these products and their mechanical properties, both at small and large deformations.

Before continuing, it should be noted that composite products are not the same as mixed products. The latter term denotes products formed by mixtures of different protein and/or polysaccharide molecules. As long as during product preparation the different macromolecules stay mixed at the molecular scale, the term "mixed products" should be used exclusively. When phase separation takes place, both "mixed products" and "composite products" can be used. For products formed by mixtures of macromolecules and particles, only the term composite products will be used. For gelled systems, nearly all protein gels in food products are mixed gels. One exception is gelatin gels. Examples of mixed gels are acid- and rennet-induced skim milk gels which consist of a mixture of casein molecule, heat-set milk protein gels (containing both different casein and whey proteins), soy protein gels (containing various soy proteins), and gels of meat proteins. The presence of dispersed particles makes such gels composite products as well. The so-called filled gels are composite products consisting of a continuous gel matrix containing dispersed particles (gel-like and non-gel particles as emulsion droplets and air bubbles) that are much larger than the pores in the network.

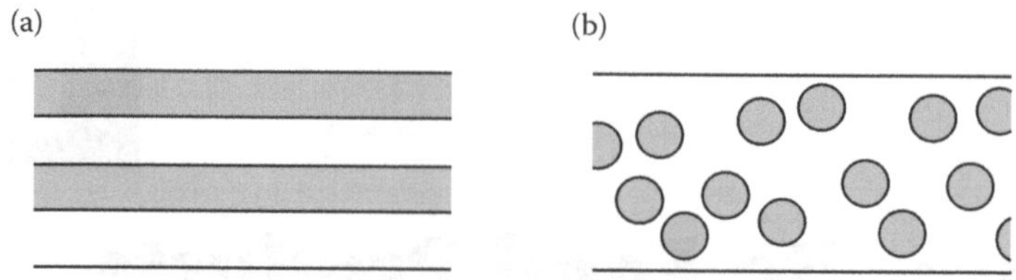

Figure 14.1 *Schematic models of composite products consisting of (a) mutually alternating adherent layers of two different materials and (b) continuous solid(-like) phase containing dispersed particles.*

The simplest spatial models of composite products are those in which mutually alternating adherent layers of two different materials form the products (layered composite products) and systems consisting of a continuous solid(-like) phase containing dispersed particles (filled composite products) (**Figure 14.1**). Because the reaction on the application of a stress differs strongly between the two types of composite products, they will be discussed separately in Sections 14.1 and 14.2, respectively.

14.1 Layered composite products

Examples of layered food products are pastry products consisting, for instance, of two or more alternating layers of cake and marmalade, or another gel-like product, and dough for puff pastry preparation that consists of layers of wheat dough alternated by layers of margarine. Schematically, they can be deformed in the direction of the alternating layers or perpendicular to these layers (**Figure 14.2a and b**, respectively).

First, we consider a material or product composed of straight layers that are mutually rigidly fixed at both sides. When this material is extended in the direction of the alternating layers (**Figure 14.2a**), the strain in the different layers will be the same, regardless of whether this is due to applying a tensile stress or strain. Moreover, assuming that the material is made up of two

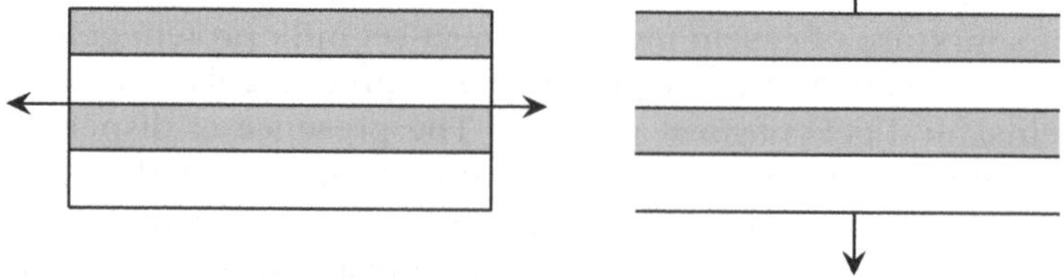

Figure 14.2 *Products consisting of alternating layers of materials x and y. Product is deformed because of a stress applied (a) in the direction of layers and (b) perpendicular to this direction (exerted stress indicated by arrows).*

components, each present as separate layers, the stress σ_c to uniaxially stretch or compress the material is given by

$$\sigma_c = \varphi_x \sigma_x + \varphi_y \sigma_y, \tag{14.1}$$

where φ_x and φ_y are the volume fraction of components x and y, respectively, and σ_x and σ_y are the stress felt by components x and y, respectively. Equation 14.1 holds regardless of the number of layers making up the structure of the material. Because the strain is the same in all layers and $\sigma = E\varepsilon$, the Young's modulus E_c of the composite product can be calculated easily if the volume fraction φ_x and φ_y of the two components and their Young's moduli E_x and E_y are known by

$$E_c = \varphi_x E_x + \varphi_y E_y \tag{14.2}$$

For this calculation, it does not matter if the material consists of two layers or more. The same equation can be used for the calculation of the modulus of a material consisting of a continuous matrix containing fibers with a homogeneous cross section, which extend over the whole length of the product. Note that if the modulus of the layers differs, as will usually be the case, the tensile stress in the layers will be different. As long as the strain in both layers is the same, it does not matter if the layers are mechanically connected to each other over their whole length or not.

For materials where the tensile deformation or stress is perpendicular to the direction of the layers, the situation will be completely different (**Figure 14.2b**). It is easily understandable that the weakest layer will be (much) more strongly deformed. On deformation, the tensile stress will be the same in all layers, so $\sigma_c = \sigma_x = \sigma_y$. However, the strain will differ. The overall strain ε_c of the material is given by

$$\varepsilon_c = \varphi_x \varepsilon_x + \varphi_y \varepsilon_y \tag{14.3}$$

Because $\varepsilon = \sigma/E$ and σ is the same in each layer for these systems, the Young's modulus of the composite material is given by

$$\frac{1}{E_c} = \frac{\varphi_x}{E_x} + \frac{\varphi_y}{E_y} \tag{14.4}$$

Again, for the calculation, the number of layers does not matter; only their volume fraction does. By the derivation of this equation, it is implicitly assumed that the materials in the different layers are rigidly connected at their interface. If this is not the case, the composite material will behave as if there is a layer of air in between these layers, and since the tensile stress of air is negligible, the composite material will be separated at that point.

The difference in mechanical behavior between the two materials depicted in **Figure 14.2** can be nicely visualized by considering the rise of a puff pastry during baking. Puff pastry dough is prepared by making first dough from wheat flour, water salt, fat (10 wt.% on flour), and some additives. After preparing the dough, a piece of fat (~75% on flour) is folded in the dough and next, the whole is sheeted, left to relax, folded, and sheeted again, etc., resulting in a product consisting of alternating dough and fat layers, although the fat layers will be locally interrupted by connections between the dough layers and the other way around. On baking, the products will strongly rise in the direction perpendicular to the layers (sheeting direction), whereas usually some shrinkage occurs in the direction of the layers.

Of course, for many layered food products the deformation during its preparation will not be perfectly in the direction of the layers or perpendicular to them, but somewhat in between, also implying bending deformations of the layers. If this is the case, the resistance against shear deformation of the interface between the layers will also play a part. If the shear connections are weak, bending of the product occurs much more easily than when these connections are stiff. This is an important aspect for construction materials but is much less important in food products. Therefore, it will not be discussed further.

Another aspect is the behavior of layered composite products during large deformations including fracture. As long as tensile deformations are applied, large deformation and fractured behavior are primarily governed by the stiffest/strongest material when deformation is in the direction of the layers, and by the weakest material when it is in the direction perpendicular to the layers. The situation becomes different when the deformation is in uniaxial compression. Then, when the deformation is in the direction of the layers and these are rather thin, it will primarily be the bending resistance of the stiffest layers, their thickness, and the presence of shear connections between the various layers. The latter will be important when the difference in stiffness between the layers is relatively small. When deformation is primarily uniaxial compression perpendicular to the layers, it is the weakest component that determines the overall deformation. When the difference in stiffness against large deformations is large, the weakest component is often expelled from in between the stronger layers. This aspect may be the cause of problems when one is eating millefeuille (known as Napoleon in the United States, vanilla slice pastry in Australia, etc.), which traditionally consists of three layers of puff pastry, alternating with two layers of cream patisserie, whipped cream, or jam (the precise makeup varies between different countries). A similar problem may occur when one is eating larger deep-fried products that have a solid crust and a soft, semisolid core such as some croquettes (e.g., Dutch-type ones).

14.2 Filled composite products

A second category of composite products are those consisting of a continuous solid(-like) phase containing dispersed particles. The continuous phase may vary from a hard solid matrix to a soft gel, whereas the dispersed particles may vary from gas bubbles to hard particles (e.g., cocoa particles). Examples of filled composite products are many dairy desserts containing emulsion droplets, chocolate consisting essentially of a fat phase containing fat, sugar crystals, and cocoa particles, and bread dough after mixing essentially consisting of a protein gel containing starch granules and air bubbles. The mechanical properties of these materials mainly depend on the rheological properties of the continuous phase, the volume fraction and shape of the dispersed particles, their deformability, and the interactions between the dispersed particles and the continuous phases.

It should be noted that products such as breads, cakes, toast, potato crisps, and cookies are primarily sponge-like products, even though the continuous matrix may be a filled composite product. In a sponge-like structure, there is an open connection between the pores and the environment, primarily air for food products. During production, these products can normally be characterized as foams. In this chapter, foams will only be discussed when the volume fraction air is below the transition from dispersed gas cells to a system consisting of close packed gas cells as in beer foam and in bread dough at the end of the proofing phase and during baking, that is, at concentrations below the maximum volume fraction for spheres. The mechanical properties of concentrated foams are discussed in Section 15.2.

Below, the discussion will focus on products consisting of a semisolid continuous phase containing dispersed spherical particles such as emulsion filled gels. Extremes are soft solids containing much harder particles and solids containing particles with a negligibly low modulus (e.g., large gas bubbles). Sponge structures will be discussed in Section 16.1. The discussion will focus on elasticity modulus because that has been studied best and theory has been developed. We will refrain from presenting the derivation of the theoretical equations since the mathematical level involved in it is too advanced for this book. In Section 14.2.1, a more qualitative overview will be given of factors affecting large deformation and fracture behavior. For this topic, theory has not been developed far enough to provide good predictions of effects, although explanations of observations can be given.

Deformation of a filled soft solid, containing spherical particles, will locally involve both shear and elongational strains, regardless of whether it concerns a shearing or elongation type of deformation. Thus, for the calculation of shear and Young's moduli of the filled material, the Poisson ratio (Equation 3.11) of the continuous (matrix) phase as well as of the filler material should be known. For many solid food products containing solid particles, both ratios can, as a first approximation, be set equal to 0.5. For fluids, Poisson's ratio is not defined. It is only defined for materials that exhibit a dominant elastic contribution (shear and Young's modulus) on application of a stress or strain. Because emulsion droplets and air bubbles covered by an interfacial layer react elastically on deformation, one can define a Poisson's ratio for them. And since the volume of emulsion droplets and very small air bubbles will not change substantially, as a first approximation, their Poisson's ratio can also be set equal to 0.5.

Various rather advanced theories exist that result in analytical expressions for the moduli of filled gels as a function of volume fraction dispersed particles. An overview of these theories is given by Yang et al. (2011). An often used equation that suffices for most filled food products is that according to Kerner (1956). For the shear modulus, it reads as

$$\frac{G_c}{G_m} = \frac{\dfrac{\varphi G_f}{(7-5\mu)G_m + (8-10\mu)G_f} + \dfrac{(1-\varphi)}{15(1-\mu)}}{\dfrac{\varphi G_m}{(7-5\mu)G_m + (8-10\mu)G_f} + \dfrac{(1-\varphi)}{15(1-\mu)}}, \tag{14.5}$$

where G_c, G_m, and G_f are the shear modulus of the composite material, matrix, and filler, respectively; μ is the Poisson's ratio of the matrix, and φ is the volume fraction filler. Basic assumptions in the derivation of this equation are that the filler particles are spherical, perfectly bound to the matrix, do not mutually interact, and are homogeneously distributed through the continuous phase so that the composite material is macroscopically homogeneous and isotropic. For food products, the effect of Poisson's ratio of the filler material can usually be neglected as is done in the derivation of Equation 14.5. This equation is precise up to φ not exceeding 0.2. The difference between the modulus calculated according to more advanced theories and Equation 14.5 stays below 25% for $\varphi < 0.3$ for rigid filler particles and for $\varphi < 0.4$ for $G_f/G_m < 30$. For larger φ, a more elaborate theory should be used. For high volume fractions, at least an empirical correction for the maximum volume fraction should be incorporated in Equation 14.5. A relatively simple correction reads as

$$\varphi_{eff} = \varphi + \varphi^2(1-\varphi_m)/\varphi_m^2, \tag{14.6}$$

where φ_m is the maximum volume fraction, its value depending on filler characteristics.

For a product where the Poisson's ratio of the matrix is 0.5, Equation 14.5 reads as

$$\frac{G_c}{G_m} = \frac{7.5\varphi\ G_f + (1-\varphi)(4.5G_m + 3G_f)}{7.5\varphi\ G_m + (1-\varphi)(4.5G_m + 3G_f)} \tag{14.7}$$

Equation 14.7 will suffice for most food composite products, possibly in combination with Equation 14.6, within the limits mentioned above.

As indicated above, Equations 14.5 and 14.7 have been derived assuming that the filler material is perfectly bound to the matrix material (the so-called bound, active, or interacting filler particles) (**Figure 14.3a**). The particles will behave as multiple cross-links and in that way effectively increase the amount of material forming the elastic network in the product. For gels, the addition of active filler particles has been shown to decrease, for instance, the critical protein concentration for heat-set whey protein gels and the temperature at which a gel is formed for heat-set soy protein gels. When there are no bonds/

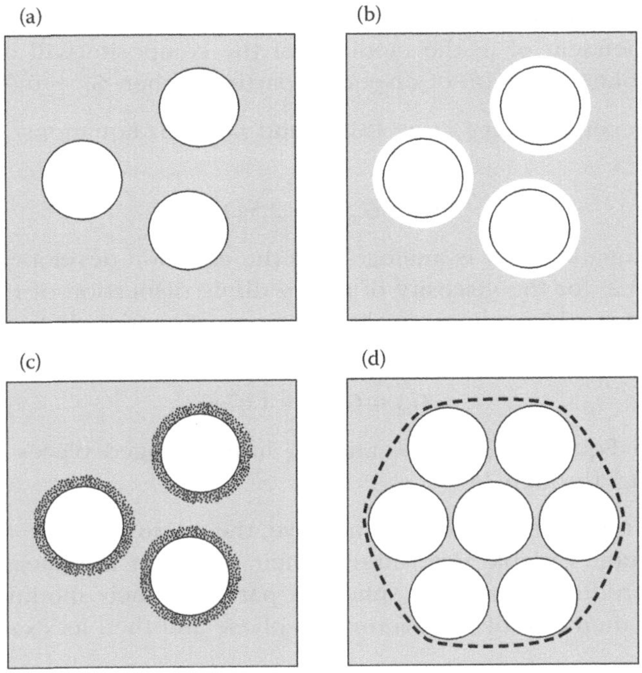

Figure 14.3 *Highly schematic representation of the structure of filled gels (not drawn to scale). Shaded area denotes gel matrix. (a) Bound particles; (b) unbound particles; (c) bound but with intermediate layer (dotted); (d) bound and aggregated. (Adapted from van Vliet, T.,* Colloid Polym. Sci., *266, 518–524, 1988.)*

interactions between the filler particles and the matrix, they will behave as particles with modulus $G_f = 0$ (unbound particles). Bonding will occur when the matrix material adsorbs onto the particle surface or when it forms chemical or physical bonds with the material covering the particles, for example, with the protein molecules providing stability to emulsion droplets and gas bubbles. If the gel material does not adsorb, but the matrix is quite stiff, a little surface roughness of the particles may cause effective bonding by friction. At large deformations, these effective bonds may become disrupted; on the other hand, the matrix becomes strongly deformed and, as a consequence, may be pressed locally at the particle surface, resulting in increasing friction. The particles will behave as (partly) bound ones. If the continuous phase consists of a gel matrix made up of nonadsorbing macromolecules, a thin (water) layer depleted of them may be formed around the particles, resulting in unbound particles (**Figure 14.3b**).

In case the modulus of the filler particles approaches zero, Equations 14.5 and 14.7 can be simplified to the next equation

$$\frac{G_c}{G_m} = \frac{(7-5\mu)G_m(1-\varphi)}{15(1-\mu)\varphi\, G_m + (7-5\mu)G_m(1-\varphi)} = \frac{4.5G_m(1-\varphi)}{7.5\varphi\, G_m + 4.5G_m(1-\varphi)} \quad (14.8)$$

The expression at the right side holds when the Poisson's ratio of the matrix is 0.5. Independent of μ, the modulus of the composite will decrease with increasing volume fraction of dispersed particles when $G_f \rightarrow 0$ (**Figure 14.4**).

For G_f/G_m becoming very large (>10^3) and $\mu = 0.5$, Equations 14.5 and 14.7 reduce to

$$G_c = G_m (1 + 2.5\varphi)_{\varphi \rightarrow 0} \tag{14.9}$$

for $\varphi \rightarrow 0$. Equation 14.9 is analogous to the equation developed by Einstein (Equation 10.2) for the viscosity of a very dilute dispersion of rigid spherical particles in a liquid medium. In the same way, Equation 14.8 can be simplified to

$$G_m = G_c (1 + 1.67\varphi) \tag{14.10}$$

Note that in Equation 14.10, G_c and G_m have changed places compared to Equations 14.5 through 14.9.

As is clear from the equations given above, the modulus of the filler particles is an important variable determining their effect on the shear modulus of composite products. For hard spherical particles, their modulus is usually much larger than that of the continuous phase and then its exact value does

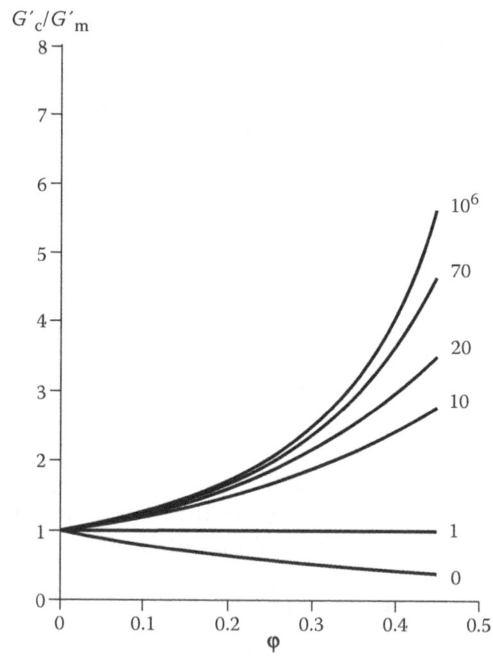

***Figure 14.4** Theoretical calculated effect of the modulus of filler particles G_f (G_f/G_m indicated) on the modulus of the composite product G_c divided by those of gel matrix G_m as a function of volume fraction dispersed particles. (Adapted from van Vliet, T.,* Colloid Polym. Sci., *266, 518–524, 1988.)*

not have to be known (**Figure 14.4**). However, for filled gels containing emulsion droplets or gas bubbles, the dispersed particles are often not considered very stiff compared to the gel matrix. When dispersed emulsion droplets and gas cells are subjected to a shear deformation, they will become somewhat ellipsoidal. This gives rise to an extra pressure difference over their interface, besides the Laplace pressure difference. Dividing this extra pressure by the accompanying relative elongation of the dispersed particles results in the next simple expression for G_f.

$$G_f = \frac{2\gamma_{int}}{R_p} = \frac{4\gamma_{int}}{d} \tag{14.11}$$

where γ_{int} is the interfacial tension (van Vliet 1988). For a protein-covered droplet with a diameter $d = 1$ μm and $\gamma_{int} = 10$ mN m^{-1}, the modulus is 40 kPa, that is, it is quite high. Conversely, for droplets of 1 mm or higher, G_f will usually be lower than G_m and thus causes a lowering of G_c with increasing dispersed particle concentration. Equation 14.11 shows that both the droplet diameter and the interfacial tension will affect G_f and, in this way, their effect on G_c. The droplet diameter effect has been confirmed experimentally various times for different types of gels containing emulsion droplets.

To summarize, of the factors determining the effect of dispersed particles on the modulus of composite products, the following ones are accounted for in Equation 14.5: (1) volume fraction dispersed particles, (2) particle modulus, and (3) indirectly, particle bonding via particle modulus. Other factors—not discussed so far—that influence the effect of filler particles on G_c are (4) particle shape, (5) the formation of intermediate gel layers between the gel matrix and the dispersed particles, and (6) the formation of particle aggregates during gel formation. The last factor, in particular, is often important for the properties of composite food products. These factors will be discussed in the following paragraphs.

(4) Particle shape. The modulus of products containing nonspherical particles can roughly be calculated by estimating an equivalent volume fraction of dispersed particles. Another very simple method, which is sometimes used, is by calculating an upper and a lower bound of the modulus using Equations 14.2 and 14.4. This can be done when the moduli of the matrix and the particles do not differ too much ($E_f/E_m < 5$); otherwise, the difference between the upper and lower bounds is so large that the results do not provide any useful information. More advanced theories on the modulus of composite products require a more precise characterization of the morphology of the particles, their mechanical properties, and those of the matrix than is usually available for food products.

(5) Formation of intermediate gel layers. Even for a matrix consisting of adsorbing flexible polymers, their interaction with the adsorbed layer around the droplet will often be less intensive than that with other dissolved molecules.

Moreover, a kind of depletion layer may often be formed around the particles. These effects will lead to an intermediate layer around the particles with a lower modulus than that of the matrix as is schematically depicted in **Figure 14.3c**. This will lead to a lower increase in G_c than theoretically predicted for bound particles. The data points for the polyvinylalcohol (a random coil polymer) gels containing bound particles in **Figure 14.5** are somewhat below the theoretical line, because the particle modulus was lower than infinite and it is likely that an intermediate layer was formed (van Vliet 1988). For very weak intermediate layers, products will behave as gels containing unbound particles (**Figure 14.5** "unbound" particles).

(6) Particle aggregation. Gel formation by protein is often a relatively slow process during which aggregation of dispersed particles may also occur, especially when they are covered with adsorbed protein molecules. Aggregation of emulsion droplets has been shown to occur by microscopic observation, for example, for acid-induced milk gels and heat-set whey and soy protein gels. It will lead to an effective volume fraction of dispersed particles that is

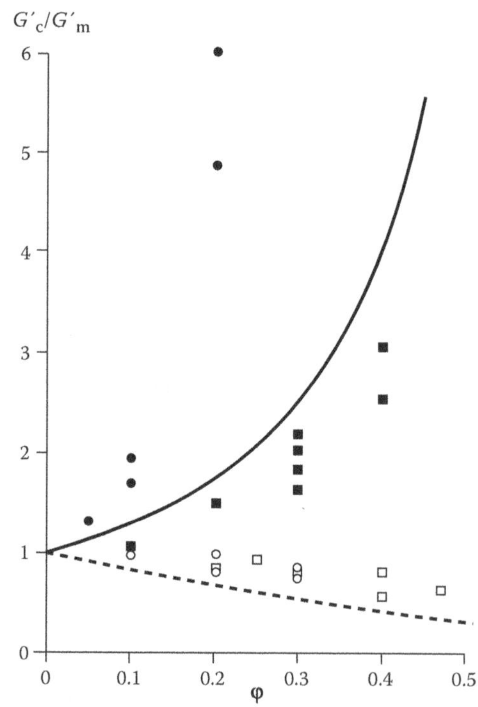

Figure 14.5 *Modulus of filled gels G_c divided by those of gels without dispersed particles G_m as a function of volume fraction dispersed particles φ. (□,■) Polyvinyl gels; (○,●) acid-induced milk gels. Filled symbols denote bound particles; open symbols denote "unbound" particles. Full curve calculated according to the van der Poel theory for rigid particles and dashed curve for particles with $G_f = 0$. (Reprinted with permission from van Vliet, T.,* Colloid Polym. Sci., *266, 518–524, 1988.)*

higher than φ (**Figure 14.3d**). Generally, it causes a much stronger increase in G_c than expected based on φ particles present (**Figure 14.5**, data for acid-induced milk gels, bound particles). By taking the φ_{eff} of the aggregates 2.3 times that of the primary emulsion droplets, the data points for acid-induced milk gels in **Figure 14.5** will fit with the theoretical line. However, another complicating factor is that, in addition to leading to a higher φ_{eff}, aggregation also results in a lower modulus of the aggregates than that of the primary particles. This will cause a smaller increase in G_c/G_m than would be expected based on the increase in φ_{eff} alone.

A last remark is that one should make a distinction between gels filled with emulsion droplets (filled gels) and emulsion gels. The first systems are discussed above. The latter, made up of aggregated emulsion droplets, are discussed implicitly in Section 13.3.

14.2.1 Large deformation behavior of filled composite products

Although of great importance for their usage properties, large deformation and fracture behavior of filled composite food products have been studied much less thoroughly than small deformation properties. Moreover, a much less readable applicable theory is available that allows estimation of this behavior for these products based on some simple determinable parameters such as φ, G_m, G_f, and mutual particle matrix bonding. In addition to these characteristics of the materials, the following factors will also play a role in determining the large deformation and fracture properties of filled products.

1. *Large deformation behavior of the continuous phase.* Obviously, fracture stress and fracture strain of the continuous phase will directly affect these properties of the filled composite product. In addition, G or E as a function of γ or ε, respectively, will play an important role since they directly affect the ratio of the filler modulus over that of the matrix. Materials may exhibit a linear increase of σ versus strain, strain hardening, or strain weakening, the latter implying an increase or decrease in G or E as a function of γ or ε, respectively (**Figure 4.9**). For those products where G_f/G_m is not very large or small ($0.01 < G_f/G_m < 100$), such a change in the ratio may substantially affect the influence of the filler on the modulus. Literature data for filled gels show that usually the observed effect of incorporated emulsion droplets on the shape of the stress-strain curve is small, except at larger strains, where it is likely that already small cracks are formed as a result of local stress concentration (**Figure 14.6**).
2. *Stress concentrations due to the presence of defects.* Stress concentration leads to a lowering of the stress at which defects present in a material start to grow, resulting in a lower fracture stress. The presence of filler particles may affect stress concentrations in two ways. First, unbound particles will act as defects in the continuous matrix if

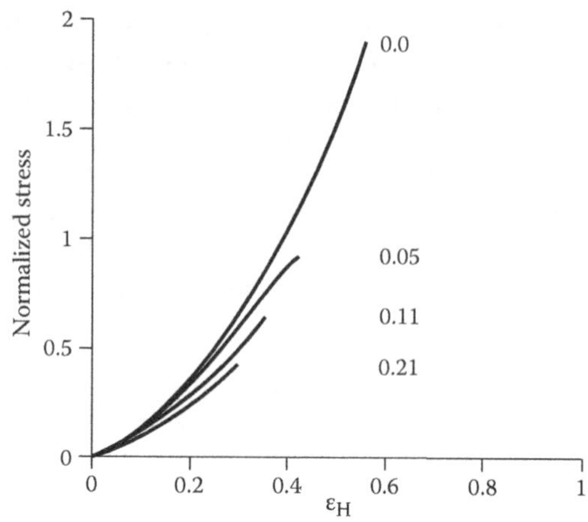

Figure 14.6 *Effect of volume fraction φ bound emulsion droplets on normalized stress versus strain curves of κ-carrageenan gels. Stress normalized by dividing by Young's modulus. φ oil indicated. (Adapted from Sala, G. et al.,* Food Hydrocolloids, *23, 1381–1393, 2009.)*

their particle size is larger than the size of the inherent defects already present in the matrix. More small particles may have a bigger effect than less big particles at the same φ. Remember that stress concentration due to spherical particles is three times, independent of their size (Equation 5.1). The presence of bound particles will always lead to inhomogeneous deformation of the matrix. If one considers a tensile deformation of a piece of material containing two spherical particles (**Figure 14.7**), the part between the two particles will be stretched out less, to the same extent, or more, depending on whether G_f/G_m is <1, 1, or >1, respectively. In the first case, the particles will act as a kind of defect, whereas in the latter case there is stress concentration in the region between the particles (**Figure 14.7**). This may result

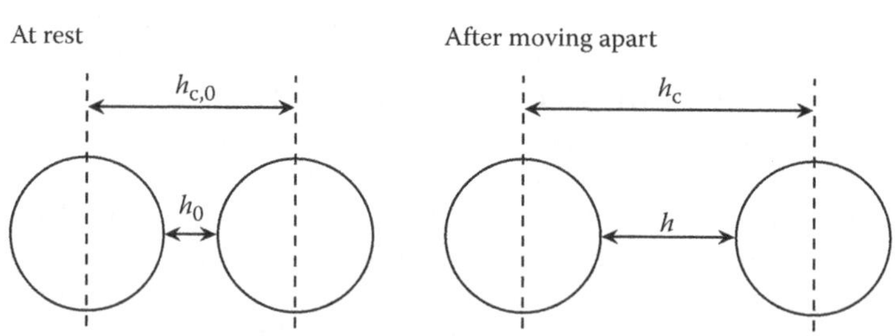

Figure 14.7 *Illustration of stronger increase in strain of the material between two stiff droplets ($G_f/G_m > 1$) when they are moved further apart compared to the material surrounding two particles. Hencky strain [$\ln(h/h_0)$] measured at shortest distance between particles is much larger than Hencky strain [$\ln(h_c/h_{c,0})$] measured in the matrix over a distance equal to that between centers of particles.*

in a lower overall fracture stress. An alternative is that the bonding layer between "hard particles" and the matrix may yield. This diminishes stress concentration and in addition gives energy dissipation, and in this way, increases toughness. In material science, it is well known that material toughness can be improved by making use of the debonding mechanism between hard particles and the matrix during deformation. The importance for the behavior of food products is not known.

3. *Changes in the energy balance governing fracture propagation.* As discussed in **Chapter 5** (Section 5.2.3), many aspects of fracture behavior of materials can only be understood if fracture is considered to be determined primarily by an energy balance that reads as (previously introduced as Equation 5.13 in Section 5.2.3)

$$W = W_e + W_{d,v} + W_{d,f} + W_{d,b} + (W_{fr}), \tag{14.12}$$

where W is the amount of supplied energy, W_e is the elastically stored energy, $W_{d,v}$ is the energy dissipation due to viscous (plastic) deformation of the material, $W_{d,f}$ is the energy dissipated due to friction between structural elements as a result of the inhomogeneous deformation of the material, $W_{d,b}$ is the energy dissipation due to debonding between dispersed particles and the matrix, and W_{fr} is the fracture energy. The term W_{fr} is enclosed by parentheses because it is derived from W_e during the fracture process. For an ideal elastic material, fracture is governed by the terms W, W_e, and W_{fr}. Crack propagation will occur when the differential energy release during crack growth (stemming from W_e) surpasses the differential energy (W_{fr} per unit area necessary for crack growth) required. Energy dissipation will always result in such a way that more energy has to be applied before crack propagation occurs. As noted above, debonding of bonds between filler particles and the matrix will lead to energy dissipation. The same holds for friction between unbound filler particles and the matrix. As a result of energy dissipation, more energy has to be supplied before fracture takes place, and it will result in a higher fracture stress and strain. The same holds for more intensive viscous flow of the matrix because of the presence of filler particles, for example, as a result of local yielding of the matrix (not due to debonding). Energy dissipation during deformation not due to the formation of cracks can be determined by measuring the amount of recoverable energy by loading–unloading experiments (Section 5.2.3, **Figure 5.11**). A short overview of various mechanisms that affect large deformation and fracture behavior of filled composite products is given in **Table 14.1**.

As **Table 14.1** illustrates, various mechanisms play a role in determining the effect of filler particles on large deformation and fracture properties of filled products. Their effect on the modulus, fracture stress, and strain can be positive or negative. This makes it very difficult or almost impossible to predict

Table 14.1 Mechanisms Involved in Effect of Dispersed Filler Particles on Large Deformation and Fracture Properties of Filled Composite Products

	Affected Parameter					
	Bound Droplets			Unbound Droplets		
Mechanism	G_c	σ_r	γ_f	G_c	σ_{fr}	γ_{fr}
Filler effect on modulus	+++[a]	++		−−−	−−	
Stress concentration		−−−	−−−		−−[b]	−−[b]
Strain induced viscous flow matrix	+[c]	+	+	+[c]	+	+
Friction filler matrix				+[c]	++	++
Debonding filler matrix interface		+	+			

Note: G_c, shear modulus composite product; σ_f, fracture stress; γ_f, fracture strain; + denotes an increase in the parameter with increasing volume fraction dispersed particles; − denotes a decrease. The stronger the effect, the more + or − used.

[a] For $G_f > G_m$.

[b] For size particles larger than inherent defect length matrix.

[c] Concerns modulus measured at large deformations.

whether the incorporation of filler particles will lead to a higher or lower σ_{fr} and γ_{fr} without profound knowledge of the system, and even then, it will still be difficult. The effect on σ_{fr} and γ_{fr} will strongly depend on the relative importance of the various mechanisms involved. A short overview of some results published on the fracture properties of emulsion filled gels is given in **Table 14.2**.

Gelatin gels filled with bound particles were fully elastic up to a deformation of 25%, so no energy dissipation was observed, whereas a clear increase in W_{diss} with increasing φ was observed for unbound particles. Moreover, the carrageenan and whey protein isolate gels showed an increase in W_{diss} with increasing φ of the filler particles. The measured effects on E of incorporating emulsion droplets in the gels are, as expected, in agreement with the theory

Table 14.2 Effect of Incorporation of Emulsion Droplets (Volume Fraction 0.05–0.20) on Young's Modulus E, Fracture Stress σ_{fr}, Fracture Strain ε_{fr}, and Relative Amount of Dissipated Energy W_{diss} on Deformation of Gels up to a Strain of 25%

Type of Gel	E	σ_{fr}	ε_{fr}	W_{diss}
Gelatin bound particles	+++	−[a]	−−−	0
Gelatin unbound particles	−−−	−−−	−−	+++
κ-Carrageenan bound particles	++	−−[a]	−−−	+
κ-Carrageenan unbound particles	−−	−−−	−−	+
Whey protein isolate gels bound particles	+++	++	−	++
Heat-set soy protein gels bound particles	+++	+++	+	ND

Source: Sala, G. et al., *Food Hydrocolloids*, 23, 1381–1393, 2009. With permission.

Note: ND, not determined

[a] Effect larger at higher strain rate.

discussed in Section 14.2. **Table 14.2** illustrates that σ_{fr} and ε_{fr} may increase as well as decrease on incorporating emulsion droplets. For the gelatin and carrageenan gels, the effect of filler particles on σ_{fr} can be explained by a combined effect of the higher modulus and larger stress concentration due to the added emulsion droplets. For the coarse soy and whey protein gels, higher stress concentration is likely to be much less important. In addition, for the whey protein gels, strain-induced viscous flow likely also plays a part. The decrease in fracture strain with increasing φ is likely mainly due to the larger stress concentration. For the gels containing unbound particles, it is possible that friction effects also play a role in counteracting the effect of stress concentration. For the whey protein gels, the stronger viscous flow will also lead to a lower effect of increasing φ on ε_{fr}.

Because the magnitude of all the terms in the right-hand side of Equation 14.12 depends on deformation speed, except for ideally elastic materials, the effect of dispersed particles on large deformation and fracture behavior of filled composite products will also depend to some extent on it. However, for most composite food products, the effect of deformation speed on the large deformation and fracture properties of the continuous phase (gel matrix) has been found to be dominant.

In summary, large deformation and fracture behavior of filled composite food products is still not a well-studied field. In recent years, various mechanisms have been identified that play an important role in determining these properties for gel systems containing emulsion droplets. This allows us to explain the observed effects of dispersed spherical particles in qualitative or, at best, semiquantitative terms, but it does not allow researchers to make predictions on a theoretical basis. Nevertheless, the information provided can help in product development in such a way that less experimentation is needed in order to achieve a desired quality.

References

Kerner, E.H. 1956. The elastic and thermo-elastic properties of composite media. *Proc. Phys. Soc. B* 69: 808–813.

Sala, G., T. van Vliet, M.A. Cohen Stuart, G. van Aken, and F. van de Velde. 2009. Deformation and fracture of emulsion-filled gels: Effect of oil content and deformation speed. *Food Hydrocolloids* 23, 1381–1393.

van Vliet, T. 1988. Rheological properties of filled gels. Influence of filler matrix interaction. *Colloid Polymer Sci.* 266: 518–524.

Yang, X., N.R. Rogers, T.K. Berry, and E.A. Foegeding. 2011. Modeling the rheological properties of Cheddar cheese with different fat contents at various temperatures. *J. Texture Stud.* 42: 331–348.

15

Gel-Like Close Packed Materials

15.1 Gels of swollen starch granules

Starch is produced by most higher plant species as an energy store. It is present in the form of roughly spherical semicrystalline granules, with diameters ranging from 2 to 100 μm. The granules do not dissolve in water at normal ambient temperatures. It consists of two components, amylose and amylopectin. Amylose is a linear polymer of 1→4 linked anhydroglucose monomers with a degree of polymerization ranging from several hundreds to about 10^4. It mostly makes up 20% to 30% of the starch. The remaining part is amylopectin, a highly branched polymer. About 4–5% of its monomers also have a 1→6 linkage. The degree of polymerization is in the range of 10^5–10^8. The semicrystalline regions in starch granules stem from ordered linear segments of amylopectin, whereas the amylose and branching regions of amylopectin form amorphous glassy regions. Starch granules also contain some lipids, protein, and minerals, in total about 1%.

At room temperature, starch granules will contain some water (for potato starch, up to 0.5 g/g starch when dispersed in water), with the amount depending on the relative humidity at which they are stored and the type of starch. On heating, granules will take up water, resulting in a volume increase by 30–40%. Some of the amylose may leach from the granules. At higher temperatures, melting of the semicrystalline region will take place although entanglement of the amylopectin chains will persist. With increasing temperature, much stronger swelling of the granule will take place to several times their original size. Moreover, phase separation of the amylose and amylopectin will occur, resulting in leaching out of virtually all amylose. At still higher temperatures, the swollen granules are broken up into far smaller fragments, the extent of which depends on temperature and the application of shear forces. At T below 100°C and without appreciable shear forces, the granules will maintain their integrity.

In most commercially available instruments used for fast characterization of the rheological properties of starch suspension on heating and after cooling, the suspensions are heated and stirred at the same time. Such a procedure imitates, to some extent, the heating of starch suspensions during industrial practice. This implies that during heating, starch granules will be disrupted already at relatively low heating temperatures (e.g., for potato starch at >60–65°C), which makes interpretation of the results in terms of relations between structure and mechanical properties more difficult. Because the latter is the topic of this section, measurements of starch pasting properties will not be discussed.

Amylose is poorly soluble in water at room temperature. It readily forms helices, of which at least part are double helices, which tend to align forming microcrystalline structures. On cooling dilute solutions of amylose, a precipitate may be formed, whereas in more concentrated solutions a gel will be formed. The rheology of this type of system is discussed in Section 13.2.2.

At high enough concentrations (about 0.5% for potato starch and 3–4% for wheat starch depending on conditions), starch dispersions that contain swollen starch granules and amylose, which is leached out of the granules during heating, will form gels on cooling consisting of swollen granules in an amylose gel (**Figure 15.1a**). Such a material in which a continuous amylose gel phase is interspersed with filler particles may be regarded as a compos-

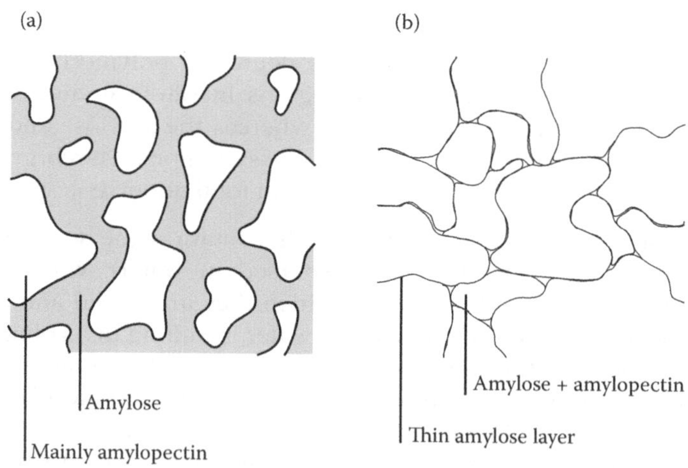

***Figure 15.1** Schematic presentation of starch gel structures: (a) fully swollen granules acting as a filler in an amylose matrix; (b) partly swollen granules that are tightly packed with a thin amylose gel layer in between. (Reprinted from Keetels, C.J.A.M. et al.* Food Hydrocolloids, *10, 343–353, 1996a. With permission.)*

ite material. The mechanical properties of this type of material have been discussed in **Chapter 14**.

On heating, systems containing still higher starch concentrations (e.g., ≥5 w/w% for potato starch and ≥15 w/w% for wheat starch systems), the swollen granules will already fill the whole available volume and a gel-like structure is formed (**Figure 15.2**). The rheology and fracture mechanics of these systems will be discussed on the basis of published data for potato and wheat starch systems, but the conclusions drawn have a more general validity.

On heating 15 w/w% wheat and potato starch systems, the storage moduli G' start to increase strongly at a temperature of about 55°C and 60°C, respectively, and reach a maximum at 82°C and 69°C, respectively. For most systems, G' decreases on prolonged heating, more stronger when the heating temperature is higher. The temperature at which the maximum level in G' was found, decreased from 70°C to 65°C and from 88°C to 64°C, with starch concentration increasing from 10% to 30% for potato and wheat starch, respectively. The initial increase in G' coincides with the first stage of crystallites melting and strong swelling of the granules. It is likely that the granules become tightly packed at temperatures close to the temperature at which the maximum in the modulus versus time plot is found. This will be the case at lower temperatures when the starch concentration is higher.

The decrease in G' on further increase in temperature coincides with the melting of the remaining crystallites in the temperature range up to about 80°C, depending to some extent on the type of starch. This implies softening of the starch granules and, in that way, a lower G'. The ongoing decrease in G'

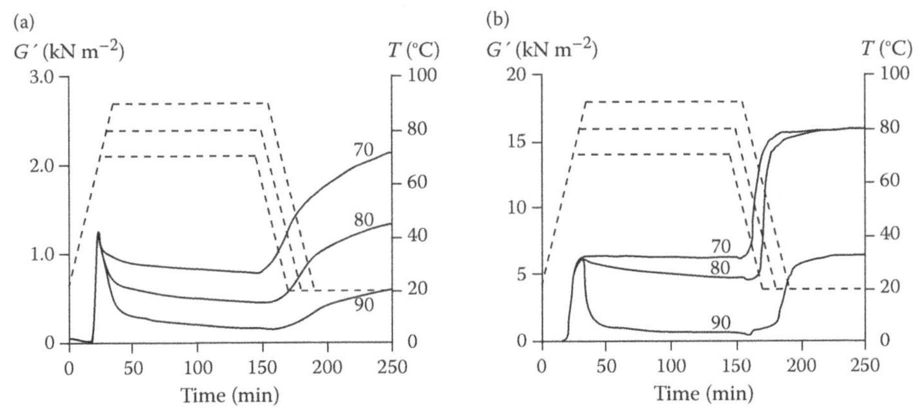

***Figure 15.2** Storage modulus G′ of concentrated starch suspensions as a function of time during a heating and cooling cycle; (a) 15 wt.% potato starch and (b) 15 wt.% wheat starch. Maximum heating temperature T is indicated. Note the difference in scale. Dashed lines show temperature against time. (Reprinted from Keetels, C.J.A.M. et al.,* Food Hydrocolloids, *10, 343–353, 1996a. With permission.)*

on further heating is thought to be due to disentanglement of the amylopectin present in the granules. Demixing of amylose and amylopectin cannot be an important cause, because a similar decrease is observed for waxy starch that does not contain amylose. For highly cross-linked potato starch (i.e., with covalent cross-links between the molecules in the granules) suspensions, no decrease in G' is observed during ongoing heating after reaching the "maximum" in G'. For these starches, no disentanglement of the amylopectin will occur.

On cooling, G' strongly increases. The increase continues also after reaching room temperature although at a lower rate. Generally, the short-time change is ascribed to heat-irreversible gelation of solubilized amylose in the continuous phase and the long-time changes to heat-reversible crystallization of the amylopectin in the granules. For concentrated starch suspensions, no full-phase separation of amylose and amylopectin will occur, but still a thin layer with amylose will be formed between the tightly packed swollen granules (**Figure 15.1b**). The concentration of amylose will be high enough to form a gel, gluing the swollen granules together. Moreover, on cooling and further storage, recrystallization of amylopectin will occur. The relative effect of the latter will increase with increasing storage time.

For starch concentrations high enough to form a gel containing tightly packed starch granules, those containing wheat starch exhibit higher G' than those containing potato starch granules directly after cooling (**Figure 15.2**). This is also the case for 30% starch systems. The G' of these concentrated systems is mainly determined by the stiffness of the swollen granules, implying that wheat starch granules are much stiffer than swollen potato starch granules directly after cooling the gels. However, a crossover is observed for G' versus time after ca. 1.5 days. For 30 wt.% potato starch systems, Young's modulus was 28 and 240 kPa directly after cooling and after 100 h at room temperature, respectively, whereas for wheat starch Young's modulus was 75 and 115 kPa, respectively. This crossover accounts for the much faster retrogradation (recrystallization in a different conformation than in the native starch granules) of potato starch amylopectin than of wheat starch amylopectin, leading to a much faster stiffening of the swollen granules in time.

For low concentration systems, where no gel is formed, potato starch systems give a much higher apparent viscosity than wheat starch systems also directly after applying a heating cycle. This is attributable to the much stronger swelling capacity of potato starch granules compared to wheat starch granules, resulting in a much larger volume fraction dispersed granules and, with that, in a much higher apparent viscosity. This much stronger swelling capacity of potato starch granules is also the cause for the lower starch concentration required for gel formation compared with wheat starch.

The difference in softening of the starch granules during heating and in the rate of retrogradation also results in very strongly different large deformation and fracture properties for both types of starches (**Figure 15.3**). For both types

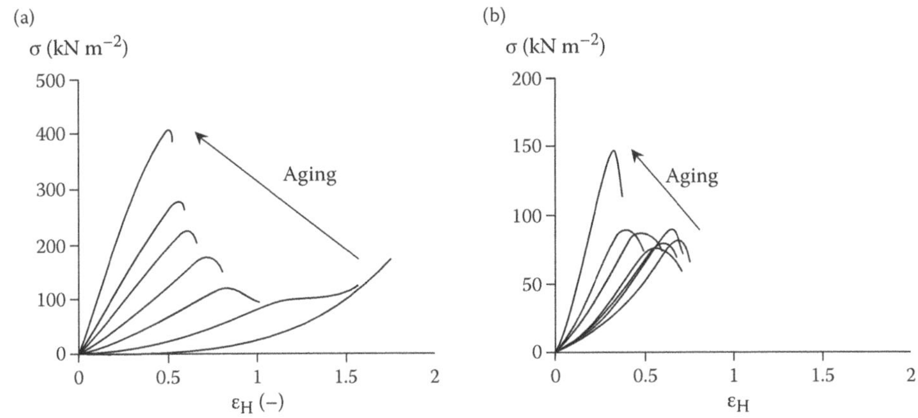

***Figure 15.3** Stress versus strain curves of 30 wt.% starch gels: (a) potato starch gels; (b) wheat starch gels. Measurements were performed after storage at 20°C for 0 and 4 h, and 1, 2, 6, 16, and 130 days. Fracture occurred at the maximum in curves. Note the difference in scale of vertical axis. (Reprinted from Keetels, C.J.A.M. et al.* Food Hydrocolloids, *10, 355–362, 1996b. With permission.)*

of gels, fracture stress increases and fracture strain decreases during storage, but the changes are much stronger for the potato starch systems. The fracture stress was found to increase linearly with the Young's modulus for both starches, although the increase is more pronounced for potato starch gels. Directly after cooling, potato starch gels are transparent, whereas wheat starch gels are opaque. After storage for some hours, the potato starch gels become also opaque. Freshly made potato starch gels are much more deformable than wheat starch gels, but after aging the difference becomes much smaller. Wheat starch and aged potato starch gels were found to fracture around the starch granules by microscopic observation, whereas for fresh potato starch gels the swollen granules were damaged on extension. This latter observation indicates that the amylose gel (the glue) in between the granules is weaker than the granules except for fresh potato starch gels. In agreement with this, the size of the inherent defects in 30 wt.% aged potato gels was found to be ~0.13 mm, which is roughly equal to the size of the largest swollen granules.

The simplest model of a system fully packed with deformable particles is a system of interacting cubes (**Figure 15.4a**). For such a system, Young's modulus will primarily depend on the Young's modulus of the cubes, whereas as long as the glue is weaker than the cubes, the fracture stress is primarily determined by the fracture stress of the glue. The fracture strain is roughly the strain of the cubes when the stress on the glue equals its fracture stress. However, amylose retrogradation proceeds quickly, usually in 1 day, so this model cannot account for the observed increasing fracture stress over longer aging times. A model more in agreement with microscopic observation is shown in **Figure 15.4b**. It is based on irregularly shaped, partly swollen granules that are hooked into each other, almost like a three-dimensional jigsaw

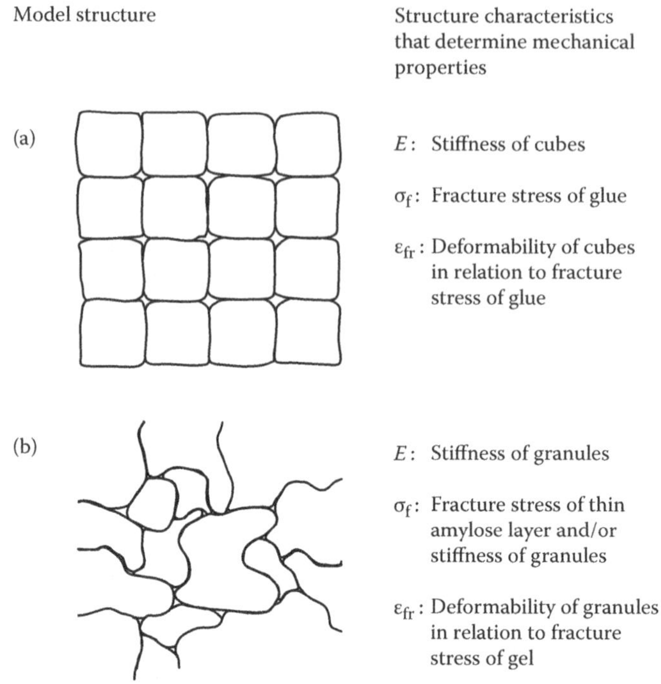

Figure 15.4 *Highly schematic presentations of: cubes that are glued together (a) and partly swollen granules that are tightly packed with a thin amylase gel layer in between (b). (Reprinted from Keetels, C.J.A.M. et al.* Food Hydrocolloids, *10, 355–362, 1996b. With permission.)*

puzzle. For such a system, the stiffer the granules are, the larger the fracture stress will be, because the hooks are more rigid. Consequently, the fracture stress of concentrated starch gels will depend both on the fracture stress of the amylose layer in between the granules and on the stiffness of the swollen granules. Of the two factors, the latter is more important, as is also clear from the observed linear relation between the fracture stress and the Young's modulus. The decrease in fracture strain will be due to the granules becoming less deformable as a result of the reordering (retrogradation) of amylopectin.

Based on the model shown in **Figure 15.4b**, it is to be expected that, for highly concentrated starch systems containing closely packed swollen granules, their large deformation and fracture properties will depend more and more on the stiffness of the granules, whereas the strength of the thin amylase layer will become less important with increasing concentration and storage time. This has indeed been observed. Heating temperature and time will affect the mechanical properties by affecting the mechanical properties of the granules after heating (likely via further disentanglement of the starch molecules and more extensive separation of the amylose and amylopectin at higher heating temperatures and longer heating times).

Measurements on cross-linked and waxy starches have shown the general applicability of the models described above. However, depending on the starch systems (type of starch, concentration, etc.), the balance between the different factors may vary.

15.2 Close packed foams and emulsions

In Section 10.2, the viscosity of concentrated dispersions has been discussed. Equations describing the relation between the dispersed volume fraction φ of particles and the viscosity η predict that η will go to infinity as φ approaches the maximum packing fraction of particles (e.g., Equations 10.10 through 10.12). Moreover, it was indicated that this will not be the case if the particles are deformable, which was illustrated for swollen starch suspensions and concentrated dispersions of fruit cells. In this section, this aspect will be worked out further for close packed foams and emulsions. The discussion will focus on systems where $\phi \to 1$. In such systems, the gas bubbles in foams and the emulsion droplets in emulsions will deform each other, the extent of which depends on the volume fraction. There is a transition of dispersed spherical bubbles to polyhedral structures, implying the formation of flat films in between the bubbles (**Figure 15.5**). This aspect can easily be visualized for concentrated foams present on well-foaming beer some time after foam formation. Although less visible, it is also often the case for products such as aerosol whipped cream, shaving foams, and mayonnaises containing 80 wt.% oil (i.e., ~82 vol.%). All mentioned products are characterized by a yield stress and flow behavior at larger shear stresses with both the yield stress and the apparent viscosity increasing with increasing ϕ. Below, some general aspects of the relation between the mechanical properties of these products and their structure will be discussed in

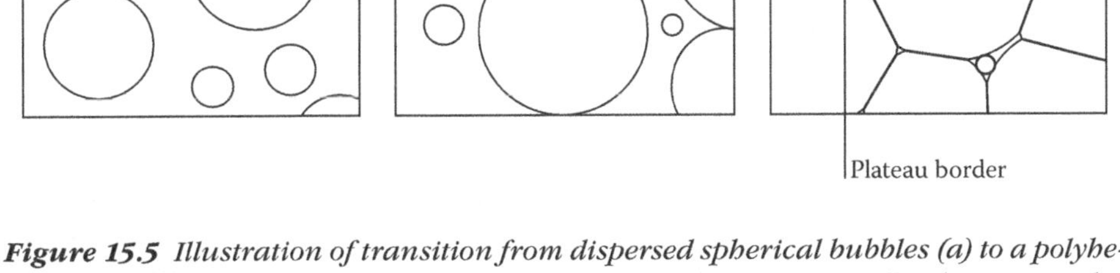

Figure 15.5 *Illustration of transition from dispersed spherical bubbles (a) to a polyhedral structure (c). Thickness of films between bubbles is too small to be seen on this scale. (Reprinted with permission from Walstra, P.,* Physical Chemistry of Foods, *Marcel Dekker, New York, 2003.)*

a qualitative or semiquantitative manner. Polyhedral structures are much more common in foods for foams than for emulsions, so the discussion will focus on these systems.

First, the mechanical properties will be considered of foams for which the volume fraction gas bubbles $\phi = 1$. In a "two-dimensional" foam, cells of equal volume will generally show a regular array of close-packed hexagons, as in a honeycomb. The thin films between the bubbles will always meet at angles of 120°, provided that the surface tension is constant (**Figure 15.6a**). Then, a balance of forces exists. On application of a shear stress on the foam in the *x* direction, the gas cells will be deformed (**Figure 15.6b**), implying an extension of the films marked by e and a shortening of those marked by f. When the angle of the lines connecting the centers of the gas cell DAB has been changed from 60° to 120°, the two intersection points of the films between the gas cells ABCD merge into a single point (**Figure 15.6c**). This configuration is unstable, which resolves itself by the generation of a new film from the center to restore the original unstrained configuration (**Figure 15.6d**). The configuration depicted in **Figure 15.6c** has a total surface area that is about 15% higher than that in the unstrained case. The end result of the shear deformation is that the upper gas bubble layer has moved to the right by one gas cell width, relative to the stationary bottom layer.

As indicated above, during shearing deformation, the total interfacial area increases leading to an opposing force proportional to two times the interfacial tension γ_{int}. To calculate the reaction stress of the foam on shearing deformation, in addition, the relative extension of the interfacial area as a function of the shearing deformation, the change in the angle ψ between the films marked by f and the horizontal plane, and the width of the unit cells have to be taken into account. The latter parameter is equal to $l\sqrt{3}$, where l is the length of each side of the honeycomb at rest. Replacing l by the radius R of a sphere with the same volume as a hexagon gives the next expression for the stress

$$\sigma \propto C_\sigma \frac{2\gamma_{int}}{R} \cos\psi, \tag{15.1}$$

where C_σ is a numerical constant. The shear strain is given by the displacement of the line *AB* in **Figure 15.6** divided by the distance between the lines *AB* and *CD*. The shear strain γ and the change in the angle ψ are related in a roughly linear way. The combination of this aspect and Equation 15.1 results in the following simple expression for the shear modulus *G* (Princen 1983; Stamenovic 1991)

$$G = C_G \frac{2\gamma_{int}}{R}, \tag{15.2}$$

where C_G is a numerical constant of about 0.25. Equation 15.2 shows that *G* is proportional to the Laplace pressure of a sphere with the same volume as the gas cells forming the close packed foam. A similar equation to Equation 15.2 can be deduced for the yield stress of the foam where the numerical constant is somewhat larger.

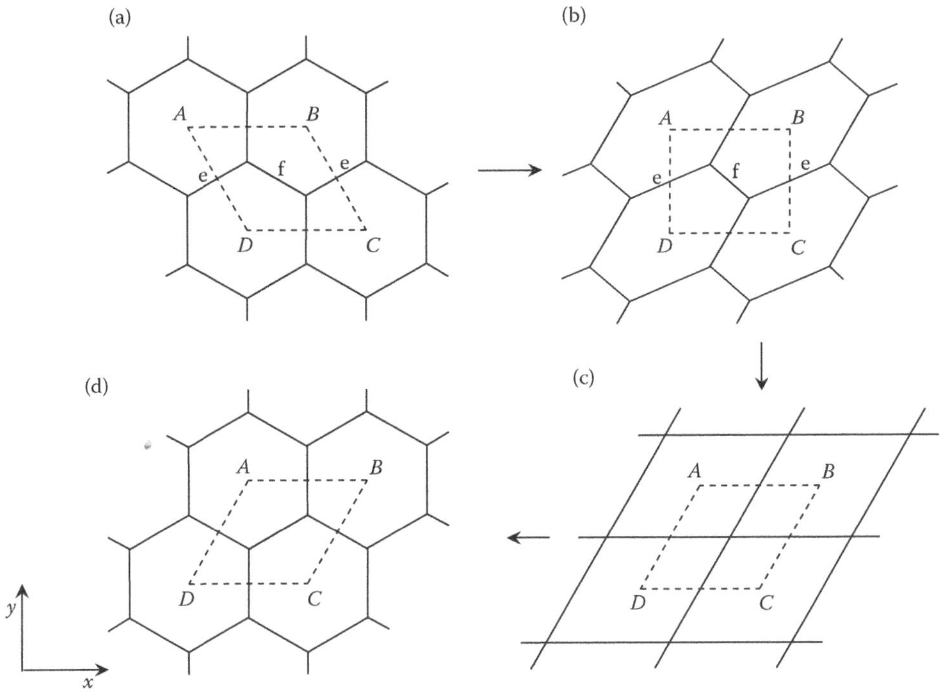

***Figure 15.6** Shear deformation of a two-dimensional foam with a honeycomb structure in x direction. Dotted lines connect centers of four depicted gas cells: (a) undisturbed configuration, angle between lines DA and AB, 60°; (b) intermediate configuration strained below stability limit; (c) configuration strained to stability limit, angle between lines DA and AB, 120° (see text); (d) new unstrained configuration involving gas bubbles reorientation. (Adapted from Princen, H.M.,* J. Colloid Interface Sci., *91, 160–175, 1983.)*

In reality, all gas cells will not have the same volume, but will be characterized by a size distribution. The parameter R should then be replaced by the volume surface average radius R_{32}, which is defined as

$$R_{32} = \frac{\sum n_i R_i^3}{\sum n_i R_i^2} = 3\frac{\varphi}{A}, \qquad (15.3)$$

where A is the surface area of the equivalent spherical droplets per unit volume of the dispersed phase.

For volume fractions below 1, the presence of another structural element, the so-called plateau border (**Figure 15.5c**), has to be taken into account. Plateau borders form a kind of channel having three concave surfaces. During deformation of the foam, the process of straining will be the same as that described above for a foam with $\varphi = 1$. However, at some point, adjacent plateau borders will meet each other and merge into one. The larger the plateau borders will be, that is at lower ϕ, the sooner this transition will occur. This will affect the stress versus strain relation, leading to yield stresses strongly decreasing with decreasing φ. A third factor affecting the stress versus strain behavior is the presence and size of a contact angle. The presence of short-range colloidal interaction forces such as van der Waals forces (Section 10.3.1) will result in a sudden change from the flat film between the droplets to the curved surfaces away from the film. This sudden transition is characterized by a contact angle. Its importance has clearly been shown for highly concentrated emulsions. Both effects together give the following dependence of the shear modulus on φ (Princen and Kiss 1986)

$$G \approx C_G \frac{2\gamma_{int}}{R_{32}} \varphi^{1/3}(\varphi - \varphi_{max}) \tag{15.4}$$

where C_G is a constant of about 0.9 and φ_{max} is the maximum volume fraction when the gas cells would be undeformable. For the yield stress, one obtains

$$\sigma_y \propto \frac{2\gamma_{int}}{R_{32}} \varphi^{1/3} f(\varphi) \tag{15.5}$$

where $f(\varphi)$ is an empirical function that increases sharply with increasing φ to about 0.25 at $\phi = 1$.

At shear stresses higher than σ_y the close packed foam will flow at a certain shear rate $\dot{\gamma}$ (**Figure 15.7**). For stresses that do not surpass σ_y to a large extent, it is reasonable to assume that the resistance to flow can be calculated simply by adding up the elastic and viscous term leading to an equation similar to the Herschel-Bulkley model (Equation 4.14). For the elastic term the yield stress is taken, whereas the viscous term is assumed to be related to the energy dissipation involved in the flow of the continuous liquid in and out of the plateau border on stretching and shrinkage of adjacent films. It leads to the following semi-empirical equations for the shear stress and apparent viscosity (Princen and Kiss 1989)

$$\sigma = \sigma_y + C_\eta(\varphi - \varphi_{max})\left(\frac{\gamma_{int}\eta_c\dot{\gamma}}{R_{32}}\right)^{1/2} \tag{15.6}$$

$$\eta_{app} = \frac{\sigma_y}{\dot{\gamma}} + C_\eta(\varphi - \varphi_{max})\left(\frac{\gamma_{int}\eta_c}{R_{32}\dot{\gamma}}\right)^{1/2}, \tag{15.7}$$

where C_η is a numerical constant of about 32.

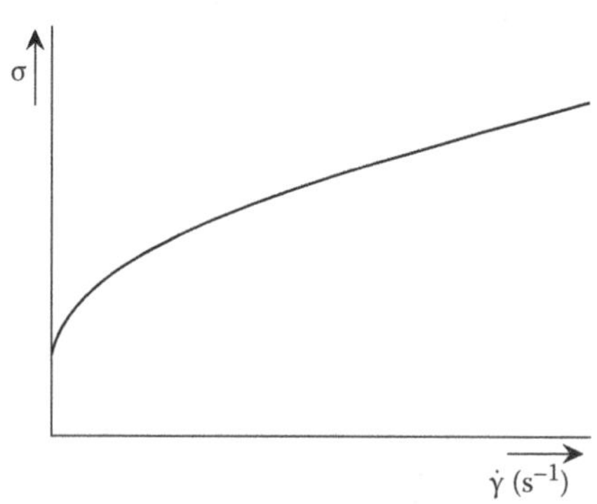

***Figure 15.7** Schematic plot of shear stress σ as a function of shear rate $\dot{\gamma}$ for a highly concentrated emulsion ($\varphi \gg \varphi_{max}$).*

One should be aware that for high shear stresses, breakup of the gas cells may occur, leading to the disruption of the foam.

Most remarks made above for foam systems also hold for highly concentrated emulsions. However, because the size of emulsion droplets is usually orders-of-magnitude lower than that in foams, the stress levels involved will be much higher. This is so despite the generally lower interfacial tension of oil–water than air–water interfaces. The Laplace pressure differs roughly by a factor $10–10^3$. In addition, other energy-dissipating mechanisms may also play a role aside from those considered during the derivation of Equations 15.6 and 15.7, leading to lower or higher power-law indexes than 1/2.

References

Keetels, C.J.A.M., T. van Vliet, and P. Walstra. 1996a. Gelation and retrogradation of concentrated starch systems: 1. Gelation. *Food Hydrocolloids* 10: 343–353.

Keetels, C.J.A.M., T. van Vliet, and P. Walstra. 1996b. Gelation and retrogradation of concentrated starch systems: 2. Retrogradation. *Food Hydrocolloids* 10: 355–362.

Princen, H.M. 1983. Rheology of foams and highly concentrated emulsions: I. Elastic properties and yield stress of a cylindrical model system. *J. Colloid Interface Sci.* 91: 160–175.

Princen, H.M., and A.D. Kiss. 1986. Rheology of foams and highly concentrated emulsions: III. Static shear modulus. *J. Colloid Interface Sci.* 112: 427–437.

Princen, H.M., and A.D. Kiss. 1989. Rheology of foams and highly concentrated emulsions: IV. An experimental study of the shear viscosity and yield stress of concentrated emulsions. *J. Colloid Interface Sci.* 128: 176–187.

Stamenovic, D. 1991. A model of foam elasticity based upon the laws of plateau. *J. Colloid Interface Sci.* 145: 225–259.

Walstra, P. 2003. *Physical Chemistry of Foods*. New York: Marcel Dekker.

16

Cellular Materials

Cellular materials are generally characterized by connected fairly rigid cell walls, enclosing a fluid material. The fluid material may be liquid-like (wet cellular materials) or gas (dry cellular materials). Most vegetable and many fruit tissues belong to the first category, whereas most manufactured cellular materials belong to the second category. They contain gas-filled cells. Examples are bread and related products, expanded products formed by extrusion or by deep frying in oil, and biscuits.

The cells forming the structure may be closed or open. The former type is like foam. The matrix can geometrically be compared with the continuous phase of foam, which consists of relatively thin lamellae between the cells (the cell walls) and places where they meet (beams, struts or ribs). When the cells are filled with gas, the material can be called *solid foam*. When the lamellae contain holes, the cells are open; this type of material is called a *sponge*. The gas-filled products mentioned above are sponges although they also contain a small volume fraction closed cells. In these products, it is the matrix that provides the stiffness of the material. Most plant tissues are built up of closed cells filled with an aqueous liquid. Their stiffness is determined both by the matrix and the turgor pressure in the cell, that is, the extent by which the osmotic pressure in the cell is higher than that outside. For many fruits and vegetables, the turgor pressure is the main factor.

Below, first the mechanical properties of dry cellular materials will be discussed and second, in qualitative way, those of wet cellular materials.

16.1 Dry cellular materials

The most important characteristics of dry cellular materials are the volume fraction air, the mechanical properties and structure of the solid matrix, and whether the cells are closed or open. Below, the discussion of the relation between structure of these materials and the mechanical properties will start with considering a two-dimensional array of hexagonal cells of equal size (honeycombs), as depicted in **Figure 16.1**. The mechanical properties of the sponge will be expressed relative to those of the matrix material. In this way, the obtained relations have a more universal applicability. First, relations for Young's modulus *E* will be discussed. Much of the discussion below is based on the work of Gibson and Ashby (1997) as reviewed in their book on cellular solids, structure, and properties.

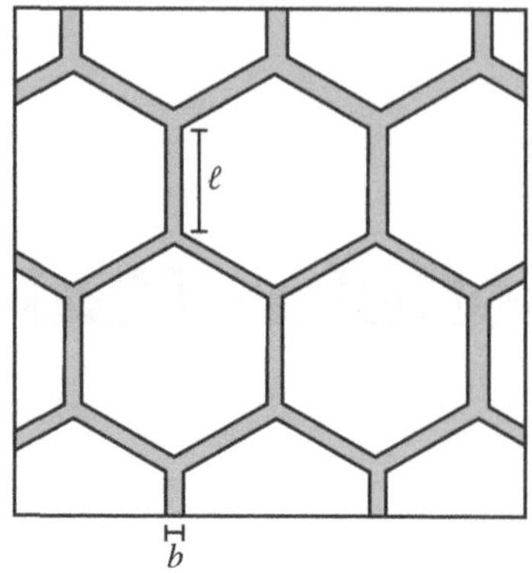

Figure 16.1 *Structure of an undeformed two-dimensional honeycomb, l length cell walls/beams, and b their thickness.*

For sponge structures in which nearly all materials are in beams, deformation of the structure will initially lead to bending of the beams, as illustrated in **Figure 16.2**. The extent of bending will depend on the stress applied, its direction relative to the beam, beam thickness *b* and width, beam length *l*, and the Young's modulus of the beam E_m. To illustrate the derivation of the various relations, cellular structures will be considered as composed of ideal open or closed cells (**Figure 16.3**). Open cell structures consists of solid struts surrounding a void space containing air and closed cell structures of solid plates (walls) also surrounding a void space. Application of a remote compressive stress σ on the sponge will exert a force $f \propto \sigma l^2$ on the beams. This will cause them to bend as shown in **Figure 16.2**. Standard theory gives for

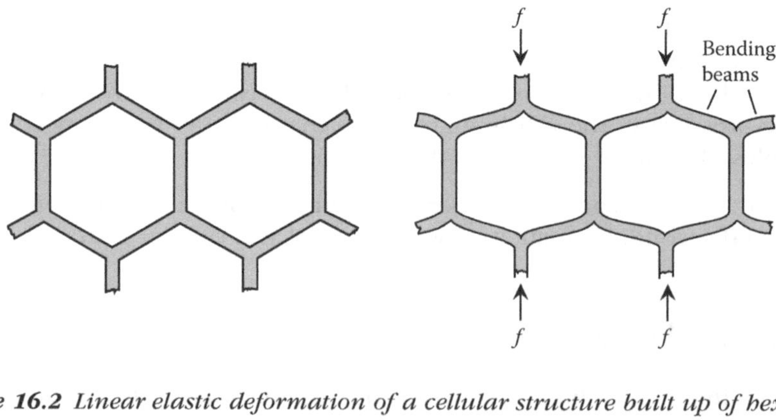

Figure 16.2 *Linear elastic deformation of a cellular structure built up of hexagonal cells of equal size due to a compressive force. Cells bend due to force f.*

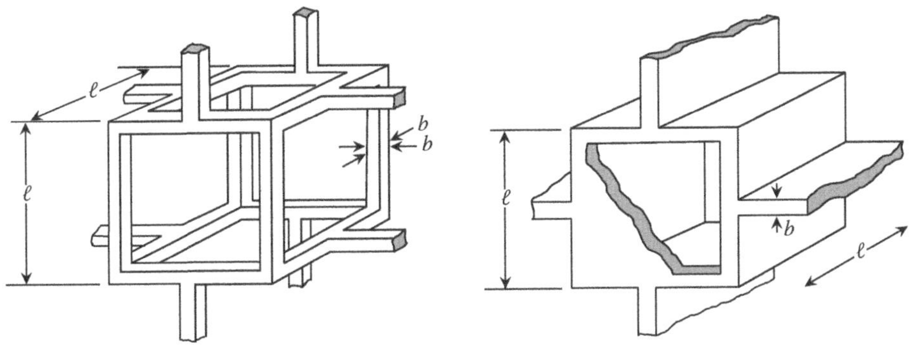

Figure 16.3 *Ideal open and close cells. (Reprinted with slight modifications from Ashby, M.F.,* Metall. Trans. A, *14, 1755–1769, 1983. With permission.)*

deflection δ of the free end of a single beam, which is fixed at the other end because of a force f at the free end,

$$\delta \propto \frac{fl^3}{E_m I}, \tag{16.1}$$

where I is the second moment of inertia of the area, which is equal to $b^4/12$ for an open-cell beam and $b^3l/12$ for a closed cell side. The strain ε_m in the beam—and with that, ε of the sponge structure—will be proportional to δ/l, and the overall stress σ to f/l^2. From this, it follows directly—taking into account Equation 16.1—that Young's modulus for the considered cellular structures is given by

$$E = \frac{\sigma}{\varepsilon} = C\frac{E_m I}{l^4}, \tag{16.2}$$

where C is a dimensionless constant. Equation 16.2 can be simplified by considering the ratio of density ρ of the cellular material to density ρ_m of the matrix material (relative density cellular material). It can be deduced easily that, for the open-cell structure,

$$\rho/\rho_m = (b/l)^2 \tag{16.3}$$

and for the closed-cell structure

$$\rho/\rho_m = b/l \tag{16.4}$$

The combination of Equation 16.2 with the expressions for the second moment of inertia of the area and Equations 16.3 and 16.4 gives, for the Young's modulus of cellular structures with open and closed cells, respectively:

$$E_o \propto E_m \frac{b^4}{l^4} \propto E_m \left(\frac{\rho}{\rho_m}\right)^2 \tag{16.5}$$

$$E_c \propto E_m \frac{b^3}{l^3} \propto E_m \left(\frac{\rho}{\rho_m} \right)^3 \tag{16.6}$$

Both for open- and closed-cell structures, it is the relative density of the cellular material over that of the solid matrix together with E_m which determines Young's modulus of the cellular material.

At larger stresses, cellular materials will exhibit nonlinear stress–strain behavior. **Figure 16.4** shows the typical stress versus strain curves for dry cellular materials subjected to uniaxial compression. At low stresses, it deforms linear elastically (region 1), but above a critical stress a kind of plateau stress is observed (region 2), and only at a much larger strain can a strong increase in stress be observed (region 3). This behavior can be understood by considering the cell walls as beams or plates that deform linear elastic to their elastic limits and at higher stresses may exhibit elastic buckling, plastic yielding by the creation of plastic hinges at the intersections of cell walls or brittle fracture (which will also primarily occur at the intersections of cell walls). The cellular material exhibits ongoing collapse with increasing strain until most of the beams/cell walls have collapsed. At still larger strains, the stress increases rapidly because of strong compaction of the material. This is a purely geometric effect. The opposite sides of the cells are forced into contact. Below, the different failure mechanisms of cellular materials will be discussed.

Elastic buckling of beams/walls implies a sideward movement (bending) of these structural elements as a result of a compressive load (**Figure 16.5a**). It occurs more easily when the structural element being considered is more slender (larger l/b ratio). It occurs already at compressive stresses lower than

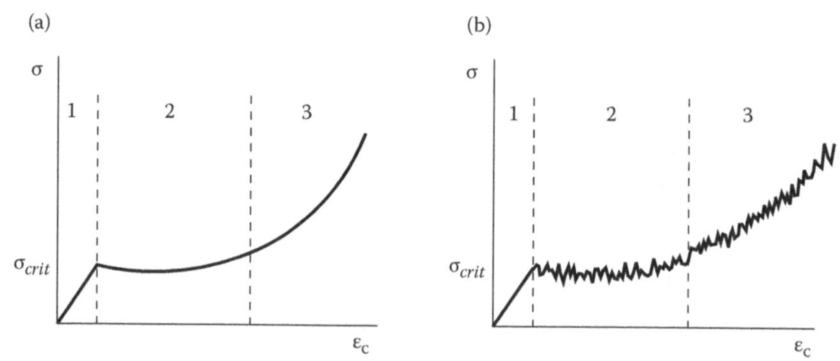

Figure 16.4 *Schematic representation of the typical shape of stress–strain curves of dry cellular solids subjected to uniaxial compression. (a) For material that shows plastic yielding; (b) for material that shows brittle fracture. Region 1, elastic bending of the beams/cell walls; region 2, collapse of structure due to plastic yielding (a) or brittle fracture (b); region, 3 compaction/densification.* σ_{crit} *denotes onset of buckling, plastic deformation, or fracture.*

the compressive stress that the material can withstand in the absence of buckling. The critical force f_{crit} at which a column of length l, Young's modulus E_m, and second moment of inertia I buckles is given by the Euler equation:

$$f_{crit} = \frac{\pi^2 E_m I}{K^2 l^2}, \tag{16.7}$$

where K is a factor that describes the degree of constraint for moving, at the end of the column (it is higher when the constraint is less; $K = 1$ for columns hinged at the ends). Using Equation 16.7 for hinged ends and $I \propto b^4$ for open sponge structures gives for the critical stress σ_{crit} on a column

$$\sigma_{crit} \propto \frac{f_{crit}}{b^2} \propto \frac{\pi^2 E_m}{(l/b)^2} \tag{16.8}$$

Equation 16.8 illustrates the importance of the slenderness ratio of columns for their sensitivity to buckling. If the critical force is reached for an entire layer of cells, they will buckle and so the foam will collapse elastically. The overall stress σ_{el} at which the cellular material will collapse is proportional to f_{crit}/l^2, from which it follows that

$$\sigma_{el,o} \propto E_m \frac{b^4}{l^4} \propto E_m \left(\frac{\rho}{\rho_m}\right)^2 \tag{16.9}$$

$$\sigma_{el,c} \propto E_m \frac{b^3}{l^3} \propto E_m \left(\frac{\rho}{\rho_m}\right)^3 \tag{16.10}$$

for open and closed cells, respectively.

Plastic yielding may occur if the matrix material yields above a certain stress. In this case, the beams/walls stay roughly straight, and deformation occurs primarily in the plastic hinges (**Figure 16.5b**). The bending moment M of a square beam with thickness b is given by fl and is proportional to $\sigma_y b^3$ for open cells and to $\sigma_y b^2 l$ for closed cells. The stress σ_{pl} (f/l^2) at which the cellular material will collapse due to plastic deformation is then proportional to M_{crit}/l^3, from which follows

$$\sigma_{pl,o} \propto \sigma_y \frac{b^3}{l^3} \propto \sigma_y \left(\frac{\rho}{\rho_m}\right)^{3/2} \tag{16.11}$$

$$\sigma_{pl,c} \propto \sigma_y \frac{b^2}{l^2} \propto \sigma_y \left(\frac{\rho}{\rho_m}\right)^2 \tag{16.12}$$

for open and closed cells, respectively.

Brittle structures will collapse by beam or cell wall fracture. Fracture is observed to occur predominantly near the cell edges just as for plastic yielding.

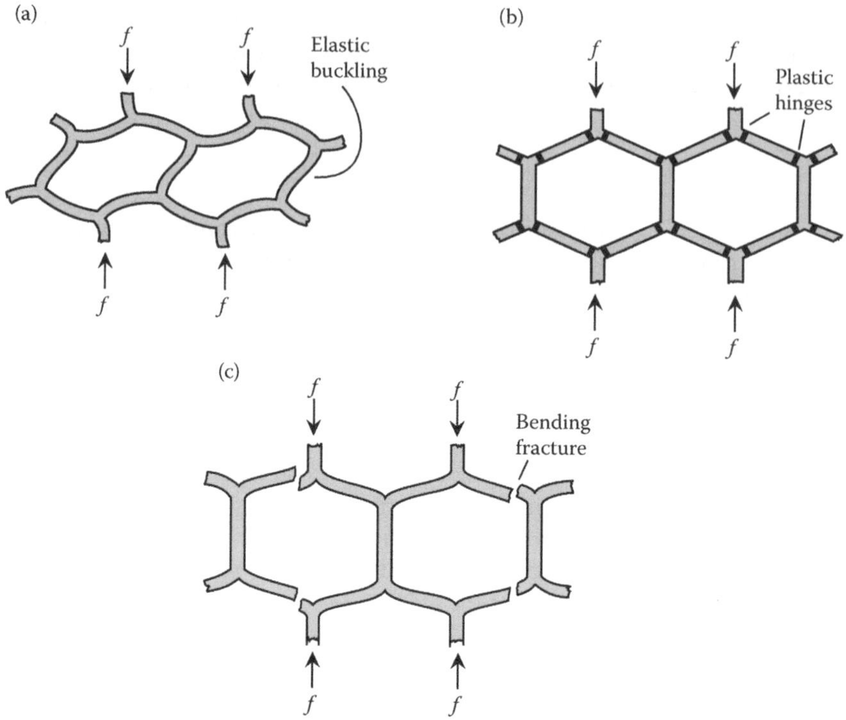

Figure 16.5 *Ways of failure of two-dimensional sponge structures due to a compressive load: (a) elastic buckling of beams, (b) plastic yielding of hinges at beam ends, and (c) fracture of beams. (Composed of Figures 7, 9, and 11 from Ahsby, M.F.,* Metall. Trans. A, *14, 1755–1769, 1983.)*

Therefore, for brittle fracture of the beams the same relations between the fracture stress σ_{fr} of the cellular material and the fracture stress $\sigma_{fr,m}$ of the matrix material are found as the relations given by Equations 16.11 and 16.12 for plastic yielding. Just as for plastic yielding, it is the force moment acting on the beams or walls that determines when they will fracture in a brittle manner.

In the discussion above, it has been implicitly assumed that the relative density of the original cellular materials was low, implying that $l > b$. The onset of densification starts usually when the instantaneous relative density is about 0.5, implying a volume fraction φ void space of about 0.5. At still higher relative densities, it is better to consider the material as a homogeneous solid material containing a limited volume fraction of noninteracting holes. Then, Young's modulus can be calculated using relevant equations given in Section 14.2, taking into account that for dry solids the Poissons's ratio will be lower than 0.5. It is often assumed to be about 0.3. In general, the scaling laws hold well for relative densities between 0.1 and 0.5 (φ = 0.5–0.9).

Figure 16.6 gives an example of published results. The effect of the water content (which is higher for storage at higher relative humidity) of the product illustrates the importance of matrix properties. This is further illustrated in **Figure 16.7** for model bread studied 4 h and 7 days after baking. The fresh bread did not exhibit a clear buckling point, although the second compression indicates that some destruction of the structure has occurred. The bread stored for 7 days not only has a much higher resistance to deformation but also shows a clear yielding/fracture point. The hysteresis between the compression and decompression curve is very large. Moreover, the second compression curve confirms that extensive irreversible structural breakdown has occurred during the first compression. The bread behaves more like a brittle product. The bread was stored under conditions that minimize water loss. Therefore, the difference in behavior due to storage is likely caused by bread staling, which involves recrystallization of the starch. It makes the matrix material much stiffer and more brittle.

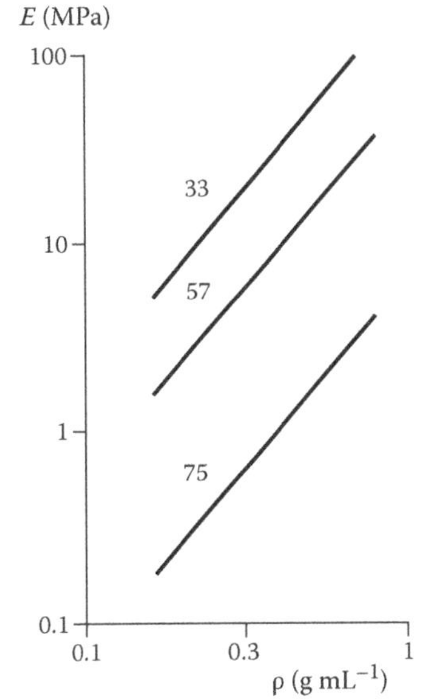

***Figure 16.6** Young's modulus E as a function of product density ρ for wavers kept in air at various relative humidity rates (indicated). (After results by Attenburrow, G.E. et al.* J. Cereal Sci., *9, 61–70, 1989.)*

The preceding equations were derived for the uniaxial compression of cellular materials containing isotropic cells. Formulas similar to the equations given above for the Young's modulus can be derived for the shear modulus. For anisotropic materials, there will be a clear effect of the loading direction on the value of the moduli and fracture parameters measured. The effect is largest for the Young's modulus. However, the scaling relations given above are not affected.

According to Equations 16.5, 16.6, and 16.9 through 16.12, there is no effect of the cell size on the mechanical parameters considered. It is only the relative density of the cellular material and the mechanical parameters of the matrix that matters. However, the above equations do not imply the absence of an effect of cell size distribution. For instance, some large cells may act as the presence of a part with a lower density than the rest of the material and, with that, they may be regarded as having lower mechanical stiffness and failure parameters. It is likely that, for Young's modulus, this will primarily affect the proportionality constant for the density dependence. However, in view of the large effect of defects (structural inhomogeneities) on fracture stress

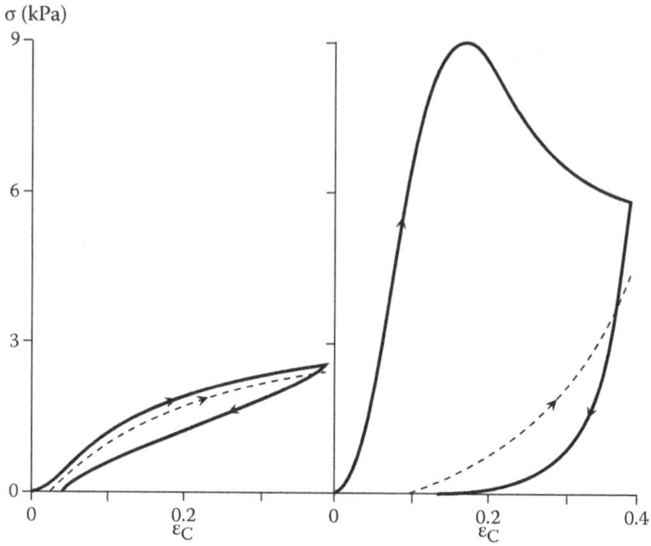

Figure 16.7 *Stress σ versus Cauchy strain ε_C curves measured during uniaxial compression and unloading of model starch bread ($\rho/\rho_m \approx 0.35$). Solid line denotes first compression and decompression; broken line denotes second compression. (Left panel) 4-h-old bread and (right panel) 1-week-old bread after storage at room temperature. (After results by Keetels, C.J.A.M. et al.* J. Cereal Sci., *24, 15–26, 1996.)*

and strain, changes in the width of the cell size distribution may affect the scaling exponent. In particular, an effect is expected as a result of a change in product density with the relative portion of cells that are much larger than the average cell size.

Another point is that structures built up of closed cells often behave like open cells. The reason for this is that for closed-cell foams, generally, most of the solid material is accumulated in the cell wall edges, causing them to become much thicker than the center part of the cell walls. Therefore, for materials that apparently consist of (approximately) closed cells, power laws are often found in between those for open and closed cells.

Finally, loading during processing, handling, and consumption of cellular solids often involves local loading only (e.g., during biting) and may involve different modes of deformation, for example, shear and uniaxial compression loads. For indentation of a brittle cellular material, the force experienced by the indenter will depend on the number of beams/cell walls with which it is in contact. For a brittle material, a structural element that fractures will no longer contribute to the force experienced by the indenter since it falls off and drops away from the contact area (**Figure 16.8**). As a result, the number of beams in contact with the indenter may vary substantially when the contact area is not much larger than the cell sizes, leading to a highly fluctuating measured force

(**Figure 16.4b**). For brittle products, this may be a product characteristic during biting and chewing. The measured force fluctuations will be smaller when the contact area is larger and when the broken parts do not drop well away, which will be the case for closed cell materials and products with a high relative density.

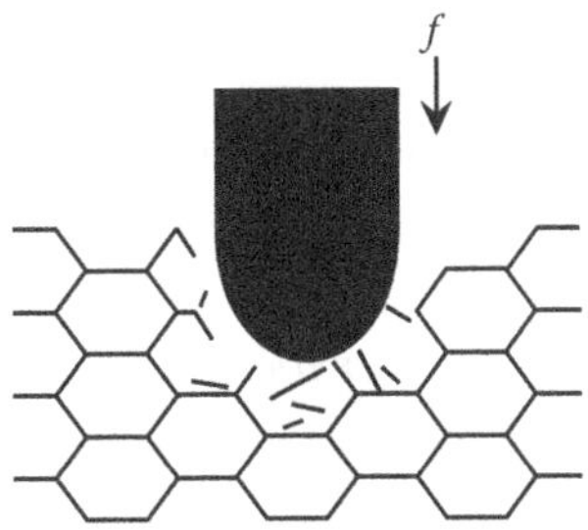

***Figure 16.8** Schematic layout of contact area between a brittle cellular material and an indenter on penetration by a relatively small indenter.*

16.2 Wet cellular materials

Most fruits and vegetables belong to the category of wet cellular materials. These materials are built up of closed cells filled with an aqueous solution. A main factor that highly complicates the establishment of general relations between the structure of fruit and vegetable tissues is that they are made up of numerous structural elements of strongly different length scales that are arranged in a hierarchical order. For instance, an apple consists of different tissues that consist of cells with intercellular spaces in between. Each tissue is an assembly of cells, consisting of a cell wall and a liquid content in which cell organs, macromolecules, and salts are dispersed. Cells are often anisotropic; in apple flesh, they are elongated in the radial direction, whereas in potato tubers, for example, they are roughly isotropic. Cell walls consist of a matrix of polysaccharides and proteins in which cellulose fibers are embedded. They are multilayer composites, which are frequently anisotropic because of the orientation of the cellulose fibers. To conclude, in apples—but also in other fruits and vegetables—the next hierarchical levels can be distinguished as molecules, cell walls and organs, cells, tissue, and fruit or vegetable.

It is considered to be outside the scope of this book to discuss the relation between the different structures of the large diversity of fruit and vegetables and their mechanical properties. Moreover, we will refrain from discussing the relation between the structure of the cell walls and their mechanical properties extensively. In addition, the structural aspects of the content of the cells will be neglected. The discussion will focus on some general aspects of parenchyma tissue.

Edible fruit and vegetables consist generally to a large extent of weak, nonstructural parenchyma cells that are mainly used by the plants for storage of food. Parenchyma cells are thin-walled cells of the ground tissue that make up the bulk of most nonwood structures. Cell walls of living tissues are usually stretched to some extent because of the presence of a *turgor pressure* in the cell. This is the main pressure of the cell content against the cell wall in plant cells. Its value is determined by the water content of the vacuole, which results from osmotic pressure, that is, the hydrostatic pressure produced by a solution in a space (cell) and separated by a semipermeable membrane from a solution at its outside with a lower osmotic pressure. It is kept at a high level by the cell metabolism. The cells are stuck together by adhesive pectins in the so-called middle lamella or because they are confined within a limited epidermis.

There are three main factors contributing to the stiffness of such wet cellular material:

(1) The turgor pressure in the cell
(2) The rigidity of the cell walls including cell material properties and the geometry of the cell wall
(3) The stiffness of the middle lamella

For various fruits (e.g., tomatoes) and vegetables, the stiffness and strength of the skin will be especially important for their mechanical properties.

The importance of the turgor pressure for cell stiffness is the main factor that distinguishes wet cellular materials from dry cellular materials. In addition, one has to make a distinction between factors (2) and (3). The importance of turgor for compressive (and with that, for bending) stiffness can easily be noticed when leaf vegetables such as lettuce and spinach wilt. Tensile stiffness of the tissue will be less sensitive to decrease in turgor pressure than stiffness against compression as long as the cells adhere. Shear stiffness also depends strongly on the degree of adhesion of cells. Its importance can be seen when considering large deformations.

In general, two main mechanisms of tissue failure can be distinguished:

(a) Cells break open on deformation and their content is released.
(b) Cell-to-cell separation occurs whereby neighboring cells are separated at the middle lamella without rupture of the cells walls. The cells do not break open.

In practice, a combination of both mechanisms may also take place. Which failure mechanisms will be dominant is determined by the three factors determining material stiffness and by the deformation mode. A higher turgor pressure gives a higher tensile deformation or stress in the cell wall at rest. Thus, the cell wall is likely to rupture at a lower applied stress than at a lower turgor pressure. On the other hand, a high turgor pressure will increase the extent of cell–cell interaction, given that for cell-to-cell debonding a higher stress is required. Bruising of apples, which involves rupture of the cell walls, under impact loading can be reduced by a slight reduction in hydration (2–3% mass loss).

The strength of the middle lamella is considerably affected by biochemical processes during storage of fruit. These cause the breakdown of the pectin in the middle lamella. Among others, it greatly diminishes the resistance against shearing deformations and favors mechanism (b) over (a). Its extent is an important factor in determining whether apples become mealy during storage, especially when they become overripe. Other factors are the presence of large air spaces between the cells, which contributes to easy separation of parenchyma cells, and the thickness of the cell walls. On the other hand, release of the cell content is a prerequisite for the juicy character of fruits and vegetables.

The importance of the presence of large air spaces for the material properties (inherent defects in fracture mechanics terms) can easily be observed during biting in various apples. If you remove the skin and the core of a large piece of a fresh, crispy apple and cut two approximately cube-shaped pieces of the same size, then bite slowly into each piece with your front teeth at a slow speed in the tangential or radial direction of the material in the original apple, you will observe that on biting in the radial direction the piece of apple fractures at lower indentation by the teeth. Already at relatively small indentation, a free running crack is formed. This will not be the case when one is biting the apple in the tangential direction (Vincent 1993).

The direction of loading of a plant tissue built of anisotropic cells may also determine the failure mode. Again taking an apple as example, during compression of cylindrical test pieces between two flat plates in the tangential direction, they often fracture in shear at 45° to the direction of the applied force. On the other hand, when compressed in the radial direction, failure occurs by a sudden and unstable collapse of just one or two layers of cells, usually in the middle of the specimen at right angles to the applied force. On further compression, another layer of cells will collapse, etc. The collapsed cells release their content. Those outside the collapse zone remain visually undamaged. On compression in the tangential direction, less and shorter cell walls are compressed, resulting in a lower stiffness of the parenchyma tissue and less energy storage. Failure occurs finally by the breaking of inclined cell walls due to shear forces acting at an angle of 45° to the direction of the force (Khan and Vincent 1993). Release of the content of the cells takes place depending on the relative strength of the cell wall compared to the middle lamella.

References

Ashby, M.F. 1983. The mechanical properties of cellular solids. *Metall. Trans. A* 14: 1755–1769.

Attenburrow, G.E., R.M. Goodband, L.J. Taylor, and P.J. Lillford. 1989. Structure, mechanics and texture of a food sponge. *J. Cereal Sci.* 9: 61–70.

Gibson, L.J., and M.F. Ashby. 1997. *Cellular Solids, Structure and Properties*, 2nd ed. Cambridge: Cambridge University Press.

Keetels, C.J.A.M., K.A. Visser, T. van Vliet, A. Jurgens, and P. Walstra. 1996. Structure and mechanics of starch bread. *J. Cereal Sci.* 24: 15–26.

Khan, A.A., and J.F.V. Vincent. 1993. Compressive stiffness and fracture properties of apple and potato parenchyma. *J. Texture Stud.* 24: 422–435.

Vincent, J.F.V. 1993. Mechanical and fracture properties of fruit and vegetables. In *Food Colloids and Polymers: Stability and Mechanical Properties*, ed. E. Dickinson and P. Walstra, 191–203. Cambridge: Royal Society of Chemistry.

17

Hard Solids

Only few foods are macroscopically homogeneous, hard solids. Most hard foods are cellular materials having a sponge or foam structure as discussed in Section 16.1. Macroscopically homogeneous, hard solid foods generally have an amorphous glassy structure (e.g., hard sweet candies) or a crystalline structure (e.g., large sugar or salt crystals).

A main disadvantage of macroscopically homogeneous, hard food products is that they are very difficult to break down by biting and chewing during mastication. The fracture stress of such products will be of the same order as that of the teeth or higher, so we can eat them only when they dissolve readily in the mouth or if they contain large inhomogeneities. The latter is the case in hard cellular solids such as biscuits and toast, where the sponge structure allows them to be broken down easily during mastication. Small pieces of hard solids will, in general, readily dissolve or contain large enough structural defects that humans can fracture them between the teeth (e.g., sugar crystals). Hard sweet candies are usually kept in the mouth and left to dissolve. Products such as dark chocolate bars consisting of cocoa solids and butter and sugar are not homogeneous solids because of the presence of some uncrystallized cocoa butter and amorphous and crystallized sugar. These products contain large inherent defects, which allow humans to bite a piece of a chocolate bar without endangering their teeth.

Hard solid materials will behave elastically on deformation and fracture at small strains (usually <0.01). Their Young's modulus will generally be about 10^9–10^{10} Pa, to some extent depending on the presence of defects in the structure. The fracture behavior can be characterized as linear elastic or brittle fracture, as discussed in Section 5.2.1. The main factor of importance is the size of the inherent defects present in the material and the amount of energy elastically stored in it on deformation. Because the latter factor will not vary much, the inherent defect length of the material is, in fact, determining fracture behavior. For a further discussion of the relation between the structure of hard solid materials and large deformation and fracture behavior, see Section 5.2.1.

V

Relationships among Food Structure, Mechanical Properties, and Sensory Perception

18

Texture Perception

For a long time, research on sensory perception was focused on taste, flavor, and some simple mechanical properties of a product such as viscosity, firmness, and hardness. In the past decades, it has become increasingly clear that many sensory attributes cannot be related to a single physical or chemical property of the food in a simple way. This was found to apply for taste, odor, as well as texture perception. For texture, it is often the interplay between several physical and—in certain cases—chemical parameters that determine its perception. One should realize that during sensory perception of food, people use all their senses. The main ones are

- Vision. Appearance is an important factor in determining the acceptability of a food product. It comprises color, shape, size, and gloss. It may affect texture perception in a subliminal manner. It is probable that when we see a food, our mind has a "texture program" as a reference, and this guides our oral processing and evaluation.
- Hearing. Sound perception is an essential part of the sensory grading of crispy and crunchy products and may affect other texture characteristics in a subtle way.
- Touch. This also includes the sense of movement and position (kinesthetic). It is sensed both outside and inside the mouth.
- Smell. This concerns the perception of volatile stimuli in the olfactory region of the nose. These stimuli may reach the nasal region via the nose as well as via the back of the mouth.
- Taste. This involves the perception from dissolved nonvolatile stimuli. The main taste components that are distinguished are salty, sweet, bitter, sour, and umami. They are perceived on the tongue, the palate, and the throat. Odor and tastants are combined in the term "flavor."

In addition, there are receptors for temperature, pain, and irritants (e.g., capsaicin from chili peppers and carbon dioxide in carbonized drinks).

The importance of texture characteristics for the acceptability of foods varies widely with the type of food. Consumers find texture—along with flavor—the most important attribute of foods. In general, it is more important for solid foods than for liquid ones. For products such as potato chips, celery, and meat, texture is a critical quality attribute, whereas it is of minor importance for thin liquid products that are homogeneous down to the molecular scale, as is the case for many beverages, at least as long as their viscosity is low. When food texture does not meet the expectations of consumers, it leads to a negative impression of the food or even to complete rejection.

A common definition of *texture* is "all the mechanical (geometrical and surface) attributes of a food product perceptible by means of mechanical, tactile and where appropriate, visual and auditory receptors" (International Organization for Standardization 1992). It implies that texture is a multiparameter attribute and not a single entity. This means that it is mostly necessary to use a number of different experimental configurations and different test principles in order to fully specify the texture of a food product. The choice must depend on the food. Instrumental determination of a textural attribute of a hard solid food, a soft solid food, and a liquid requires to a large extent the determination of different physical parameters and, consequently, different measuring techniques.

As mentioned above, texture is a multiparameter sensory attribute. Moreover, in contrast to other sensory attributes (e.g., taste and color), there are no single and specific receptors for it because of its multiparameter character. For most foods, the perception of texture in the mouth involves the transformation of the product structure into a stage that it can be swallowed. The extent and the way of transformation depend on product characteristics. Thus, the combination of three main areas of research is necessary to advance our understanding of texture perception. These are sensory research on signal processing in the brain, oral physiology, and physics/chemistry. The latter should involve the study of rheological and fracture characteristics, microstructure, ingredient interaction, and other relevant food characteristics in relation to texture perception.

This chapter will focus on the relation between rheological and fracture characteristics of food and texture perception. By establishing such relations, one should realize that texture perception is fundamentally a human response to a food, and its only true measure is sensory analysis. However, for industry, it is essential to be able to predict, at least to some extent, what the texture perception of their products will be. The use of panels is a good, but expensive way of determining this. Therefore, there is an urgent need for relatively simple, reliable, repeatable, and inexpensive methods to estimate the texture of food products. A problem in developing such methods is that foods are very complex systems and texture is a multiparameter attribute. As a result, application of the principles of rheology and fracture mechanics leads, in many cases, to difficult experimental and theoretical problems. An additional problem is that mechanical tests will most likely not determine all the texture attributes that humans experience during eating. In spite of these problems, intelligent use of principles of rheology and fracture mechanics can lead to tests that can estimate certain texture attributes and can contribute to linking these attributes to specific elements of food structure. To develop such tests, it is essential to have a good understanding of the way humans judge these attributes both in and outside the mouth, implying a good understanding of sensory perception and oral physiology.

Because the way of judging is strongly dependent on the mechanical properties of the food product, a subdivision has to be made between all foods in various classes, depending on these properties. Based on the mechanical properties, one can distinguish four main groups of foods whereby one should realize that there is no absolute transition between the different groups, but that all transitions are gradual. These groups are

- Liquids—these foods flow under a stress. They vary from simple Newtonian liquids (e.g., water, wine, and clear fruit juices) to strongly non-Newtonian liquids (e.g., condensed milk, stirred yogurt, and tomato juice). During drinking, liquids are primarily transported by the tongue, which pushes small portions of it to the throat, followed by swallowing.
- Semisolids—these products have elastic properties at rest and under low stresses, but when subjected to higher stresses, they first yield and then flow. Semisolids include many dairy desserts, sauces such as tomato ketchup, and mayonnaise. They are processed in the mouth by squeezing between the tongue and the palate.
- Soft solids—usually they require biting off and further processing in the mouth by squeezing between the tongue and the palate or by chewing between the molars. Soft solids include bakery products (e.g., bread, cakes, and pies), soft fruits (e.g., strawberries and bananas), meat and meat products, ice cream, and cheese. Part of the particles formed in the mouth will aggregate to a so-called food bolus at the back of the oral cavity, which is eventually swallowed.
- Hard solids—these products must fracture in a brittle way during biting and chewing, or they (e.g., hard candies) must dissolve before they are ready for swallowing. Most hard solids have a cellular structure characterized by connected fairly rigid cell walls that enclose a fluid material. This may be liquid-like (wet cellular materials) or gas (dry cellular materials) (**Chapter 16**). Various vegetables and fruits belong to the first category, whereas most manufactured cellular materials belong to the second category. They contain gas-filled cells. Examples are various bakery products and expanded (starch) products formed by extrusion or by deep frying in oil. On consumption, these products fracture in a brittle manner, whereby fracture is usually accompanied by acoustic emission. Particles formed will aggregate to a food bolus and are eventually swallowed.

Below, the relation between food structures, mechanical properties, and texture perception will be discussed for these four groups. The discussion will be limited to some general principles and to some texture attributes. When studying the relation between mechanical properties and oral evaluation of the texture of a food product, one has to make a clear distinction between studies focused on establishing a correlation between a texture attribute and a (set of) instrumental characteristics and those directed on understanding the relation between the texture attribute and mechanical properties of the product. Both aims have their own value. In the literature, many publications are directed on establishing a correlation. The danger of establishing a correlation, without understanding how it works, is that a change in product properties or in measuring methods could make this correlation invalid. Below, the focus will be on how to devise methods that allow at least a relatively good understanding of the relation obtained between the texture attribute studied, the microstructure of the product, and its measurable mechanical properties.

A series of measuring techniques has been discussed extensively in **Chapters 7 and 8** that can be used to determine the fundamental mechanical parameters of various food materials. In addition, a short, but far from complete, overview was given of empirical tests. A more extensive overview of empirical tests, especially in relation to texture perception, is presented by M.C. Bourne in his book, *Food Texture and Viscosity* (2002). It also provides an extensive review of the determination of texture attributes by panel tests and instrumental methods.

A general point, which is too often neglected, is that the meaning of terms used in sensory science and in rheology and fracture mechanics may differ greatly. For instance, hardness may be defined as the stress (force) needed to deform a material to a set strain (e.g., 0.1 or 0.5) as well as the stress to fracture the material. In the context of the often-applied texture profile analysis (Section 8.4.2.3), it is defined as "the force necessary to attain a given deformation." In the literature dealing with rheology and fracture mechanics, hardness is usually related to permanent deformation, implying yielding or fracture. This clearly shows that the description of how a term is evaluated (e.g., force needed to fracture the food on first chew) is more important than the term itself. The latter always has a cultural bias (A.E. Foegeding, personal communication).

18.1 Liquids and semisolid products

The important texture characteristics of liquid foods are thickness, sliminess, creaminess, fattiness, roughness, and the so-called after-feel texture characteristics primarily due to a coating formed on the oral mucosa during oral processing. During oral processing, the liquid and dispersed molecules and particles will interact with the saliva present in the oral cavity and with the oral surfaces leading to a coating of primarily the tongue. Properties such as coating formation do not depend primarily on the mechanical properties of the liquid, but mainly on the adhesion properties of the molecules/particles involved. In this section, discussion will be limited to one important texture attribute of liquids—their "thickness." This attribute is an important characteristic of many liquid products on its own. In addition, it contributes very significantly to a multiparameter attribute, creaminess. This attribute has been found to depend on such texture characteristics as thickness, smoothness, and roughness of the liquid, but also on taste and, likely, flavor attributes.

Thickness is the most studied characteristic of liquids. It is generally thought to be related to the viscosity of the products under the flow conditions being associated with their oral manipulation. In several studies, the relation has been determined between the sensory rating of thickness (or viscosity) by a panel and shear stress versus shear rate curves for various Newtonian and non-Newtonian liquids. An extensive study has been performed by Shama and Sherman (1973a). Information on the relevant shear rates during oral processing can be obtained, for instance, by providing panelists each time with six pairs of four different food samples. The shear stress–shear rate curve of one of these four samples intersected and crossed over the curves for the other three samples (**Figure 18.1**). The panelists had to compare the liquids pair by pair and to indicate which one appears more viscous (or has a higher perceived thickness) in the mouth. In this way, shear stress–shear rate bounds could be determined in which a particular set of four liquid products are orally evaluated. By repeating this procedure for a wide range of products, Shama and Sherman and other researchers were able to deduce bounds for shear stresses and shear rates associated with oral evaluation of "viscosity" of liquids and semisolids that behave as a liquid after yielding (**Figure 18.2**).

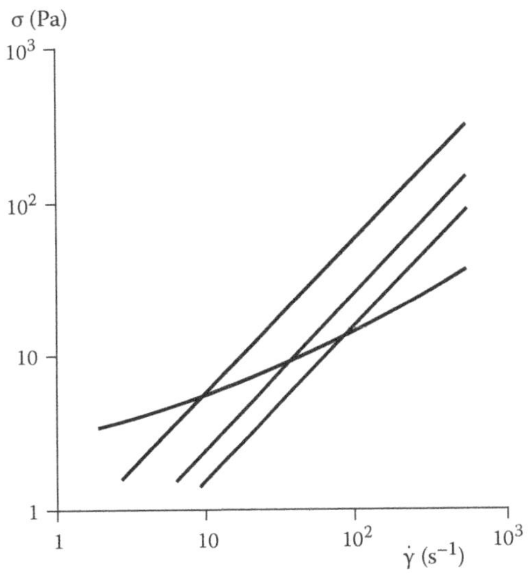

***Figure 18.1** Shear stress versus shear rate curves for four liquids that allows, in principle, users to obtain information on average shear rate at which "viscosity" is evaluated orally. For further explanation, see text. (From Shama, F. et al.,* J. Texture Stud., *4, 102–110, 1973.)*

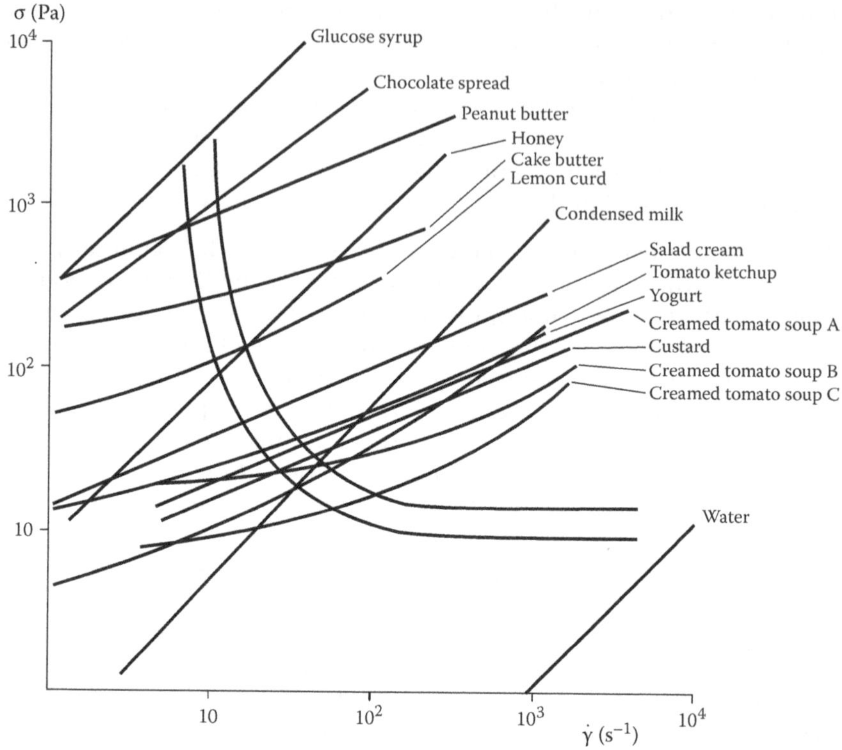

Figure 18.2 *Bounds of shear stress–shear rate associated with oral evaluation of viscosity, and superimposed on these bounds shear stress–shear rate data for various liquid and semisolid foods. (Reprinted from Shama, F., Sherman, P.,* J. Texture Stud., *4, 111–118, 1973a. With permission.)*

Figure 18.2 shows that sensory evaluation of "viscosity" does not occur at constant shear rate or shear stress, but that their effective values depend on the viscosity of the products. It suggests that low viscosity products (viscosity roughly below 0.1 Pa s) are evaluated at a constant shear stress of about 10 Pa, whereas products with a viscosity exceeding 10 Pa s are evaluated at a constant shear rate of about 10 s^{-1}. Data by other researchers are roughly in agreement with these data. These results suggest that the evaluation of low viscosity liquids is based on an appraisal of the ease with which the materials flow between the upper surface of the tongue and the palate. Low viscosity samples will spread faster (higher shear rate) and over a longer distance. Products with a high viscosity or a yield stress above about a few Pascals (1 Pa is equal to the pressure exerted by a water column of 0.1 mm high) will not spread significantly under gravity during the usual oral evaluation time. It is likely that these products are evaluated by the pressure (stress) required to produce significant flow. In agreement with this, a relation has been found between the yield stress of semisolid foods and sensory thickness perception

(Dickie and Kokini 1983). Moreover, in various studies, a positive relation has been observed between the pressure exerted on the tongue or palate as measured with pressure sensors and the thickness/firmness of semisolid and soft solid foods.

To some extent, it is surprising that the correlation observed between the instrumentally measured viscosity in shear flow under the conditions shown in **Figure 18.2** and the orally evaluated viscosity is so good. One can make a series of critical remarks based on basic principles of rheological measurements (van Vliet 2002).

- Short evaluation times. The residence time of liquids and semisolid products in the oral cavity will be short, usually of the order of seconds, and will depend on the viscosity/thickness of the product. For Newtonian liquids the viscosity is independent of shearing time, but for many non-Newtonian liquids and semisolid products the viscosity will decrease as a function of shearing time. These products exhibit thixotropic behavior. This implies that thixotropic products should be evaluated instrumentally for viscosity/thickness over similar times as the average residence time in the mouth. For semisolid products, one option is to take the maximum in the shear stress versus shearing time on shearing. This may work well for their thickness evaluation.

  Another consequence of thixotropic behavior is that every disturbance of a product before and during the loading in the mouth (and instrument) will affect the viscosity or thickness perceived.
- Occurrence of turbulent flow. For low viscosity liquids, turbulence may occur at the high "shear rates" they are subjected to in the mouth. After mixing with saliva, turbulence will be damped to some extent because of the macromolecules present in it. The occurrence of turbulence will lead to higher energy dissipation and so to an effective higher viscosity. Whether it is really a factor of importance for thickness perception is not known.
- Occurrence of extensional flow next to shear flow. In the literature, it is mostly assumed that the prevailing flow pattern between the tongue and palate is shear flow. One assumes that the liquid is squeezed between the tongue and then palate, and then sheared during backward and forward or sideways movements of the tongue. However, a squeezing action between two plates will also involve an extensional flow component, with its relative importance depending on the extent of friction between the liquid (material) and the plates (Section 8.3.6). There are no conclusive studies on the relative importance of both flow fields for the oral evaluation of viscosity and thickness. For Newtonian liquids, the relative importance of the elongation flow component is likely not so important because the ratio between the elongation viscosity η_e and the shear viscosity η is constant and rather low ($\eta_e/\eta = 3$, Section 3.2). However, for non-Newtonian liquids, the ratio η_e/η may

be much higher and will depend on the prevalent elongation and shear flow rates. As a result, the occurrence of elongation flow may affect oral perception to a larger extent than for Newtonian liquids and, therefore, the relation between measured shear viscosity and oral evaluated viscosity. An indirect indication of the importance of elongation flow during oral perception is the popularity of flow cups for quality control of products such as stirred yogurts (Section 8.4.1.1).

- Occurrence of range of strain rates. It is clear from the shape of the oral cavity and from the movements of the tongue that, depending on the place in the mouth, the type of flow and the flow rate will differ during oral processing of liquids and semisolids. In addition to the type of flow, the strain rate will also greatly vary from place to place in the mouth. Whether this variation in strain rate with the place in the mouth is an important factor for the oral evaluation of the viscosity (or perceived thickness) has not been studied thoroughly.
- Shear rate thinning effects. Apart from thixotropy, likely, shear rate thinning character of non-Newtonian liquids may also affect their perceived viscosity. It has been suggested in the literature that the shear rate thinning behavior of non-Newtonian liquids correlates to sliminess perceived in the mouth. However, it is likely a combination of factors that determine sliminess, whereby viscosity, shear rate thinning behavior, and possibly behavior in extensional flow plays a role.
- Mixing with saliva. Mixing with saliva will increase the viscosity of low viscosity liquids and decrease the viscosity of liquids and semisolids with a high viscosity. However, for residence times in the mouth of 2 s, only small effects on the viscosity of liquids have been observed due to mixing with saliva. In addition, enzyme action may affect the viscosity (e.g., by the action of amylase in starch-containing products). These effects may be substantial, as shown for custards, for example.
- Temperature. No effect of measuring temperature (range 20–40°C) has been found on the correlation between experimentally determined viscosity and oral perceived viscosity. Over such a temperature range, viscosity changes are relatively small (for water viscosity decrease 30% between 20°C and 37°C). The effect of temperature will be much larger if a phase change occurs in the temperature range between product temperature before consumption and the temperature in the mouth (e.g., due to melting of fat or of a gelatin gel).

18.2 Soft solids

A huge variety of food products belong to the category soft solids. They differ from semisolids insofar that on consumption usually first a part is bitten off from a larger piece, although they can also be brought directly into the oral cavity by using a spoon or a fork. The latter holds for certain food products such as cooked potatoes and soft fruits (e.g., strawberries). Further processing takes place by squeezing between the tongue and palate and/or by chewing between the molars. The distinction with semisolid products, in particular, is rather vague and for various products, a matter of taste. Important texture attributes of soft solids include firmness, soft, creamy, crumbly, sticky, juicy, spreadability, thin, and thick. Thereby, one has to make a distinction between mouth-feel and after-feel properties. The latter ones concern the texture properties that are perceived after the bulk of the food has been swallowed. The intensity of after-feel properties is clearly positively related to the extent of mouth coating by the product due to adhesion between food particles/food components and the oral mucosa.

On eating a solid food, it will first be further processed in the oral cavity before swallowing. In general, the following stages can be distinguished: (1) ingestion/biting by the front incisors, (2) chewing of the harder foods by the molars and deforming of the softer ones between the tongue and palate, (3) wetting by saliva, (4) manipulation by the tongue to form a bolus at the back of the oral cavity, (5) swallowing, and finally (6) cleaning of the mouth. For the formation of a coherent bolus, the food has to be broken down into sizes of roughly several cubic millimeters. In addition, the particles have to be wetted by saliva, which may result in sticking together of the particles. For a further discussion of oral processing, see Chen (2009) and van Vliet et al. (2009) and references therein.

When establishing a relation between the mechanical characteristics of soft food products and their texture perception, a complicated factor is that these characteristics often depend on the rate of deformation (strain rate). The first systematic study on the effect of the applied deformation rate during an instrumental test on the correlation between sensory evaluation of hardness and the force required to compress a food product was conducted by Shama and Sherman (1973b). They performed uniaxial compression experiments on Gouda and White Stilton cheese at various compression speeds: 10, 20, 50, and 100 cm min^{-1}. A trained sensory panel evaluated the Gouda cheese as harder than the White Stilton cheese. When the cheeses were compressed at 5 cm min^{-1}, the force required to compress the White Stilton cheese was larger than that for the Gouda cheese over the whole relative deformation range (Cauchy strain 0–0.8). They only obtained agreement between sensory evaluation and instrumental data for compression speeds ≥20 cm min^{-1} and Cauchy strains between about 0.35 and 0.65 (range somewhat increasing with deformation speed). In addition, one should remember that also the fracture strain might depend on the deformation speed. (e.g., **Figure 5.1** for young Gouda cheese and **Figure 5.12** for potato starch gels). The above examples illustrate the importance of the selection of an appropriate compression speed (strain rate) and degree of compression (maximum strain) of foods if one wants to relate results to texture perception.

Another complication is that, in most instrumental tests, the food is deformed only one or two times, whereas several texture attributes are only graded after multiple chews. This implies that information on the breakdown rate and pattern is not measured in most instrumental tests, whereas it is measured during oral processing. The dependency of measured food properties on the applied maximum strain implies that its selection for the first compression may be essential for the results obtained both during the first and later compressions. Therefore, in the still very popular Texture Profile Analysis test, the choice of the maximum degree of deformation during the first compression should depend on the type of foods studied.

A last complication is that during mastication, saliva lubricates particles, assists in bolus formation, and provides enzymes that may lead to the breakdown of certain food components (primarily starch). These aspects will only be discussed explicitly where appropriate.

Below, we will discuss the relation between texture perception and product properties in more depth for two texture attributes: creaminess and crumbliness. Several other cases are discussed by van Vliet et al. (2009), Foegeding (2007), and Foegeding et al. (2011).

18.2.1 Creaminess of soft solids

Creaminess is a clear example of a multiparameter sensory attribute. An extensive description of this attribute is "range of sensations typically associated with fat content, such as full and sweet taste, compact, smooth, not rough, and not dry, with a velvety (not oily) coating." The food should disintegrate at a moderate rate. This description combines texture (e.g., smooth and not rough) and taste (or possibly flavor) sensations. In addition, it concerns both bulk (e.g., compact) and surface characteristics (e.g., velvety and not oily coating) of the food oral bolus.

Of the characteristics mentioned in the description of creaminess, the most important depends on the product. Limiting ourselves to texture attributes, according to the given description, the main ones are smooth and not rough. In addition, the basic texture attribute thickness has been observed to be important. For instance, for a range of dairy food products, creaminess was found to be strongly related to the sensory characteristics thickness and smoothness (Kokini and Cussler 1983): creaminess = thickness^a × smoothness^b. The exponents were estimated to be about 0.54 and 0.84, respectively. In this relation, smoothness was assumed to be inversely proportional to the contact friction force between the tongue and the surface of the food. The texture attribute thickness was assumed to be directly related to the viscosity or yield stress of the bulk of the product. However, for other products, the given relation for creaminess was less strong, indicating that other factors were also important. This may be attributable to various factors.

First, creaminess is negatively affected by texture attributes such as roughness (nonslippery). It can be instrumentally measured with friction measurements and is primarily related to the surface properties of the food bolus.

Second, a bulk characteristic such as heterogeneous mouth-feel (not smooth; note that here smooth has a different meaning than was assumed by Kokini and Cussler) also affects creaminess negatively. The importance of these characteristics of the food bolus follows also from the observation that the method of oral processing strongly affects texture attributes such as creamy, thick, heterogeneous, and smooth. For thickness grading, a simple up-and-down movement of the tongue is enough, but for creamy a smearing action of the tongue is required to obtain maximum sensation.

Third, for thicker products the mechanical breakdown of the product structure is also important. For mayonnaise and custards, rheological parameters characterizing their mechanical breakdown (for mayonnaise steep decrease in shear stress after yielding) on shearing are the most relevant rheological attributes relating to creamy scores. The importance of the mechanical breakdown properties follows also from the observation that reduction of amylase activity caused an increase in creamy perception of gelatinized starch-containing products such as custards. This is likely caused by affecting the perceived thickness and heterogeneity of these products.

Fourth, factors other than texture attributes also play a dominant role. For many products, fat content and dairy flavor are important factors.

The prediction of creamy after-feel perception based on mechanical characteristics is not as good as that for creaminess. It is likely that characteristics such as adhesion of the product to the oral mucosa will also play a role.

18.2.2 Crumbliness

In the study discussed below, crumbly was defined by the sensory panel as "sample falls apart in (small) pieces upon compression between the tongue and palate." This definition implies that the sample has to fall apart in pieces, implying fracture, and that it has to be broken down to small pieces, which still can be noticed by the person evaluating the product. The latter means that the result of the fracture process is also important. The study was performed using mixed whey protein isolates–polysaccharide gels with different microstructures. Both the protein/polysaccharide ratio and the type of polysaccharide were varied (van den Berg et al., 2008). Panelists could significantly discriminate among the gels with high and low crumbly scores. No correlation was observed between the crumbly scores and the type of polysaccharide used. In addition, there was no direct relation with the mesoscopic microstructure of the gels as determined by confocal scanning laser microscopy.

In view of the definition given by the panel, correlations were expected between fracture behavior and sensory crumbly scores. However, observed correlations between sensory crumbly scores and classical characteristics of fracture behavior such as fracture stress, fracture force, fracture stress after correction for the change in cross section of the test piece, strain at fracture, and energy supplied up to fracture were low (correlation coefficients,

$R^2 \leq 0.50$). In contrast, a relation could be seen between the crumbly scores and the shape of the force versus strain curves after the gels were fractured (**Figure 18.3**). Gels were more crumbly if the decrease in force after fracture was relative stronger. If the force decrease was only relatively minor, indicating a more yielding type of fracture, crumbly scores were low.

Fracture behavior was further studied using a wedge test. At low speeds, the gels did not fracture during wedge penetration but were deformed and/or cut by the penetrating wedge (tip angle 30°), whereas at high speeds all gels fractured by a self-propagating crack. A negative correlation was observed between the crumbly score of a gel and the speed at which 50% of the test pieces fractured by a self-propagating crack. To obtain a self-propagating crack, sufficient energy has to be stored elastically in the gel and released during crack growth (Section 5.2). This, in combination with the negative correlation observed, indicates a relation between crumbly perception of a product and its viscoelastic behavior.

Viscoelasticity of materials outside the linear region can be determined by measuring the recoverable energy after deforming them to a large strain (Section 5.2.3; **Figure 5.11**). Gels were first compressed to 60% of their

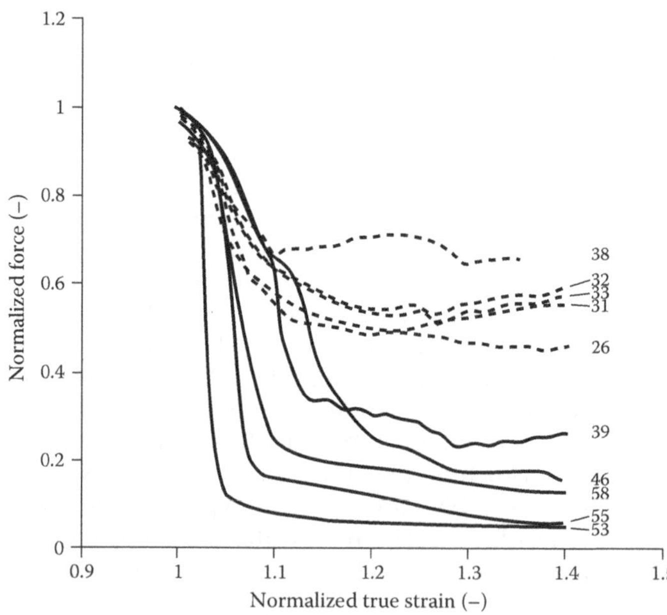

***Figure 18.3** Normalized force versus normalized Hencky strain curves after the fracture point of whey protein gels and mixed whey proteins–polysaccharides gels subjected to uniaxial compression at a strain rate of 0.8 s⁻¹ to 10% of gels' initial height. Bold curves correspond to gels with high crumbly scores, indicated. (Redrawn from van den Berg, L. et al.,* Food Hydrocolloids, *22, 1404–1417, 2008.)*

original height, and subsequently the stress was released at the same speed as the compression speed. The recoverable energy is the ratio of the energy released as mechanical work during unloading over the energy invested during compression of the gels. For viscoelastic materials, this ratio will depend on the deformation speed. A very clear correlation was observed between sensory crumbly scores and the recoverable energy determined at a deformation speed of 20 mm s^{-1} ($R^2 = 0.87$), but not at a deformation speed of 1 mm s^{-1} (**Figure 18.4**). The high deformation speed is of the same order as the biting and palating speed applied by humans. Thus, under conditions similar to those in the mouth, crumbly gels have a high recoverable energy of 70–85% (implying that reaction on applied deformation has a high elastic component), whereas gels with low crumbly scores have a low recoverable energy.

The effect of deformation speed can be understood by considering that the formation of a self-propagating crack in a gel during deformation is governed by the following energy balance (Section 5.2.3);

$$W = W_e + W_d + (W_{fr}), \qquad (18.1)$$

where W is the amount of energy supplied to the material during deformation, W_e is the part of the deformation energy that is elastically stored, W_d is the part that is dissipated, and W_{fr} is the fracture energy. (Note that Equation 18.1 is a simplified version of Equation 5.13.) A free-running (self-propagating) crack will only be formed when the differential energy (dW_e/dl) released due

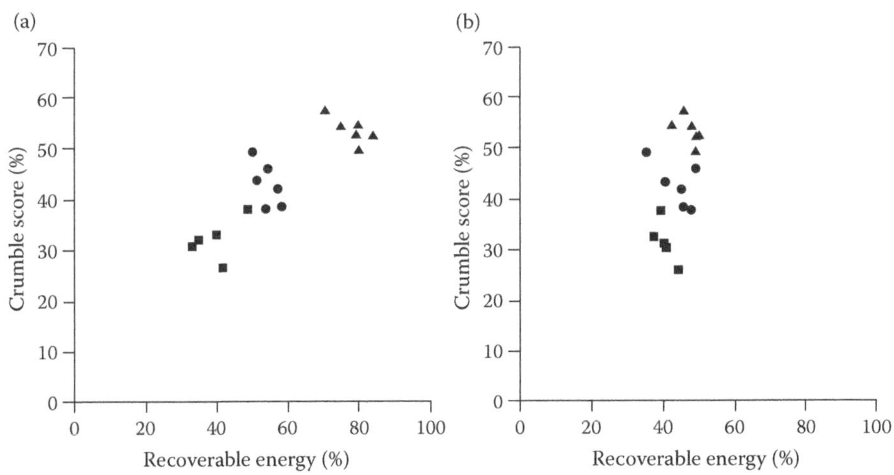

***Figure 18.4** Crumbly scores of whey protein and mixed whey protein–polysaccharide gels as a function of the energy recovered during compression–decompression test at 20 mm s^{-1} (left panel) and 1 mm s^{-1} (right panel). Gels were compressed to 60% of their original height. (Redrawn from van den Berg, L. et al.,* Food Hydrocolloids, *22, 1404–1417, 2008.)*

to stress relaxation in the material in the vicinity of the newly formed crack surpasses the differential energy (dW_{fr}/dl) required for further crack growth.

Energy dissipation other than that resulting from fracturing (W_d) directly affects the amount of energy available for crack growth and hence the formation of a self-propagating crack. For the systems of which results are shown in **Figures 18.3 and 18.4**, energy dissipation was mainly caused by serum transport through the gels, resulting in serum release. The extent of serum release was positively related to the speed at which 50% of the test pieces fractured by a self-propagating crack. In addition, serum release was related to the microstructure of the gels. As a result, there was only an indirect relation between microstructure and crumbly perception. Apparently, formation of a self-propagating crack in the gels under conditions similar to those in the mouth is a prerequisite for getting a high crumbliness score during oral evaluation.

Finally, in order to obtain a crumbly sensation, the gel particles formed during the fracture process should not be too soft. A positive relation between the modulus of gels and their crumbly perception has been observed for emulsion filled gels. Although not studied, there must be an optimum in particle and product stiffness for a product to be crumbly. Very hard elastic solids that fracture by a free-running crack will not be considered crumbly, but hard or crispy, although the recoverable energy at deformation below fracture will be about 100%. Then the formation of hard sharp particles on fracture may contribute to the usually negative texture attribute splinter formation.

The examples discussed above show that, in order to predict texture attributes such as creaminess and, even more clearly, crumbliness of soft solids, one has to determine not only the large deformation properties of the foods up to yielding or fracture, but also their behavior after that. The determination of classical product characteristics such as yield stress and fracture stress may be sufficient for estimating sensory hardness, but will be insufficient for the estimation of various other texture characteristics.

18.3 Hard solids

Important texture attributes for these types of products include hardness/firmness, brittleness, crispiness, crunchiness, splinter formation, and rate of dissolving (the latter for, e.g., hard candies). Hard solid foods encompass a large variety of products, which range from breakfast cereals, dry toast, potato crisps, to several vegetables and fruits such as celery, fresh carrots, and apples. Products such as (just) deep-fried chicken nuggets and (fresh) baked French baguettes consist of a soft solid inner part and a brittle crust with a low water content. These crusts have a similar mesoscopic structure to dry cellular solids and, at low water content, similar mechanical properties. However, their mechanical properties will be affected to some extent by the soft layer below the hard crusts. Here, these effects will not be discussed separately. For a general discussion of layered products, see Section 14.1. Below, we will discuss briefly two examples of established relations between sensory texture attributes and instrumentally determined parameters. For a more extended discussion, see van Vliet and Primo-Martín (2011, and references therein). With these examples, we intend to illustrate that understanding fracture mechanics may lead to increased understanding of the sensory attributes involved and of the mechanisms determining the intensity of these attributes.

The relation between sensory and instrumentally determined hardness of fruit and vegetables has been studied by Vincent (2004). In this study, sensory hardness was defined as "the amount of force required to break a sample." Fracture properties were determined by 3-point bending of notched specimens using the so-called single-edge notch bending test (Section 8.3.3). The critical stress intensity K_{IC} was calculated using Equation 8.70, whereby for f the force (load) at the onset of crack propagation has been taken. An excellent correlation was observed between the instrumentally determined K_{IC} and sensory hardness (**Figure 18.5**). This correlation means that hardness of the studied fruits and vegetables is closely related to the stress required to start a crack running through the sample. It also implies that, for these products, the main fracture events in the food must take place close to the tip of the crack as they exhibit brittle fracture.

Together with hardness, crispiness and crunchiness are considered as the most important texture attributes of hard solid products. By establishing a relation between these texture attributes and instrumentally determined

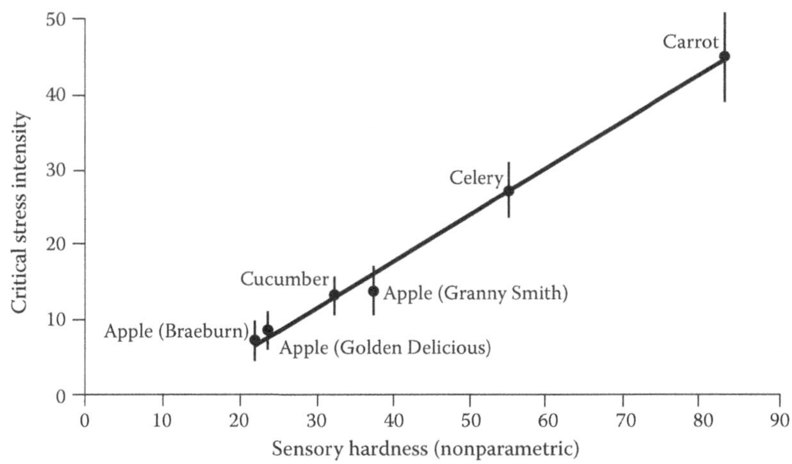

Figure 18.5 *Relationship between critical stress intensity factor, K_{IC}, and sensory hardness for some fruits and vegetables. (Reprinted from Vincent, J.F.V.,* Eng. Failure Anal., *11, 695–704, 2004. With permission.)*

parameters, one has to realize that these attributes are complex concepts that combine a wide range of perceptions such as fracture and emitted sound characteristics, density, and geometry. The definitions of both terms may even vary from person to person. In the literature, various definitions of crispiness and crunchiness have been given. In view of this, the discussion below will be limited to crispiness of dry cellular foods.

Most authors accept that crispy behavior requires multiple fracture events accompanied by acoustic emission at relatively low work of fracture. Acoustic emission requires crack growth speeds to be above one-fourth to one-half of the maximum speed of stress waves in a material, the so-called Rayleigh speed (v_{Rayleigh}), which is given by

$$v_{\text{Rayleigh}} = \sqrt{E/\rho} \tag{18.2}$$

where E is Young's modulus of the solid material and ρ is its density (Section 5.2.1.3). For crispy foods, this implies that the critical crack speed is about 300–500 m s^{-1}. The low work of fracture implies that the fracture strain and energy dissipation on deformation and during fracture must be low, that is, brittle behavior. The presence of high crack speeds required for acoustic emission can only be understood well by considering the fracture process being governed by Equation 18.1, shown above.

Fast crack growth (fracture propagation) will occur when the differential energy (dW_e/dl) released due to stress relaxation in the material in the vicinity of the newly formed crack greatly surpasses the differential energy (dW_{fr}/dl) required for further crack growth. Energy dissipation other than that resulting from fracturing (W_d) directly affects the amount of energy available for crack growth and hence the speed of crack propagation and the critical crack length l_c for fracture propagation.

The very high crack growth speeds involved imply that a homogeneous product of 1 cm length will fracture in less than 0.1 ms. In addition, humans will perceive a short passage in music as one single acoustic event when the pause in between is less than about 10 ms. Something similar to this holds for fracture events. This information, together with the requirement that consumers must hear several successive sound events, implies that while a crispy product is being consumed, fast-growing cracks must be initiated at different places at different times. Moreover, the cracks must stop after some time. Hence, the presence of crack stoppers at distances of the order of a few hundred microns is essential for a product to be perceived as crispy. Likely, the large pores in the cellular structure, typically for dry crispy products, act as crack stoppers.

It is well known that an increase in water content of crispy products causes a decrease of their crispiness. At high water content, crispiness is completely lost. The shape of the force and sound pressure versus deformation curves changes drastically with increasing water content (**Figure 18.6**). Various

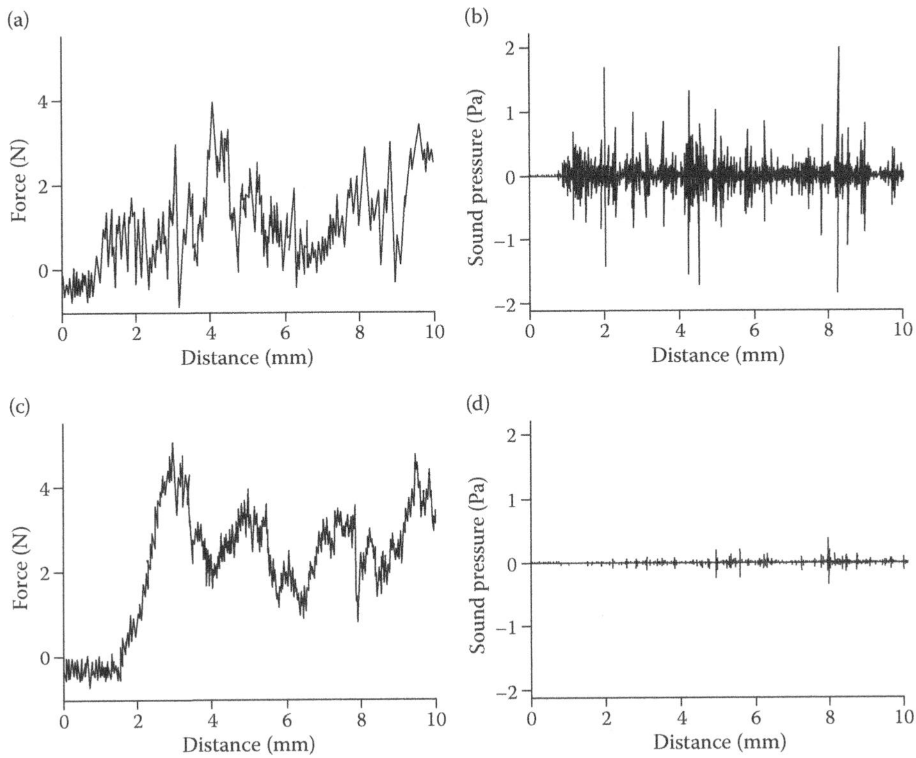

Figure 18.6 *Force and sound pressure versus cutting distance using a scalpel of toasted rusk rolls conditioned at a relative humidity of 30 (panels a and b) and 70% (panels c and d). Deformation speed, 4 cm s*$^{-1}$. *(Reprinted from Castro-Prada, E.M. et al.,* J. Texture Stud., *40, 127–156, 2009. With permission.)*

characteristics of these curves are used to characterize the decrease in crispiness with increasing water content of the products. These include numbers of force peaks, linear length of the force versus deformation curve, jaggedness of the force versus deformation curve, number of large force peaks, number of sound peaks, total sound intensity, and average sound intensity. Most of these characteristics show a good correlation with crispiness on comparing different products and between each other.

The observed correlations between instrumentally measured characteristics and oral evaluation of crispiness do not mean that all these characteristics are directly related to the way humans perceive crispiness. As mentioned above, humans need a pause of about 10 ms between two sound peaks before they are able to distinguish these as two separate sound events. In addition, humans also have a latency time of the muscles involved in biting and chewing before they react on a drastic change in load on the teeth. Latency times for the reflex reaction on a load change of these muscles is about 10 ms, whereas the conscious reaction time is about 150 ms. For normal biting speeds of 20–40 mm s^{-1}, biting crispy products of 1 cm thickness in two halves will only take 250–500 ms. This implies that humans can at best distinguish two to three force drops consciously and possibly somewhat more unconsciously and 25–50 sound events. More likely, they notice only some large force drops or only a clear variation in the required force for fracturing. Something similar will hold for sound events. Another possibility is that they primarily notice total sound intensity and the variation in sound pressure.

The consequence of incorporating human physiological aspects in the consideration of the relation between crispiness and instrumentally measured characteristics is that this relation must be indirect for determined parameters such as numbers of force and sound peaks (often a few hundred to a few thousand), linear length of the curves, and curve jaggedness. This does not detract from the usefulness of these characteristics as long as one is aware that, for other measuring conditions, the relation may be different. Therefore, it is likely that the determination of the number of large fracture or sound events in combination with mean sound pulse intensity is usually to be preferred for estimating sensory crispiness.

As mentioned above, human biting speed is of the order of 20–40 mm s^{-1}. However, in general, instrumental characterization of crispiness is performed at much lower speeds (e.g., 0.1–1 mm s^{-1}). For comparing different products, usually this does not matter very much. However, it has been observed that deformation speed clearly affects the range in water activity over which a transition is instrumentally observed from crispy to noncrispy behavior. For toasted rusk rolls stored at a relative humidity (RH) of 70%, instrumentally no sound emission was measured at a deformation speed of 0.4 mm s^{-1}, whereas there was clear sound emission when the deformation speed was 40 mm s^{-1}. Moreover, the force versus deformation curve was more jagged when the test piece was deformed at a higher speed. At a low deformation speed, the transition from crispy to noncrispy behavior was between RH 40 and 56%, whereas at high speed it was between RH 50–60% and 74%. Oral evaluation by a panel showed loss of crispness between RH 48 and 75%.

A complication in the instrumental determination of the force versus deformation curve is the inertia of the pressure transducer. This is of the order of 1 ms. In addition, several smaller peaks (oscillations) will follow the main force drop event in a force versus deformation curve as a result of damped vibrations of the transducer around a new equilibrium position due to the sudden stress release. They may be as big as real force drops resulting from smaller fracture events. This limits the use of parameters such as the total number of measured force peaks as a characteristic of crispiness.

As argued above for obtaining acoustic emission crack growth, speeds must exceed 300 mm s^{-1}. For beams/cell walls of 300 μm thickness, this implies a cracking time of 10^{-6} s. This time is much shorter than the reaction time of available load cells. This implies that when deforming crispy products at a speed comparable to the biting speed, there will be an extensive overlap between the measured force peaks. In view of the latency times for the reaction of the muscles involved in biting and chewing, this will also be the case for the fracture events that humans notice. The same will apply for the sound events that humans notice during oral evaluation.

Another consequence of the overlap of force peaks, especially at high deformation speeds, is that the jaggedness of force deformation curves of brittle foods and the measured number of force peaks provides only information on the relation between product structure and fracture behavior when the curve is determined at a low deformation speed. In addition, the signal has to be cleaned with respect to the effects of vibrations of the pressure transducer.

Finally, a correlation is often observed between hardness and crispiness both by sensory and instrumental measurements. This may suggest that they are correlated sensory terms. However, in contrast with this notion, it has been observed various times for dry products that with increasing water content a decrease in crispness may go together with an increase in hardness (the so-called moisture toughening). Only at somewhat higher water contents than that at which crispness starts to decrease, does hardness also decrease. It is likely that, in order to be crispy, products need to have a certain hardness; however, this does not mean that these factors have a 1:1 correlation.

In conclusion, the combination of the principles of fracture mechanics and the oral evaluation of crispy products leads to a better understanding of the relation between food structure, mechanical properties, and perception of crispiness. However, because crispiness is a multiparameter attribute, there is no single parameter that can be used for estimating instrumentally oral crispiness perception for all crispy products. The parameter or combination of parameters to be measured should depend on the goal of the measurement and the crispy product studied.

18.4 Concluding remarks

The examples discussed above show that combining the principles of rheology and fracture mechanics and their relation with product structure with a basic understanding of oral processing and evaluation of food stimuli may lead to a better understanding of texture perception by humans. In the long term, such an understanding will lead to better methods of checking the texture of produced foods. To come to a better control of the texture of produced food, it is also required to understand the relation between ingredients, manufacturing (processing), and texture of the food produced. This will require knowledge discussed in various chapters of this book combined with information on the composition and (mesoscopic) structure of the product involved. However, regardless of the reason why one wants to determine the texture of a food product, it should be realized that texture is a multiparameter characteristic of food products. This implies that there is no single method that is able to predict the oral texture perception for nearly all food products, except by eating them. The latter also applies for many texture characteristics.

Some simple texture characteristics such as oral viscosity or thickness of Newtonian liquids such as water, clear fruits drinks, wine, and soda can be estimated by a simple mechanical test. However, this becomes even more complicated for non-Newtonian liquids and semisolid materials, where test conditions such as shear rate and shearing time are much more critical for predicting oral thickness and may depend on product properties. In addition, texture attributes such as sliminess and stickiness may also contribute significantly to texture perception. For soft and hard solids, many texture attributes themselves are already multiparameter attributes such as creaminess and crispiness. Thus, for a full characterization of a single texture attribute, one has to determine various instrumental parameters. However, for quality checks, the measurement of a well-chosen, single instrumental parameter will often be sufficient.

The discussion given above does not detract from the usefulness of many of the existing methods developed to measure specific texture attributes, as long as these methods are suitable for reaching the goal of the measurement. The biggest mistake one can make is to use any available method without first analyzing in a profound way what it measures, and whether the obtained results relate in a direct or indirect way to the desired texture attributes.

References

Bourne, M.C., 2002. *Food Texture and Viscosity: Concept and Measurement*. 2nd ed. New York: Academic Press.

Castro-Prada, E.M., C. Primo-Martin, M.B.J. Meinders, R.J. Hamer, and T. van Vliet. 2009. Relationship between water activity, deformation speed and crispness characterization. *J. Texture Stud.* 40: 127–156.

Chen, J. 2009. Food oral processing—a review. *Food Hydrocolloids* 23: 1–25.

Dickie, A.M., and J.L. Kokini. 1983. An improved model for food thickness from non-Newtonian fluid mechanics in the mouth. *J. Texture Stud.* 14: 377–395.

Foegeding, E.A. 2007. Rheology and sensory texture of biopolymer gels. *Curr. Opin. Colloid Interface Sci.* 12: 242–250.

Foegeding, E.A., C.R. Daubert, M.A. Drake, G. Essick, M. Trulsson, C. Vinyard, and F. van de Velde. 2011. A comprehensive approach to understanding textural properties of semi- and soft-solid foods. *J. Texture Stud.* 42: 103–129.

International Organization for Standardization. 1992. *ISO standard* 5492:1992.

Kokini, J.L., and E.L. Cussler. 1983. Predicting the texture of liquid and melting semi-solid foods. *J. Food Sci.* 48: 1221–1225.

Shama, F., and P. Sherman. 1973a. Identification of stimuli controlling the sensory evaluation of viscosity: II. Oral methods. *J. Texture Stud.* 4: 111–118.

Shama, F., and P. Sherman. 1973b. Evaluation of some textural properties of foods with the Instron Universal Testing Machine. *J. Texture Stud.* 4: 344–353.

Shama, F., C. Parkinson, and P. Sherman. 1973. Identification of stimuli controlling the sensory evaluation of viscosity: I. Non-oral methods. *J. Texture Stud.* 4: 102–110.

van Vliet, T. 2002. On the relation between texture perception and fundamental mechanical parameters for liquids and time dependent solids. *Food Qual. Preference* 13: 227–236.
van Vliet, T., and C. Primo-Martín. 2011. Interplay between product characteristics, oral physiology and texture perception of cellular brittle foods. *J. Texture Stud.* 42: 82–94.
van Vliet, T., G.A. van Aken, H.H.J. de Jongh, and R.J. Hamer. 2009. Colloidal aspects of texture perception. *Adv. Colloid Interface Sci.* 150: 27–40.
van den Berg, L., A.L. Carolas, T. van Vliet, E. van der Linden, M.A.J.S. van Boekel, and F. van der Velde. 2008. Energy storage controls crumbly perception in whey proteins/polysaccharide mixed gels. *Food Hydrocolloids* 22: 1404–1417.
Vincent, J.F.V. 2004. Application of fracture mechanics to the texture of food. *Eng. Failure Anal.* 11: 695–704.

Appendix A: List of Frequently Used Symbols

Latin

A	Area	(m^2)
b	Length statistical chain element (**Chapter 11**)	(m)
b	Thickness (**Chapter 16**)	(m)
c	Concentration	
c^*	Critical concentration for significant coil interpenetration	
c_0^*	Chain overlap concentration demarcating boundary between dilute and semidilute macromolecular solutions	
d	Diameter	(m)
D	Diffusion coefficient	($m^2\ s^{-1}$)
D_r	Rotational diffusion coefficient	(s^{-1})
D_f	Fractal dimensionality	(–)
E	Young's modulus	(Pa)
f	Force	(N)
F	Gibbs energy (Section 13.1)	(J)
g	Acceleration due to gravity	($m\ s^{-2}$)
G	Shear modulus	(Pa)
G_{coll}	Colloidal interaction free energy	(J)
$G(t)$	Relaxation modulus	(Pa)
G'	Storage (shear) modulus	(Pa)
G''	Loss (shear) modulus	(Pa)
G_C	Critical strain energy release rate	($J\ m^{-2}$)
h	Distance	(m)
H	Height	(m)
$H(\tau_{rel})$	Relaxation spectrum/spectrum of modulus densities	($N\ m^{-2}$)
I	Second moment of area	(m^4)
I	Ionic strength (**Chapter 11**)	($mol\ L^{-1}$)
J	Shear compliance	($m^2\ N^{-1}$)
k	Consistency index	($N\ m^{-2}\ s^n$)

k_B	Boltzmann constant	(1.381×10^{-23} J K^{-1})
k_H	Huggings constant	(–)
K	Compression modulus (**Chapter 3**)	(Pa)
K	Stress intensity factor (**Chapter 5**)	(N $m^{-1.5}$)
K_I	Stress intensity factor in mode I (tensile) deformation	(N $m^{-1.5}$)
K_{IC}	Critical stress intensity factor in mode I (tensile) deformation	(N $m^{-1.5}$)
l	Length	(m)
l_c	Critical notch (crack) length for crack propagation	(m)
L	Distance between supports in bending tests	(m)
$L(\tau_{ret})$	Retardation spectrum/spectrum of compliance densities	(m^2 N^{-1})
M	Molar mass	(kg $kmol^{-1}$)
n	Power law index	(–)
n	Number of segments in polymer chain (**Chapter 11**)	(–)
n'	Number of statistical chain elements in a polymer chain	(–)
N	Number of elastically effective strands per cross section gel (**Chapter 13**)	(m^{-2})
N_1	First normal stress difference	(Pa)
N_2	Second normal stress difference	(Pa)
N_{AV}	Avogadro's number	(6.022×10^{23} mol^{-1})
N_{agg}	Number of particles in an aggregate	(–)
p	Pressure	(Pa)
P	Perimeter	(m)
Q	Volume flow rate	(m^3 s^{-1})
r	Radius (variable)	(m)
r_m	Root-mean-square end-to-end distance	(m)
r_g	Radius of gyration	(m)
R	Radius	(m)
R_p	Particle radius	(m)
R_g	Gas constant	(8.315 J K^{-1} mol^{-1})
R_{agg}	Aggregate radius	(m)
R_s	Specific fracture energy	(J m^{-2})
t	Time	(s)
T	(Absolute) temperature	(K)
T	Torque (Section 8.2)	(N·m)
U	Helmholtz interaction energy	(J)
v	Velocity	(m s^{-1})

V	Volume	(m^3)
w	Width	(m)
W	Supplied energy during deformation	($J\ m^{-3}$)
W_e	Energy elastically stored	($J\ m^{-3}$)
W_d	Energy dissipated	($J\ m^{-3}$)
W_{fr}	Fracture energy	($J\ m^{-3}$)
x	Distance	(m)

Greek

α	Angular displacement	(rad)
α	Linear expansion coefficient (Section 11.1)	(–)
γ	Shear strain	(–)
$\dot{\gamma}$	Shear rate	(s^{-1})
γ_{fr}	Fracture (shear) strain	(–)
γ_{int}	Interfacial tension (Sections 14.2 and 15.2)	($N\ m^{-1}$)
ε	Tensile strain	(–)
ε_C	Cauchy strain	(–)
ε_H	Hencky strain	(–)
ε_v	Volume strain	(–)
ε_b	Biaxial strain	(–)
ε_{fr}	Fracture strain	(–)
$\dot{\varepsilon}$	Elongational strain rate	(–)
η	(Shear) viscosity	(Pa s)
η'	Dynamic viscosity	(Pa s)
η_c	Viscosity continuous phase	(Pa s)
η_e	Elongational viscosity	(Pa s)
η_{app}	Apparent viscosity	(Pa s)
$[\eta]$	Intrinsic viscosity (dispersions of particles)	(–)
$[\eta]$	Intrinsic viscosity (solution of macromolecules)	($m^3\ kg^{-1}$)
θ	Angle	(rad)
κ	Screening parameter of electric charge	(m^{-1})
μ	Poisson's ratio	(–)
υ	Kinematic viscosity (**Chapter 8**)	($m^2\ s^{-1}$)
ν	Number of chains between cross-links per volume (**Chapter 13**)	(m^{-3})
ξ	Correlation length	(m)
ρ	Mass density	($kg\ m^{-3}$)
ρ_p	Particle density	($kg\ m^{-3}$)

σ	Stress	(Pa)
σ_e	Tensile/uniaxial compressive stress	(Pa)
σ_y	Yield stress	(Pa)
σ_{fr}	Fracture stress	(Pa)
σ_R	Shear stress at wall	(Pa)
τ_{rel}	Relaxation time	(s)
τ_{ret}	Retardation time	(s)
ψ	Velocity gradient (Section 3.3)	(rad s^{-1})
ψ	Electrostatic potential (Section 10.3)	(V)
ψ_0	Electrostatic surface potential (Sections 10.3 and 11.1)	(V)
φ	Volume fraction	(–)
φ_{max}	Maximum volume fraction	(–)
φ_{agg}	Volume fraction particles in an aggregate	(–)
ω	Angular frequency	(rad s^{-1})
ω	Revolution rate (Section 3.3)	(s^{-1})
ω_f	Angular velocity fluid element	(rad s^{-1})
Ω	Angular velocity	(rad s^{-1})

Other

De	Deborah number	(–)
Re	Reynolds number	(–)
Pe	Peclet number	(–)
Tr	Trouton ratio	(–)

Appendix B: SI Units Used

Base Units

Quantity	Name	Symbol
Length	Meter	m
Mass	Kilogram	kg
Time	Second	s
Temperature	Kelvin	K
Amount of substance	Mole	mol

For this book, relevant supplementary units concern the unit for a plane angle, the radian (rad), and that for frequency, Hertz (Hz).

Derived Units

Quantity	Name	Symbol	SI Base Units
Force	Newton	N	kg m s^{-2}
Pressure, stress	Pascal	Pa	kg m^{-1} s^{-2} (N m^{-2})
Energy	Joule	J	kg m^2 s^{-2} (N m)
Frequency	Hertz	Hz	s^{-1}
Angular frequency		ω	rad s^{-1}

Numerical Prefixes

The most important numerical prefixes are

Multiplication factor	Name	Symbol
10^9	giga	G
10^6	mega	M
10^3	kilo	k
10^{-1}	deci	d
10^{-2}	centi	c
10^{-3}	milli	m
10^{-6}	micro	μ

Some Other Units

For the temperature scale, SI rules allow the use of degrees Celsius (°C), but not for temperature intervals, nor can °C be used with a prefix.

For angles, the "degree" symbol (°) is often used.

For volumes, the unit liter can be used (10^{-3} m^3) (L).

Appendix C: Derivation of Equations 7.32 and 7.33 Using Complex Notations

In complex notation, the complex strain $\hat{\gamma}$ and stress $\hat{\sigma}$ read as

$$\hat{\gamma} = \gamma_0 e^{i\omega t}$$

$$\hat{\sigma} = \sigma_0 e^{i\omega(t-\delta)}$$

where $i = \sqrt{-1}$. The term $e^{i\omega t} = \cos\omega t + i \sin\omega t$. Inserting these equations in $\sigma = \tau_{rel}(d\sigma/dt) + \eta\dot{\gamma}$ (Equation 7.31) gives

$$\sigma_0 e^{i(\omega t - \delta)} = \tau_{rel}\sigma_0 i\omega e^{i(\omega t - \delta)} + \eta\gamma_0 i\omega e^{i\omega t}$$

Using $\eta = \tau_{rel}G$ gives

$$\sigma_0 e^{-i\delta}(1 - i\omega\tau_{rel}) = i\omega\tau_{rel}G\gamma_0$$

From this equation, it follows that, for the complex modulus

$$\hat{G} = \frac{\sigma_0 e^{-i\delta}}{\gamma_0} = \frac{i\omega G\tau_{rel}}{1 - i\omega\tau_{rel}} = \frac{i\omega G\tau_{rel}(1 + i\omega\tau_{rel})}{(1 - i\omega\tau_{rel})(1 + i\omega\tau_{rel})} = \frac{G\omega^2\tau_{rel}^2 + iG\omega\tau_{rel}}{1 + i\omega^2\tau_{rel}^2}$$

Since in the complex notation $\hat{G} = G' + iG''$, this gives for the storage and loss modulus, respectively

$$G'(\omega) = G\frac{\omega^2\tau_{rel}^2}{1 + \omega^2\tau_{rel}^2} \tag{7.32}$$

and

$$G'' = G\frac{\omega\tau_{rel}}{1 + \omega^2\tau_{rel}^2} \tag{7.33}$$

Index

For terms mentioned at various places the page number printed in boldface refers to the place where it is defined. From frequent used terms as stress, Newtonian liquids, viscosity, etc only the page numbe(s) is given were they are defined and/or described extensively. Foods products that are mentioned only as an example of a mechanical behavior are not included in this index.

D

E

F

G

H

I

K

L

S

T

U

V

For Product Safety Concerns and Information please contact our EU representative GPSR@taylorandfrancis.com Taylor & Francis Verlag GmbH, Kaufingerstraße 24, 80331 München, Germany